# Haut als endokrines Erfolgsorgan
# Gestagene
# Geriatrische Endokrinologie des Mannes

17. Symposion der Deutschen Gesellschaft für Endokrinologie
in Hamburg vom 4.—6. März 1971

Schriftleitung: Prof. Dr. Joachim Kracht

Mit 84 Abbildungen

Springer-Verlag Berlin · Heidelberg · New York 1971

ISBN-13: 978-3-642-95217-3     e-ISBN-13: 978-3-642-95216-6
DOI: 10. 1007/978-3-642-95216-6



Offsetdruck: J. Beltz, Weinheim

Der *Schoeller-Junkmann-Preis*, eine Stiftung der Schering AG Berlin, wurde von
der Deutschen Gesellschaft für Endokrinologie 1971 verliehen an:


Dr. Govind S. Rao
Institut für Klinische Biochemie
der Universität Bonn

für die Arbeit:

„Steroidglucuronyltransferasen"




Der *Marius-Tausk-Förderpreis*, eine Stiftung der Organon GmbH München,
wurde von der Deutschen Gesellschaft für Endokrinologie 1971 verliehen an:


Dr. D. Scholer
Laboratoire de Physiopathologie Clinique
Hôpital Cantonal, Genf

für die Arbeit:

„Methodik der Plasma-Aldosteron-Bestimmung. Dynamik und Spezifität der Aldosteronstimulierung nach Angiotensin II, Kalium
und adrenocorticotropem Hormon."

# Deutsche Gesellschaft für Endokrinologie

Präsident der Gesellschaft und Vorsitzender des 17. Symposions:
Prof. Dr. J. Tamm, Hamburg

Vorstand der Gesellschaft: Prof. Dr. H.-L. Krüskemper, Hannover
Prof. Dr. J. Kracht, Gießen
Prof. Dr. G. Bettendorf, Hamburg
Prof. Dr. H. Breuer, Bonn
Prof. Dr. P. W. Jungblut, Wilhelmshaven
Prof. Dr. H.-J. Karl, München

Vorstand 1971/72
Präsident: Prof. Dr. H. L. Krüskemper, Hannover
Vizepräsident: Prof. Dr. H. Schriefers, Ulm
Sekretär: Prof. Dr. J. Kracht, Gießen

Mitglieder des Vorstands: Prof. Dr. G. Bettendorf, Hamburg
Prof. Dr. P. W. Jungblut, Wilhelmshaven
Prof. Dr. H. J. Karl, München
Priv.-Doz. Dr. F. Neumann, Berlin

# Inhaltsverzeichnis

Haut als endokrines Erfolgsorgan

Gestagene

Geriatrische Endokrinologie des Mannes

Symp. Dtsch. Ges. Endokrin. **17**, 1-3 (1971)


# Eröffnungsansprache des Präsidenten

## Opening Remarks of the President

J. TAMM

### II. Medizinische Universitätsklinik Hamburg

Sehr geehrter Herr Senator!
Meine sehr verehrten Damen und Herren!

Zum 17. Symposion unserer Gesellschaft möchte ich Sie sehr herzlich in Hamburg willkommen heißen. Mein besonderer Gruß und Dank gilt den Referenten der Tagung, die selbst weite Reisen nicht gescheut haben, um uns über ihre Forschungsergebnisse zu berichten, sowie den Kollegen aus Belgien, Holland, Schweden und der Schweiz, die zu diesem Symposion in unsere Stadt gekommen sind.

Wie Sie wissen, wurde die Deutsche Gesellschaft für Endokrinologie vor 18 Jahren von Professor Arthur Jores zusammen mit einer Gruppe gleichgesinnter Kollegen hier in Hamburg gegründet. Ich darf unserer besonderen Freude darüber Ausdruck geben, daß unser Gründungspräsident, dem ich vor kurzem im Namen der Gesellschaft eine Glückwunschadresse zu seinem 70. Geburtstag überreichen konnte, an diesem Symposion in gewohnter Frische teilnehmen kann. Der 18. Gründungstag ist normalerweise nichts besonderes und kein Grund, in Jubiläumsstimmung zu verfallen; jedoch die Tatsache, daß sich unsere Gesellschaft das erste Mal seit ihrer Konstituierung in ihrer Geburtsstadt versammelt, mag ein Anlaß sein, den Blick kurz auf die Vergangenheit zu werfen.

Es ist erstaunlich, daß in Deutschland erst im Jahre 1953 eine Gesellschaft für Endokrinologie aus der Taufe gehoben wurde. Dies umso mehr, als die Endokrinologie im deutschsprachigen Raum eine jahrzehntealte Tradition hat. Ich darf daran erinnern, daß eine der ersten Zeitschriften für dieses Fachgebiet, "Endokrinologie", im Jahre 1928 von Leon Asher, Artur Biedl u. a. in Leipzig begründet wurde, daß Biedl das erste deutschsprachige endokrinologische Lehrbuch 1910 und Hirsch das erste Handbuch für Endokrinologie 1929 herausgegeben haben. In vielen anderen Ländern war schon lange vor dem 2. Weltkrieg die Notwendigkeit erkannt worden, daß die über viele verschiedene Disziplinen hinweggreifende Endokrinologie nur in einer eigenständigen Gesellschaft ihre besonderen Belange vorantreiben kann. Wenn wir auf die immensen Fortschritte unseres Fachgebietes in den vergangenen zwei Jahrzehnten zurückblicken, erscheint es heute nicht mehr verständlich, daß die Gründung der Deutschen Gesellschaft für Endokrinologie keineswegs nur Zustimmung fand. Eine wesentliche Starthilfe für die Gesellschaft wurde dadurch gegeben, daß sie von Beginn an in den Kreis der sogenannten Acta-endocrinologica-Länder aufgenommen wurde. Dies war ein wesentliches Verdienst des inzwischen verstorbenen Professors Axel Westman, Stockholm, und des jetzigen Herausgebers der Acta endocrinologica, Dr. Christian Hamburger, Kopenhagen, die beide am ersten Symposion teilnahmen. Ein Jahr nach ihrer Gründung beteiligte sich die Deutsche Gesellschaft am ersten Acta-endocrinologica-Congress in Kopenhagen, und es zeigte sich, daß sich die deutschen Endokrinologen wieder sicher auf internationalem Parkett zu bewegen wußten.

Die Kontinuität in der Leitung unserer Gesellschaft, die 10 Jahre lang in Händen von Professor Jores und weitere 5 Jahre in Händen von Professor Oberdisse lag, hat wesentlich zu ihrer Konsolidierung beigetragen. Nach Ablauf dieser Periode schien es jedoch an der Zeit zu sein, die Statuten den moder-

nen Erfordernissen anzupassen, nicht zuletzt mit dem Ziel, die Mitglieder
stärker an der Arbeit der Gesellschaft aktiv zu beteiligen. Wir dürfen heute
mit Befriedigung feststellen, daß wir diesem Ziel merklich näher gekommen
sind. Die internationalen Beziehungen wurden inzwischen dadurch weiter ver-
tieft, daß die Gesellschaft Mitglied der International Society for Endocrinology wurde und einen ständigen Vertreter in deren Central Committee entsen-
det. Als Zeichen der steigenden internationalen Anerkennung der Deutschen
Endokrinologie dürfen wir schließlich auch die Tatsache werten, daß der
3. International Congress on Hormonal Steroids im vergangenen Jahr unter der
Schirmherrschaft unserer Gesellschaft hier in Hamburg stattfand.

Die große Bedeutung, die die endokrinologische Forschung erlangt hat, hat
die für die Wissenschaftsförderung verantwortlichen Gremien der Bundesrepub-
lik dazu veranlaßt, im Rahmen ihres neuartigen Modells für Forschungsschwer-
punktbildung, den sogenannten Sonderforschungsbereichen, zwei endokrinologische Sonderforschungsbereiche zu schaffen: im Jahre 1968/69 an der Universität Hamburg und 1970/71 an der Universität Ulm. Die anfänglich von manchen
Seiten laut gewordenen Befürchtungen, daß diese Art der Schwerpunktbildung
zu einer Benachteiligung anderer endokrinologischer Gruppen führen könne,
dürfte sich inzwischen als gegenstandslos erwiesen haben. Es hat nur manchmal
den Anschein, als ob man sich nicht immer völlig darüber klar ist, daß die
Vergabe von derartigen Fördermitteln sich ausschließlich an strengen wissenschaftlichen Leistungskriterien zu orientieren hat.

Was 1953 einer relativ kleinen Gruppe von Wissenschaftlern klar war, ist
heute für jeden einleuchtend: nämlich, daß es ohne eine Deutsche Gesellschaft
für Endokrinologie sehr viel schwieriger gewesen wäre, der so wichtigen endokrinologischen Forschung in unserem Lande nach dem Kriege die belebenden
Impulse zu geben, die notwendig waren für den Versuch, wieder Anschluß an
internationales Niveau zu gewinnen.

Erlauben Sie mir einige kurze Bemerkungen zum Programm. In diesem Jahr haben wir drei verschiedene Hauptthemen gewählt. Am heutigen Vormittag steht
die "Haut als endokrines Erfolgsorgan" im Mittelpunkt der Betrachtung. Dieser Problemkomplex wurde bislang auf unseren Tagungen noch nicht abgehandelt.
Das hat seinen Grund u. a. auch darin, daß man erst in jüngster Zeit nähere
Einblicke in den Stoffwechsel und Wirkungsmechanismus von Hormonen, vor allem von Androgenen, in der Haut und ihren Anhangsgebilden gewinnen konnte.
Es ist zu erwarten, daß diese Erkenntnisse dazu beitragen werden, die Patho-
mechanismen von Veränderungen der Haut bei Störungen des Endokriniums besser
zu verstehen und Ansätze für eine wirksame Therapie zu entwickeln.

Das Thema des zweiten Tages "Gestagene" war zuletzt auf dem Symposion
1959 Verhandlungsgegenstand. Die kaum noch überschaubare Anhäufung neuer Erkenntnisse auf dem Gebiet der Physiologie und Pathophysiologie des Progesterons sowie die umfangreichen Erfahrungen, die man mit der Anwendung synthetischer Gestagene gesammelt hat, waren Veranlassung, über dieses wichtige
Teilgebiet der Endokrinologie zu diskutieren. Selbstverständlich kann man
den Komplex "Ovulationshemmer" bei einer solchen Diskussion nicht völlig aus-
sparen. Hierzu ist ein besonderes Rundtischgespräch vorgesehen. Daß die
Stoffklasse der Ovulationshemmer kürzlich wieder einmal das Interesse der
breiten Öffentlichkeit finden würde, war bei der Planung dieses Programms
nocht nicht vorauszuahnen.

Der letzte Tag ist den endokrinologischen Problemen des alternden Mannes
gewidmet. Neben einer notwendigen Bestandsaufnahme der bislang bekannten Fakten über Altersveränderungen des männlichen Endokriniums und einem Überblick
über moderne Therapieverfahren bei Tumoren der Prostata stehen die jüngsten
Ergebnisse über den Stoffwechsel von Sexualhormonen in diesem Organ zur De-
batte. Diese Resultate sind nicht nur für die endokrinologische Grundlagenforschung von besonderem Interesse, sie scheinen auch sehr wohl geeignet,
der Behandlung von Prostataerkrankungen richtungweisende Impulse zu geben.

Eine besonders ehrenvolle Aufgabe des Präsidenten unserer Gesellschaft ist
es, die jährlich ausgeschriebenen Preise für besondere wissenschaftliche Lei-
stungen auf dem Gebiet der Endokrinologie zu verleihen. Die Gesellschaft ver-
zeichnet es mit besonderer Genugtuung, daß neben den schon traditionell ge-

gewordenen Schoeller-Junkmann-Preis, gestiftet von der Schering AG, der Marius-Tausk-Förderpreis der Organon GmbH getreten ist, der in diesem Jahr erstmalig vergeben wird.

Für den Schoeller-Junkmann-Preis 1971 gingen elf konkurrierende Arbeiten aus verschiedenen Ländern Europas ein. Nach sorgfältiger Prüfung, z. T. unter Hinzuziehung weiterer Fachgutachter, hat die Jury entschieden, in diesem Jahr nur einen 1. Preis zu verleihen. Den 1. Preis des diesjährigen Schoeller-Junkmann-Preises, der mit 7.000,-- DM dotiert ist, vergibt die Deutsche Gesellschaft für Endokrinologie an Herrn Dr. Govind Rao aus dem Institut für Klinische Biochemie der Universität Bonn für seine Arbeit "Steroidglucuronyltransferasen". Herr Dr. Rao hat in einer profunden Studie zahlreiche neue Erkenntnisse über diesen für den Steroidmetabolismus so außerordentlich bedeutsamen Enzymkomplex vorgelegt.

Um den Marius-Tausk-Förderpreis 1971 bewarben sich neun Autoren aus verschiedenen Ländern unseres Kontinents. Aufgrund der Statuten dürfen die Bewerber nicht älter als 30 Jahre sein. Es handelt sich also um Arbeiten von jungen Kollegen, die am Beginn einer Forscherkarriere stehen. Wie der Name des Preises zum Ausdruck bringt, soll die wissenschaftliche Ausbildung des Besten unter ihnen finanziell erleichtert und gefördert werden. Die Jury hat auch bei der Beurteilung dieser Arbeiten Fachgutachter hinzugezogen, wenn ihre eigene Kompetenz nicht ausreichte. Nach Entscheidung der Jury verleiht die Deutsche Gesellschaft für Endokrinologie den Marius-Tausk-Förderpreis 1971 an Herrn Dr. Dieter Scholer aus dem Institut für Klinische Pathophysiologie der Universität Genf für seine Arbeit "Methodik der Plasma-Aldosteronbestimmung, Dynamik und Spezifität der Aldosteronstimulierung nach Angiotensin II, Kalium und adrenocorticotropem Hormon". In einer ausgezeichnet angelegten Versuchsserie hat Herr Dr. Scholer am Menschen nachweisen können, daß Angiotensin II und Kalium spezifische Stimuli für die Aldosteron-Ausschüttung darstellen, ACTH dagegen nicht. Der Marius-Tausk-Förderpreis ist dotiert mit 15.000,-- DM. Hiervon erhält der Preisträger 3.000,-- DM zu seiner Verfügung. die restliche Summe ist für seine weitere Ausbildung bzw. für wissenschaftliche Arbeiten bestimmt.

Lassen Sie mich abschließend ein besonderes Wort des Dankes sagen. Pecunia est nervus rerum. Davon macht auch eine endokrinologische Gesellschaft keine Ausnahme. Unser Schatzmeister, Herr Professor Kracht, weiß davon ein Lied zu singen. Die Arbeit unserer Gesellschaft wäre schon längst stark reduziert worden, wenn unsere Fördernden Mitglieder uns nicht seit langem die Treue gehalten hätten. Es ist mir daher eine besonders angenehme Pflicht, den Fördernden Mitgliedern unserer Gesellschaft an dieser Stelle sehr herzlich zu danken. Damit möchte ich schließen und das 17. Symposion für eröffnet erklären.

Symp. Dtsch. Ges. Endokrin. 17, 5-6 (1971)


Schoeller-Junkmann Preis 1971
The Schoeller-Junkmann Award 1971

# Steroidglucuronyltransferasen

## Steroid Glucuronyltransferases

GOVIND S. RAO

Institut für Klinische Biochemie der Universität Bonn

## Summary

Oestrogens are excreted to a large extent as glucuronides. The enzyme UDP-glucuronyltransferase (EC 2.4.1.17) is present in several organs of the mammal in addition to the liver. The enzyme is located primarily in the microsomal fraction. Human liver microsomes conjugate the 16α-hydroxyl group of oestriol. The oestriol 16α-glucuronyltransferase of the microsomes can be "solubilised" by treatment with deoxycholate; from a study of molecular weight and phospholipid content, the enzyme appears to be a lipoprotein complex of the microsomal membrane.

Die Steroidhormone werden - ähnlich wie viele andere körpereigene und körperfremde Substanzen - aus dem Organismus in Form wasserlöslicher Verbindungen eliminiert. Unter diesen wasserlöslichen Verbindungen spielen die Glucuronide eine wichtige Rolle.

Die Östrogene werden vorwiegend als Östrogenglucuronide über die Niere ausgeschieden. Bei den Glucuroniden handelt es sich um Äther, die aus einem Aglycon einerseits und dem Glucuronsäurerest andererseits bestehen. Die Enzyme, welche die Bildung von Glucuroniden katalysieren, heißen Glucuronyltransferasen und sind in zahlreichen Organen nachgewiesen worden. Die Tatsache, daß die Östrogenglucuronide am enterohepatischen Kreislauf beteiligt sind, ist von großer Bedeutung für das Hormongleichgewicht des Körpers. Aus diesem Grunde war es notwendig, die Einzelheiten der Biosynthese von Östrogenglucuroniden in denjenigen Organen zu untersuchen, die unmittelbar am enterohepatischen Kreislauf beteiligt sind. Dazu gehören einmal die Leber und zum anderen der Dünndarm.

Während des enterohepatischen Kreislaufes der Östrogene laufen folgende Reaktionen ab: (1) In der Leber werden die Östrogene bzw. ihre Metaboliten mit Glucuronsäure konjugiert und in die Galle ausgeschieden. (2) Mit der Galle werden die Konjugate in den Dünndarm sezerniert; dort findet eine partielle Hydrolyse statt. (3) Die freigesetzten Steroide werden über die Blutbahn zur Leber transportiert und dort rekonjugiert. Die Wiederholung dieser Vorgänge führt zu einem Kreislauf der Östrogene innerhalb des Systems Leber-Galle-Dünndarm-Blut-Leber; dabei wird ein steady-state des biologisch wirksamen Hormons erreicht.

Um eine Charakterisierung und kinetische Untersuchungen der Glucuronyltransferasen durchführen zu können, müssen die an Zellpartikel gebundenen Enzyme löslich gemacht und stabilisiert werden. Die Solubilisierung der mikrosomalen Glucuronyltransferasen konnte mit Deoxycholat erreicht werden. Die solubilisierten Enzyme wurden in Gegenwart reduzierender Substanzen stabilisiert und angereichert. Die angereicherten Enzyme besitzen ein hohes Molekulargewicht und können aufgrund ihrer Eigenschaften als Lipoprotein-Kom-

plex bezeichnet werden. Dieser Befund spricht dafür, daß die Glucuronyltransferasen aus den Membran-Komponenten der Zelle stammen.

Nachdem es gelungen war, die mikrosomalen Enzyme partiell anzureichern,
konnte erstmalig die Spezifität der Glucuronyltransferasen ausführlich untersucht werden. Es zeigte sich, daß die Glucuronyltransferasen der menschlichen
Leber eine ausgeprägte Substratspezifität besitzen. Von den angebotenen
Östrogenen wird nur Östriol konjugiert, und zwar in der 16α-Stellung. Die
gleiche Enzympräparation konjugiert ebenfalls Testosteron, jedoch in geringe-
rem Umfang. Kinetische Untersuchungen zeigten, daß die beiden Steroide unabhängig voneinander an zwei verschiedenen aktiven Zentren des Enzyms glucuro-
nidiert werden. Im Gegensatz zu Testosteron und Östrogenen müssen Aldosteron
und Cortisol vor ihrer Konjugation in der Leber reduziert werden, um als
Aglycon akzeptiert werden zu können. Hierdurch wird ein Hinweis auf die Bedeutung der Lokalisierung von Glucuronyltransferasen, Hydroxylasen und Oxido-
reduktasen in der Leberzelle erhalten.

Die Glucuronidierung im enterohepatischen Kreislauf kann als ein Vorgang
betrachtet werden, mit dessen Hilfe sich der Körper vor extremen Veränderungen der Steroidkonzentrationen schützt. Diese können entweder durch eine
Überfunktion der endokrinen Organe oder durch große Gaben von Steroid-Arznei-
mitteln hervorgerufen werden. Gleichzeitig steht dem Körper damit ein Reser-
voir von Steroiden zur Verfügung, um wichtige Funktionen zu unterhalten.
Durch die Spezifität der in der Leber nachgewiesenen Glucuronyltransferasen
werden unabhängige Pools verschiedener Hormone im Organismus aufrecht erhalten.

Eine weitere Konsequenz der Glucuronidierung ist die folgende Tatsache:
Lipidlösliche Aglycone, nämlich die Steroide, werden lipidunlöslich; damit
wird der ungehinderte Eintritt der Hormone in die Zelle unterbunden. Dies
ist ebenfalls für das Hormongleichgewicht des Körpers bedeutsam.

Zusammenfassend kann gesagt werden: (1) Die Glucuronyltransferasen zeigen
strukturelle Unterschiede; in Abhängigkeit des Gewebes können sie in einer
oder in allen partikulären Fraktionen sowie im Cytoplasma der Zelle vorkommen. (2) Das an Zellpartikel gebundene Enzym kann solubilisiert, stabilisiert
und angereichert werden; es ist mit großer Wahrscheinlichkeit ein Lipoprotein-
Komplex. (3) Glucuronyltransferasen sind spezifisch für bestimmte Gruppen
von Steroidhormonen. (4) Die Glucuronidierung der Steroidhormone führt zu
Verbindungen, die leicht eliminiert werden können. (5) Die Lipidunlöslichkeit
der Glucuronide kann als ein Regulationsmechanismus betrachtet werden, der
die Konzentrierung von biologisch aktiven Steroid-Hormonen verhindert.
(6) Östrogenglucuronide sind an einem dynamischen enterohepatischen Kreislauf
beteiligt, der ein Gleichgewicht des aktiven Hormons gewährleistet; wichtige
enzymatische Grundlagen dieses Vorganges konnten aufgeklärt werden.

Symp. Dtsch. Ges. Endokrin. 17, 7–10 (1971)
© by Springer-Verlag

Marius Tausk-Förderpreis 1971
The Marius Tausk Award 1971

# Methodik der Plasma-Aldosteron-Bestimmung; Dynamik und Spezifität der Aldosteronstimulierung nach Angiotensin II, Kalium und Adrenocorticotropem Hormon[1]

## Aldosteron Determination in Peripheral Plasma of Man; Dynamic and Specific Reponse of Plasma Aldosterone after Angiotensin II, Potassium and ACTH

DIETER SCHOLER

Laboratoire de Physiopathologie Clinique
Medizinische Fakultät der Universität Genf/Schweiz

## Summary

Plasma aldosterone in man has been determined in a 2–4 ml plasma sample according to the slightly modified double isotope derivative method of Bojesen and Thuneberg (1) using $^3$H-aldosterone as recovery indicator and $^{35}$S-labeled p-toluene-sulfonic anhydride as reagent.
The short-term action of angiotensin II, potassium and ACTH on plasma aldosterone levels of the recumbent normal subject has been investigated. Baselines studies on normal sodium intake between 7 a.m. and 11 a.m. show that in the same subject the aldosterone level exhibits some variations; range for 3 subjects: 3–11 ng/ 100 ml. Angiotensin II (infusion of 7 ng/kg/min over 1–3 hours), potassium citrate (30 mEq p.o. every hour for 3 hours) and ACTH (rapid i.v.-injection of 0.5 mg $\beta^{1-2\,4}$ACTH) enhance a marked and reproducible increase of plasma aldosterone. The onset of this increase has been observed within 10 min. for angiotensin II and ACTH, both given i.v., and within 30–60 min. for potassium, given p.o. The range of plasma aldosterone elevation is about 20–25 ng/100 ml for all 3 stimuli, corresponding to a 3–5 fold increase from basal values. Angiotensin II and potassium are, in the dose used, specific stimuli for aldosterone; ACTH, however, stimulates aldosterone, corticosterone and cortisol, but to a variable degree and with a different time-course response.

Bisherige Untersuchungen der Aldosteron-Regulation am Menschen beruhen meist auf Messungen der Aldosteron-Sekretionsrate oder der Urinausscheidung einzelner Aldosteronmetabolite; diese Bestimmungen erfassen die biologisch aktive Form des freien Aldosterons indirekt und als Mittelwert einer mehrstündigen Zeitspanne.

---

[1] Diese Arbeit wurde durch den Schweizerischen Nationalfonds für wissenschaftliche Forschung unterstützt.

Ziel der vorliegenden Arbeit ist es, gewisse Aspekte der *kurzfristigen* Aldosteronveränderung mit Hilfe einer *Direktbestimmung des Plasma-Aldosterons* zu präzisieren. Angesichts der geringen Aldosteron-Konzentration im peripheren Plasma (unter Ruhebedingungen 5-10 ng/100 ml), stellt deren Bestimmung in einem relativ kleinen Plasmavolumen methodische Probleme.

## Methodik der Plasma-Aldosteron-Bestimmung nach Bojesen und Thuneberg

In einer ersten methodischen Phase haben wir eine 1967 von Bojesen und Thuneberg (1) entwickelte, aber fast ausschließlich für Tierexperimente verwendete Plasmabestimmungsmethode, leicht modifiziert (2) in unser Laboratorium eingeführt. Die Bestimmung beruht auf dem Prinzip der Doppelisotopen-Derivatmethode und umfaßt folgende Hauptetappen: 2-4 ml Plasma werden nach Zusatz von 0.05 ng $^3$H-Aldosteron mit einem organischen Lösungsmittel extrahiert und mit $^{35}$S-markiertem p-Toluol-Sulphonsäureanhydrid (spezifische Aktivität 100-150 mC/mEq) verestert. Das dabei gebildete Aldosteronderivat (Aldosteron-21-tosylester-$^{35}$S) wird durch eine Sequenz von 5 Chromatographien und durch sukzessive Umwandlung zu seinem 11,18-$\gamma$-Lacton und seinem 3-Dinitrophenylhydrazon von anderen, mitveresterten Substanzen abgetrennt und in einem Szintillationszähler gezählt.

Die Methode wurde systematisch überprüft bezüglich Leerwert, Spezifität, Genauigkeit und Präzision; die Reproduzierbarkeit der Bestimmung von 0.14 ng Aldosteron, achtfach durchgeführt, ergibt beispielsweise einen Variationskoeffizienten von 7%. Aufgrund der erhaltenen methodischen Kriterien kann Aldosteron spezifisch und genau bis in den Bereich von 0.1 ng ($10^{-10}$g) bestimmt werden. Damit ist die Bestimmung des Plasma-Aldosterons in 2-4 ml peripherem Plasma möglich, d. h. Veränderungen der Sekretion oder der Plasma-Clearancerate können via Mehrfachbestimmung im kleinen Zeitraum verfolgt werden.

## Anwendung der Plasma-Aldosteronbestimmung im Kurzzeitversuch

Vom experimentellen Gesichtspunkt aus ergeben sich mit der Bestimmung des Plasma-Aldosterons folgende Vorteile: 1. die Möglichkeit einer direkten Korrelation des Plasma-Aldosterons mit anderen Plasmaparametern, 2. die Möglichkeit, die Dynamik der Aldosteronveränderung und ev. kurzfristige, bisher unterschwellige Abweichungen mit einem gerafften Zeitraster festzuhalten, und 3. die Prüfung der Reaktivität der Aldosteron-produzierenden Nebennierenrindenschicht in dynamischen Tests; Stimulus und Inhibitor können dabei in geringerer Dosis und für kürzere Zeit appliziert werden als in früheren Langzeitversuchen, sie arbeiten damit näher den physiologischen Bedingungen und verändern die Grundsituation weniger stark durch sekundäre Prozesse (Veränderung der Na-, K-Bilanz, der Flüssigkeitsvolumina).

Eine erste Anwendung erfolgte bei der gesunden Versuchsperson während normaler Kochsalzzufuhr mit dem Ziel, *die Dynamik und Spezifität der Aldosteronantwort nach Stimulierung mit Angiotensin II, Kalium und ACTH zu dokumentieren.* In streng definierten Versuchsbedingungen (Körperlage, Uhrzeit) wurden vorerst die Spontanvariation des Plasma-Aldosteronspiegels und dann dessen Auslenkung nach Applikation der 3 auch physiologisch einwirkenden Stimuli untersucht.

In Zusammenarbeit mit P.D.Dr. M.B. Vallotton wurden gleichzeitig Bestimmungen der Plasma-Reninaktivität (3) und von Angiotensin II (4) durchgeführt, und in Zusammenhang mit Dres. M. Birkhäuser, A.M. Riondel und A.M. Tissot, Plasmabestimmungen von Corticosteron und Cortisol.

## Kontrollversuche

Die Resultate der Kontrollgruppe zeigen, daß der Plasma-Aldosteronspiegel derselben Person, nach 10-stündiger horizontaler Körperlage, zwischen 7 und

11 Uhr morgens gewissen noch unerklärten Schwankungen unterworfen ist; so
variiert die Plasma-Aldosteron-Konzentration bei 3 Versuchspersonen insgesamt zwischen 3 und 11 ng/100 ml.

Die durch Angiotensin II, Kalium und ACTH induzierten Plasma-Aldosteron-
Veränderungen unterscheiden sich jedoch bezüglich Dynamik und absoluten Werten klar von den Spontanvariationen.


## Stimulierung mit Angiotensin II

Angiotensin II (-Val$^5$-Asp-β-amid), i.v. infundiert über 1-3 Stunden in der
Dosierung von 7 ng/kg/min, bewirkt bei 3 Versuchspersonen eine innerhalb
10 Minuten einsetzende, dann konstante 3-4 fache Erhöhung des Plasma-Aldosterons. Corticosteron und Cortisol bleiben von Angiotensin II unbeeinflußt.

Diese selektive Stimulierung von Aldosteron ist angesichts der kontrover-
sen Diskussion über die Spezifität des Angiotensin II-Stimulus von Interesse.
Eine Dosis-Wirkungs-Beziehung zwischen gemessenen Angiotensin II-Werten und
Aldosteron-Werten läßt sich jedoch kaum aufstellen, da sich die venöse Angio-
tensin II-Konzentration aus verschiedenen Gründen von der Konzentration am
Erfolgsorgan unterscheidet.


## Stimulierung mit Kalium

Kalium-Citrat, verabreicht per os, 30 mEq pro Stunde während 3 Stunden, verursacht bei 3 Versuchspersonen bei unveränderter Natriämie eine variable Erhöhung des Plasma-Kaliums. Unter dieser akuten Kaliumbelastung verzeichnen
wir eine markante, wiederum selektive Erhöhung des Plasma-Aldosterons, entsprechend einem 3-6 fachen Anstieg. Eine erste Plasma-Aldosteronerhöhung
zeichnet sich nach 30-60 Minuten, d. h. nach Einnahme von 30 mEq Kalium ab.

Der minimale, mit einer Aldosteronerhöhung verbundene Plasma-Kalium-
Anstieg läßt sich in unserer Versuchsanordnung nicht genau festlegen, infolge der additiven Kaliumgabe und der zu erwartenden Latenzzeit bis zur Aldosteronstimulierung. Die Beobachtung, daß bei Veränderungen der Kaliumbilanz
im Langzeitversuch eine Aldosteronstimulierung auch bei unveränderter Kaliämie möglich ist (5, 6), schwächt die Bedeutung der Kaliämie als direkte Re-
gelgröße der Aldosteronveränderung ab. Die Erhöhung des Plasma-Kaliumspie-
gels scheint eine hinreichende, aber nicht notwendige Bedingung für die Aldosteronstimulierung zu sein.


## Stimulierung mit adrenocorticotropem Hormon

ACTH, gegeben als einmalige, rasche Injektion von 0.5 mgβ$^{1-24}$ACTH, verursacht
im Kurzzeitversuch eine ausgeprägte Aldosteronstimulierung. Aldosteron, Corticosteron und Cortisol zeigen eine ähnliche Dynamik, gekennzeichnet durch
einen Anstieg innerhalb der ersten 10 Minuten; das Ausmaß der innerhalb 60
Minuten erreichten Stimulation ist jedoch verschieden.

Die gleichzeitige Stimulierung von Aldosteron, Corticosteron und Cortisol
ist in Kenntnis früherer in vitro Studien (7) und neuerer Untersuchungen über
den Angriffspunkt des ACTH (8) nicht überraschend.


## Zusammenfassung

1. Die untersuchten 3 Stimuli, Angiotensin II, Kalium und ACTH verursachen
   bei der gesunden Versuchsperson, bei normaler Kochsalzernährung und in der
   gewählten Dosierung eine Erhöhung des Plasma-Aldosterons zu Mittelwerten
   von 20-25 ng/100 ml, entsprechend einem 3-5 fachen Anstieg.

2. Eine erste Plasma-Aldosteronerhöhung wird sichtbar:
        nach 10 Minuten für i.v. verabreichtes Angiotensin II und ACTH,
        nach 30-60 Minuten für peroral verabreichtes Kalium.
3. Angiotensin II und Kalium sind in der applizierten Dosierung aldosteron-
   spezifische Stimuli, während ACTH auch Corticosteron und Cortisol stimuliert, allerdings mit verschiedener Intensität und Zeitdauer.

Diese Ergebnisse demonstrieren die Dynamik, Spezifität und das Ausmaß der
akuten Aldosteronstimulierung.

Ausgehend von einer normalen Aldosteron-Basalsekretion, d. h. bei normaler
Kochsalzzufuhr, erzwingt eine akute Stimulierung, selbst bei wohl pharmako-
logischer Stimulusgröße, wie im Falle von Angiotensin II und ACTH, einen
Plasma-Aldosteronspiegel von höchstens 20-25 ng/100 ml. Praktisch überein-
stimmende Stimulationswerte finden sich nach Kalium. Dies belegt, daß dieser
weniger bekannte Stimulus, trotz bestimmt submaximaler Dosierung, einen den
beiden anderen Stimuli vergleichbaren Effekt auf die Aldosteronstimulierung
ausübt. Damit ist auch im Kurzzeitversuch bestätigt, daß der Begriff "Renin-
Angiotensin-Aldosteron-System" zu eng gefaßt ist und nicht alle Regelmecha-
nismen des Aldosterons erfaßt.


<u>Literatur</u>


1. Bojesen, E., Thuneberg, L.: In "Steroid hormone analysis" (ed. H. Carsten-
     sen), New York: Marcel Dekker 1967.
2. Scholer, D., Riondel, A.M., Manning, E.L.: in Vorbereitung.
3. Vallotton, M.B.: In "Immunological Methods in Endocrinology". Hormone and
     Metab. Res. Suppl. $\underline{3}$ (1971).
4. Vallotton, N.B., Page, L.B., Haber, E.: Nature (Lond) $\underline{215}$, 714 (1967).
5. Muller, A.F., Veyrat, R., Grandchamp, A.: Klin. Wschr. $\underline{23}$, 1241 (1968).
6. Veyrat, R., Brunner, H.R., Manning, E.L., Muller, A.F.: Helv. med. Acta
     Suppl. $\underline{46}$, 141 (1966).
7. Kaplan, N.M., Bartter, F.C.: J. clin. Invest. 41, 715 (1962).
8. Koritz, S.B., Kumar, A.M.: J. Biol. Chem. $\underline{245}$, 152 (1970).

Symp. Dtsch. Ges. Endokrin. 17, 11–18 (1971)
© by Springer-Verlag

# Testosterone Metabolism in Skin [1,2]

JEAN D. WILSON

Department of Internal Medicine, The University of Texas Southwestern
Medical School, Dallas, Texas, U. S. A.
With 5 Figures

<u>Summary</u>

It appears warranted to draw several tentative conclusions from these studies. First, the reduction of testosterone to dihydrotestosterone by slices of various types of human skin appears to correlate with the capacity for the skin to grow under the influence of androgens, skin from the perineal areas exhibiting the highest rates observed. Second, as the result of studies on dihydrotestosterone formation in skin biopsies from patients with the syndrome of testicular feminization and on the relation between age and the capacity of human prepuce to perform this conversion, it is possible that the ability to reduce testosterone to dihydrotestosterone may be a limiting factor in the androgen-mediated growth of these tissues. Third, the appearance of the enzyme which performs this conversion in the anlage of the perineal skin early in embryonic development appears to fulfill the requirements of an initial androgen receptor, suggesting the possibility that this reaction may be critical to the male differentiation of the genital skin. Fourth, the factors which regulate the level of the enzyme for dihydrotestosterone formation have not yet been identified. Finally, preliminary evidence suggests that dihydrotestosterone formation may also be involved in the androgen-mediated growth of sebaceous glands and hair.

This working hypothesis, namely that dihydrotestosterone may be involved critically in the androgen-stimulated growth and differentiation of human skin as well as in the accessory organs of reproduction, leaves a number of questions concerning androgen action unanswered. It does not provide an explanation for testosterone-enhanced growth of muscle or for testosterone action in tissues which lose the ability to form dihydrotestosterone with age (6, 11). The hypothesis does have considerable potential explanatory value, however, for the elucidation of testosterone action in skin and in other tissues both in normal state and in a variety of pathological conditions.

During recent years considerable data has accrued to indicate that the $5\alpha$ reduction of testosterone to 17β-hydroxy-5α-androstan-3-one (dihydrotestosterone) within certain androgen target tissues may play a crucial role in mediating some effects of the hormone (Fig. 1). Although this testosterone metabolite was observed in prostate as early as 1963 (1–3), support for a physiological role for dihydrotestosterone in hormone action has been accumu-

---

[1] This study has been supported in part by a grant (A3892) from the National Institutes of Health.
[2] Work performed in part during the tenure of a Career Development Award of the National Institutes of Health.

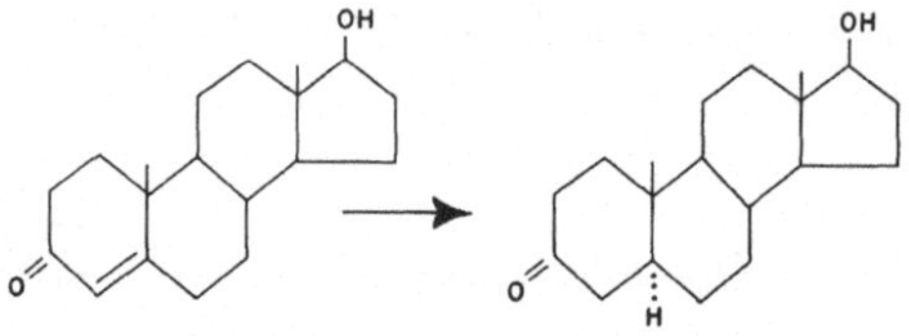

Fig. 1. The principal metabolites recovered in skin following incubation of slices of prepuce with testosterone-1,2-$^3$H.

lated more slowly as the result of a variety of different studies in this laboratory and by others. This evidence can be summarized as follows.

1. Although enzymes which perform this physiologically irreversible reduction are known to exist in other tissues such as rat liver (4), dihydrotestosterone is a principal metabolite of testosterone-1,2-$^3$H only in tissues which are major sites of testosterone action, such as the male organs of accessory reproduction (5, 6).
2. In these same target tissues it is dihydrotestosterone rather than testosterone itself which is the major steroid bound to the nuclei (7-9), a presumed site of action for the hormone (10).
3. In the prostate glands of a number of animal species, the rate of testosterone reduction correlates with the growth response of the tissue to testosterone, those prostate glands in which unregulated growth eventuates in benign prostatic hypertrophy (man and dog) exhibiting high rates of dihydrotestosterone formation throughout life (6, 11-13).
4. Both in short term *in vivo* bioassays (14-16) and in *in vitro* organ cultures of prostate (17) dihydrotestosterone appears to be more potent than testosterone as measured by the growth or maintenance of the accessory sex tissues, and even in long term studies dihydrotestosterone appears to be the more active hormone for prostatic growth in the dog (13).

Taken together, these various types of evidence suggest that dihydrotestosterone may be the active form of testosterone for certain intracellular effects of the hormone in a variety of tissues and in several species.

The skin is also a major testosterone target tissue. In man, for example, the growth of the perineal skin - the scrotum, penis, clitoris, and labia - and of at least two skin organelles, the sebaceous glands and hair in some regions of the body, is regulated by androgens. It was first observed by Gomez and Hsia that radioactive testosterone could be converted to dihydrotestosterone by minces of skin (17), and subsequently, a series of studies were undertaken in this laboratory to investigate the relationship between dihydrotestosterone formation in skin and testosterone action.

Six questions were posed in these experiments. First, do regional rates of dihydrotestosterone formation in skin correlate with the known ability of skin to grow under the influence of androgens? Second, is there any age correlation between the ability to form dihydrotestosterone and the capacity of skin to grow under the influence of testosterone? Third, could a defect in dihydrotestosterone formation provide insight into the mechanisms of the known resistance to testosterone action which exists in the syndrome of testicular feminization? Fourth, what are the characteristics of the enzyme which performs this reduction of testosterone in the accessory sex tissues? Fifth, at an embryological level is the ability of the perineal skin appendages to form dihydrotestosterone a cause or the result of the early androgen-mediated differentiation of the external genitalia? And finally, can an analysis of dihydrotestosterone formation provide insight into the androgen-enhanced growth of skin organelles such as sebaceous glands and hair?

For this purpose, the optimal conditions for assaying this reduction in slices of human skin were established, and the formation of dihydrotestoster-

one-$^3$H at an optimal substrate concentration of testosterone-1,2-$^3$H
(1 X 10$^{-6}$ M) was measured in skin biopsies obtained from various anatomical
sites in 118 normal individuals of both sexes and varying ages, in four patients with the syndrome of testicular feminization, and in one subject with
Reifenstein's syndrome (Fig. 2) (18).

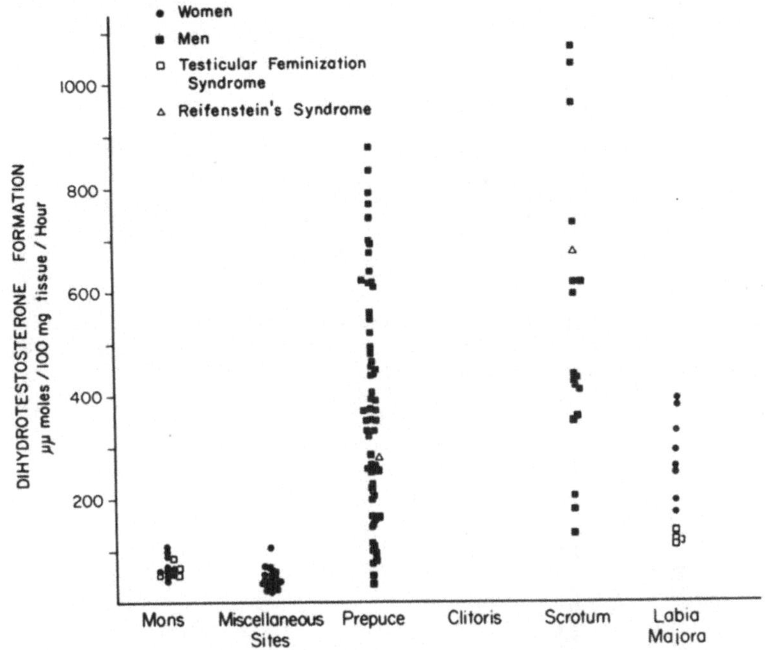


Fig. 2. Comparison of the rates of formation of dihydrotestosterone by slices of skin obtained from various anatomical sites in 118 normal men
and women, in four patients with the syndrome of testicular feminization, and one subject with Reifenstein's syndrome. Reprinted in
part from Wilson and Walker (18).

Under these conditions of assay the rate of dihydrotestosterone formation
averaged 70 μμmoles/hour in the 10 samples of control tissue obtained from
the mons veneris and 50 μμmoles/hour in the specimens of skin labeled miscellaneous sites (thigh, breast, back, inguinal area, sole of the foot,
upper abdominal wall, and chest wall). The values obtained in the samples
of perineal skin were strikingly different. The prepuce, which exhibited a
wide range of values (from 50 to 880) had a mean rate of 380 μμmoles/hour.
The single specimen of clitoris fell within this range (350). The samples of
scrotum also exhibited a wide variation with a mean of 530 μμmoles/hour, and
the labia majora exhibited slightly lower rates of dihydrotestosterone formation, averaging 280 μμmoles/hour. Thus, on an average, the rates of conversion of testosterone to dihydrotestosterone were considerably higher in
the four types of perineal skin than in the samples from the mons.
    Nevertheless, the range of variation in the prepuce, scrotum, and labia
was so great that individual samples fell within or near the normal range,
and consequently the data in Fig. 2 were plotted as a function of the age
of the patient. In the case of the mons, miscellaneous sites, scrotum, and
labia majora there was a uniform scatter of the data with age (Fig. 3). With
prepuce, however, a very interesting relation was observed between the rate
of dihydrotestosterone formation and the age of the patient (Fig. 4). In the
newborn infant (up to 5 days of age) the rate averaged 410 μμmoles/hour. In
the samples obtained from subjects between 10 days and 3 months of age the

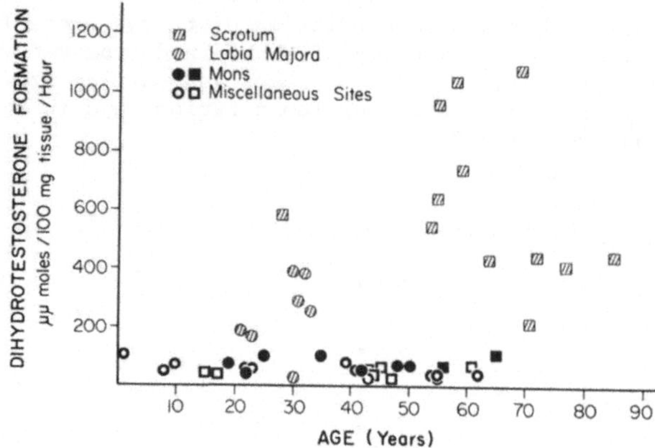

Fig. 3. Relation between patient age and dihydrotestosterone formation by
slices of skin obtained from the scrotum, the mons veneris, and from
a variety of miscellaneous sites.

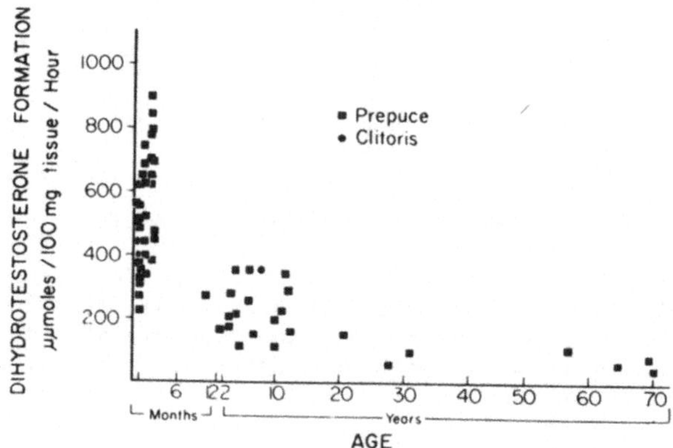

Fig. 4. Relation between patient age and dihydrotestosterone formation by
slices of prepuce and clitoris. Reprinted from Wilson and Walker (18).

mean rate rose slightly to 640. In the next age group (3-12 years) this value
fell to 220, and in the specimens from older patients the conversion rate
was only 90 μμmoles/100 mg per hour, a value not significantly different than
observed for mons veneris. To summarize these findings, the high rates of
dihydrotestosterone formation in the types of perineal skin that are known to
be under the trophic control of androgens is in keeping with the possibility
that the ability to form dihydrotestosterone may be related causally to the
capacity of these tissues to grow in response to testosterone. The apparent
decrease with age in the rate of dihydrotestosterone formation in the prepuce
also deserves comment. The ability of the penis to grow in response to testosterone is a limited one in that the penis ceases to grow at the end of puberty, and in adult, normal men testosterone administration even in pharmaco-
logical doses causes no further growth (19). It is tempting to speculate that
this phenomenon might in part be the consequence of a decrease in the ability
of the tissue to convert the hormone to a locally active derivative with
age.

14

<u>Dihydrotestosterone Formation in the Syndrome of Testicular Feminization</u>

The results obtained in studies of skin from four patients with the syndrome
of testicular feminization and one subject with Reifenstein's syndrome are
also shown in Fig. 2 (18). The four patients with testicular feminization
varied widely in age (2, 17, 44, and 56 years); each had XY karyotypes and
typical clinical findings for the diagnosis, and three had affected siblings
with the same disorder. The rates of formation of dihydrotestosterone in the
samples of skin obtained from the mons veneris were clearly within the normal range. However, the rates observed in the skin from the labia majora
were lower than the average values for normal labia majora and considerably
lower than the mean rates observed in the samples of scrotum.
   Similar results in skin from patients with testicular feminization were
reported almost simultaneously by Heinrichs et al. (20) and by Northcutt and
co-workers (21), and Mauvais-Jarvis has reported that the excretion of
5α-reduced metabolites of testosterone is decreased in this condition (22).
In contrast, in the patient with Reifenstein's syndrome (age 9) a disorder
which is inherited in an identical fashion to the testicular feminization
syndrome and in which in affected men there is a partial defect in viriliza-
tion (23), dihydrotestosterone formation in prepuce and scrotum was clearly
within the range of normal (24).
   There is now a convincing body of evidence that some peripheral defect in
androgen action must occur in the syndrome of testicular feminization (23).
Whether such a defect could be due to a deficient conversion of testosterone
to dihydrotestosterone at the time of puberty is unclear at present, for
while the values observed in the labia majora and the Wolffian duct derivatives (21) are low this deficiency might be the passive consequence of some
other defect in testosterone metabolism such as nuclear binding of the steroid. In adult patients with the disorder in whom there is a paucity of
tissues capable of androgenic response, it has been reported that the administration of dihydrotestosterone does not produce a positive nitrogen
balance (25).


<u>Characterization of the 5α-Reductase Enzyme System</u>

The characteristics of 5α-reductase enzyme system which converts testosterone
to dihydrotestosterone have been studied in rat liver by a number of investigators (4, 26-32), in skin by Voigt, Fernandez, and Hsia (33) and in the
male organs of accessory reproduction by this laboratory (5, 11, 34). In
these various systems, 5α-reductase enzyme activity shares a number of characteristics. There is an absolute requirement for NADPH as cofactor for the
reaction. The reduction is apparently irreversible under physiological conditions, and the enzyme has not yet been solubilized but is bound tightly to
subcellular particles - principally the endoplasmic reticulum in the case of
liver and skin with a variable amount of activity in the nuclei in the male
organs of accessory reproduction studied. In two regards, however, the enzyme in liver differs from that in the testosterone-target tissues. As the
result of studies on the endoplasmic reticulum from human prepuce by Voigt,
Fernandez, and Hsia (33) and on the nuclei and endoplasmic reticulum of rat
prostate by Frederiksen and Wilson (34) it has been concluded that the
enzyme in these tissues has an apparent Km of/around $10^{-6}$ rather than $10^{-4}$
as reported for liver (26). In addition, the enzyme in human skin and rat
prostate has a much narrower substrate specificity than does the liver
enzyme (33, 34); for example, steroids with substitutions on the eleven car-
bon (such as hydrocortisone) that are metabolized *in vivo* and *in vitro* via
5α-reduction in the liver are impotent as substrates for the skin and prostatic preparations, and as the result of $K_m$ and $K_i$ studies it has been concluded that there is probably only one 5α-reductase enzyme in the prostate
whereas several such enzymes may exist in liver (34). Furthermore, the metabolic regulation of hepatic 5α-reductase enzyme(s) has been investigated in

considerable detail (26, 27, 31, 32); it is subject to regulation by thyroid
hormones, by several glucocorticoid hormones, and by drugs. As yet, however,
neither the factors which are responsible for the age-related changes nor
the remarkable variations among species in the enzymatic activity within the
testosterone-target tissues have been elucidated.

## Dihydrotestosterone Formation in Fetal Tissues

Therefore, the question was posed as to whether dihydrotestosterone formation in the testosterone-target tissue is involved in the initial action of
testosterone in the embryo, namely the promotion of the differentiation of
the urogenital sinus, Wolffian ducts, and urogenital tubercle into the male
external genitalia and internal organs of accessory reproduction or whether
the ability to convert testosterone to dihydrotestosterone is the passive
consequence of some earlier testosterone effect involved in the differentiation process. To examine this problem Wilson and Lasnitzki measured the rate
of formation of dihydrotestosterone-1,2-$^3$H in tissue slices from rabbit and
rat embryos at different stages of development under circumstances in which
the concentration of the substrate testosterone-1,2-$^3$H was in the physiological range (3.5 X 10$^{-8}$ M) (35). In the earliest stages studied in both the
rabbit and rat embryo dihydrotestosterone formation was rapid only in the
urogenital sinus and urogenital tubercle.

A similar study of dihydrotestosterone formation in tissues from the
guinea pig embryo prior to and after the completion of sexual differentiation is shown in Fig. 5. In this study even at the indifferent stage

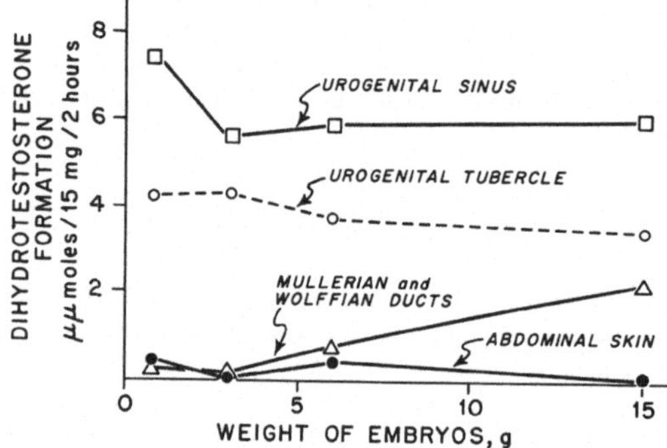


Fig. 5. Dihydrotestosterone formation by tissue slices from guinea pig embryos
as a function of the age of the embryo. Each point represents the
mean value of 2-4 experiments. The incubation conditions and
methods are described in Ref. 35.

(1.5 gm) prior to demonstrable differentiation of the internal and external
genitalia, the rate of dihydrotestosterone formation was higher in the genital tubercle and the urogenital sinus than in any of the other 17 tissues
examined. These findings suggest that the capacity to form dihydrotestosterone in these tissues (and ultimately in the prostate gland and perineal
skin of the adult) is not the result of androgen action but rather an early
property of the tissue and may fulfill in part the function of an initial
androgen receptor in these organs. In contrast, in the Mullerian and
Wolffian ducts the ability to form dihydrotestosterone appears to be ac-
quired after the initial stages of differentiation and may actually be the
result of hormonal action.

<u>Dihydrotestosterone Formation in Sebaceous Glands and Hair Follicles</u>

Testosterone metabolism in the two skin organelles which are major sites of testosterone action - the sebaceous glands and the hair follicles - has been investigated less extensively than the skin itself because of inherent difficulties in obtaining sufficient tissues for quantitative studies. However, in specialized sebaceous glands of several species (mouse and rat preputial glands, the coccygeal gland of the chicken, and the duck preen gland) dihydrotestosterone formation from testosterone-1,2-$^3$H is very active (6), and it is reasonable to infer that this conversion may also be involved in the trophic regulation of intracutaneous sebaceous glands. In regard to hair, Northcutt, Island, and Liddle have reported that hair follicles from the pubic region of normal man are among the most active tissues in the body in the conversion of testosterone to dihydrotestosterone on a per weight basis (21), and as the result of studies in isolated human scalp hair follicles, Adachi and Karo have suggested that the intracellular level of dihydrotestosterone may control the initial steps in the development of male pattern baldness (36). Clearly, the possibility deserves exploration that the androgen-mediated control of sebum formation in man and regional differences in hair growth are regulated by the local formation of dihydrotestosterone within these skin appendages.


<u>References</u>


1. Shimazaki, J., Kurihara, H., Yoshikazu, I., Shida, K.: Gunma J. med. Sci. 14, 313 (1965).
2. Farnsworth, W.E., Brown, J.R.: J. Amer. med. Ass. 183, 436 (1963).
3. Chamberlain, J., Jagarinec, N., Ofner, P.: Biochem. J. 99, 610 (1966).
4. McGuire, J.S., Jr., Tomkins, G.M.: J. Biol. Chem. 234, 791 (1959).
5. Bruchovsky, N., Wilson, J.D.: J. Biol. Chem. 243, 2012 (1968).
6. Gloyna, R.E., Wilson, J.D.: J. clin. Endocr. 29, 970 (1969).
7. Bruchovsky, N., Wilson, J.D.: J. Biol. Chem. 243, 5953 (1968).
8. Anderson, K.M., Liao, S.: Nature (Lond.) 219, 277 (1968).
9. Mainwaring, W.I.P.: J. Endocr. 44, 323 (1969).
10. Wilson, J.D., Loeb, P.M.: In: Developmental and Metabolic Control Mechanisms and Neoplasia (M. D. Anderson Hospital and Tumor Institute), Baltimore: Williams and Wilkins Co., 1965, p. 375.
11. - Gloyna, R.E.: Recent Progr. Hormone Res. 26, 309 (1970).
12. Siiteri, P.K., Wilson, J.D.: J. clin. Invest. 49, 1747 (1970).
13. Gloyna, R.E., Siiteri, P.K., Wilson, J.D.: J. clin. Invest. 49, 1746 (1970).
14. Dorfman, R.I., Shipley, R.A.: Androgens, Biochemistry, Physiology and Clinical Significance, New York: Wiley and Sons, 1956, p. 118.
15. Saunders, F.J.: Nat. Cancer Inst. Monogr. 12, 139 (1963).
16. Hilgar, A.G., Hummel, D.J.: In E.P. Vollmer (Editor), Androgenic and Myogenic Endocrine Bioassay Data, Part 11, National Cancer Institute, Bethesda, 1964, p. 15.
17. Gomez, E.C., Hsia, S.L.: Biochemistry 7, 24 (1968).
18. Wilson, J.D., Walker, J.D.: J. clin. Invest. 48, 371 (1969).
19. Turner, H.H.: The Clinical Use of Testosterone, W.O. Thompson (Editor), Springfield: Thomas, 1950.
20. Heinrichs, W.L., Karsznia, R., Wyss, R., Hermann, W.L.: Clin. Res. 17, 143 (1969).
21. Northcutt, R.C., Island, D.P., Liddle, G.W.: J. Clin. Endocr. 29, 422 (1969).
22. Mauvais-Jarvis, P., Bercovici, J.P., Crepy, O., Gauthier, F.: J. clin. Invest. 49, 31 (1970).
23. Federman, D.D.: Abnormal Sexual Development, Philadelphia, Saunders, 1967.
24. Wieland, R.G., Chen, J.C., Chambers, D.: Clin. Res. 18, 605 (1970).
25. Strickland, A.L., French, F.S.: J. clin. Endocr. 29, 1284 (1969).

26. McGuire, J.S., Jr., Tomkins, G.M.: J. biol. Chem. 235, 1634 (1960).
27. – Hollis, V.W., Jr., Tomkins, G.M.: J. biol. Chem. 235, 3112 (1960).
28. – Tomkins, G.M.: Fed. Proc. 19, 29 (1960).
29. Wilton, D.C., Ringold, H.J.: Proceedings of the Third International
    Congress on Endocrinology, Mexico City, 1968, p. 105.
30. Björkhem, I.: Europ. J. Biochem. 8, 345 (1969).
31. Shriefers, H.: Vitam. and Horm. 25, 271 (1967).
32. – Advanc. Biosci. 2, 69 (1969).
33. Voigt, W., Fernandez, E.P., Hsia, S.L.: J. biol. Chem. 245, 5594 (1970).
34. Frederiksen, D.W., Wilson, J.D.: J. biol. Chem. 246, 2584 (1971).
35. Wilson, J.D., Lasnitzki, I.: Endocrin. 89, 659 (1971).
36. Adachi, K., Kano, M.: Biochem. biophys. Res. Commun. 41, 884 (1970).

Symp. Dtsch. Ges. Endokrin. 17, 19–32 (1971)
© by Springer-Verlag

# The Effects of Steroid Hormones on the Skin of Experimental Animals

F. J. EBLING

Department of Zoology, The University, Sheffield S10 2TN, U. K.
With 8 Figures

## Introduction

Almost every component of skin – superficial epidermis, melanocytes, hair
follicles, sebaceous glands and various dermal elements – has been shown to
be influenced by steroid hormones. Some of the evidence is, however, con-
flicting. For example, while it is generally agreed that testosterone stimulates epidermal mitosis (Allen, 1957, 1958; Bullough and van Oordt, 1950;
Ebling, 1957) and accelerates the healing of wounds, (Dyson and Joseph,
1968), there have been differences of opinion about whether oestrogens
affect epidermal cell division (Bullough, 1953; Carter, 1953; Gelfant, 1960).
On the other hand, while oestrogens increase skin pigmentation in some experimental animals (Bischitz and Snell, 1960; Snell and Bischitz, 1960) there
is some doubt about the action of testosterone. It appears that, in guinea
pigs, melanogenesis can be stimulated by castration (Snell and Bischitz,
1959) whereas, in some experimental animals, androgens appear to stimulate
pigmentation (Kuppermann, 1944; Wells, 1945). Steroids may act in several
different ways on the hair follicle. In rats, oestrogens not only delay the
onset of follicular activity which initiates the moult, but they reduce the
rate of growth of the hairs and inhibit the shedding of the club hairs
(Ebling and Johnson, 1964a, b; Johnson, 1958). Similar effects have been
described in the mouse (Jensen, 1958) and in the guinea pig (Jackson and
Ebling, 1970). These facts, most of which have been more extensively reviewed elsewhere (Ebling, 1964) serve to emphasise the importance of the skin
as a steroid target. But it is the sebaceous glands which provide the most
clear cut evidence of hormonal influence; indeed, they appear to be entirely
under hormonal control. We have a much clearer picture of how they react to,
and probably transform, steroids than for any other cutaneous derivative.
This paper will be devoted mainly to them.

## Sebaceous glands

The sebaceous glands are epidermal in origin, and usually open into the
pilary canals. They are holocrine; that is to say their secretion is formed
by complete disintegration of the glandular cells which are replaced by cell
division at the periphery of the gland. In the rat there are normally one or
two lobes. All mammals, except the Cetacea (whales), have sebaceous glands,
and many species have aggregated glands of sebaceous origin, such as the
preputial glands of the rat and the supracaudal gland of the guinea pig. The
hormonal responses of these specialized glands are not necessarily identical
with those of the cutaneous sebaceous glands.

## Methods for study of sebaceous glands

The earlier studies upon the sebaceous glands of rodents involved measurements of gland size. Apart from the time-consuming nature of such procedures, the gland size is not necessarily an indication of glandular activity, for it depends not only on the output of cells but on the time they take to move through the gland (Ebling, 1963). However, two other measures of glandular activity, namely the incidence of mitosis and the rate of secretion of sebum (Ebling and Skinner, 1967), have proved particularly useful in analysing the effect of steroid hormones.

The mitotic rate can be determined by injecting the rats intraperitoneally with 0.1 mg colchicine in water/100 g body weight 5 hours before the rats are killed; this treatment arrests mitosis in the metaphase. As the incidence of mitosis varies diurnally the rats are killed at a standard time, conveniently at 15.00 hours (Ebling, 1954). The number of arrested mitoses in a given length of skin section is a direct proportional measure of the mean rate of production of cells. However, a high rate of cell division does not necessarily produce a high level of secretion, since the production of sebum depends also on the rate of intracellular synthesis (Ebling, 1957 b). A comparison of the rates of secretion and of mitosis can be a useful way of distinguishing the mode of action of different steroids.

Sebum production has been measured in two ways. One is to dip the whole animal in lipid solvents (Archibald and Shuster, 1967, 1970; Nikkari and Valavaara, 1969). Another is to estimate the changes in hair fat levels of washed animals (Ebling and Skinner, 1967). In this technique the animals are washed in warm water and sodium lauryl sulphate, and dried with a hair dryer. About a gram of hair is then immediately clipped from the left flank, weighed and extracted with five 50 ml portions of diethyl ether; the extracts from each sample are combined in tared aluminium cups, evaporated to dryness on a hot plate, and weighed. The level of hair fat is expressed in milligrams/gram of hair. 8 days later another sample is clipped from the right flank of each animal and similarly extracted. Sebum production may thus be expressed as the increase in hair fat in milligrams/gram of hair/day.

Steroids also affect the composition of sebum; under stimulation by testosterone the ratios of palmitate:stearate and of oleate:stearate are increased (Wilde and Ebling, 1969). As these ratios appear to vary in proportion to the rate of sebum production, they have been proposed as measures of glandular activity (Nikkari and Valavaara, 1970 a).

## Effect of testosterone

Testosterone increases sebum production in rats, whether it is measured by the total immersion technique (Archibald and Shuster, 1967) or by changes in hair fat levels (Ebling and Skinner, 1967). Castrated male rats treated for 24 days with implants of testosterone giving an uptake of 0.2 mg/day showed an increase of about 2.5 times (Ebling and Skinner, 1967) and even greater effects have been shown in spayed females (Ebling, 1967). The action of testosterone involves an increase in mitotic rate (Fig. 1). These findings confirm earlier experiments on the effect of testosterone on the size of the sebaceous glands (Ebling, 1948; 1957 a, b). Enlargement of the sebaceous glands by treatment with testosterone has been shown in many other species, including rabbits (Montagna and Kenyon, 1949), hamsters (Hamilton and Montagna, 1950) and mice (Lapière, 1953). Specialized sebaceous homologues such as the preputial glands (Huggins, Parson and Jensen, 1955; Korenschevsky and Dennison, 1934, 1936), the supracaudal gland of the guinea pig (Martan, 1962; Martan and Price, 1967), and the abdominal sebaceous gland pad of the gerbil (Glenn and Gray, 1965) react in a similar fashion.

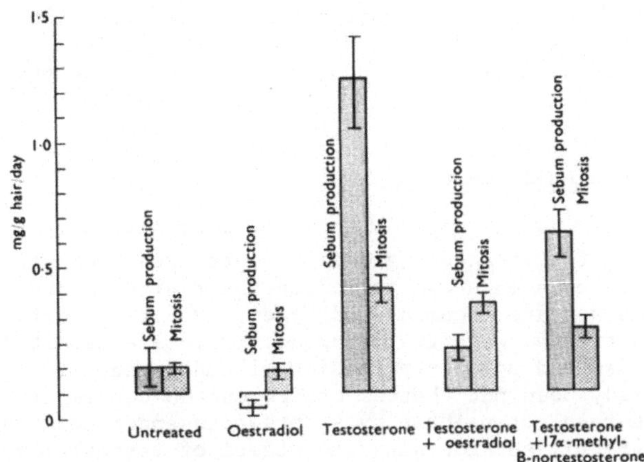


Fig. 1. Sebum production (left hand columns) in female rats, expressed as
increase in hair fat/g. hair/24h measured over an 8-day period after
washing. Proportional levels of the incidence of mitoses are shown
in the right hand columns. Means and S.E. are for groups of 10 rats.
Data from Ebling, 1967.

## Effect of progestogens

Some disagreements have been expressed about the possible effects of progesterone. Ebling (1948) could find no evidence that the size of the sebaceous
glands in intact immature female rats was affected by a dose of 0.5 mg progesterone daily for 20 days. However, Haskin, Lasher and Rothman (1953) and
Lasher, Lorincz and Rothman (1954) put the view that its action was comparable with that of testosterone. Ebling (1961) was unable to show any effects
of progesterone on gland size or on sebaceous mitosis in a range of experimental conditions, and Groot, Lely and Kooij (1965) repeated these experiments with similar results. However, they believed that 10 mg/day of progesterone produced a significant effect on the size of the sebaceous glands,
as well as enlarging the preputial glands and prostate, though not the seminal vesicles. Recent studies on sebum production have failed to reveal any
effect of progesterone even when given in doses of 10 mg/day for 24 days
(Ebling, Ebling and Skinner, 1969 b).

## Effect of oestrogens

Oestrogens inhibit sebaceous activity (Fig. 1). Implants of 1:9 oestradiol/
cholesterol giving an absorption of 2 - 4 µg oestradiol/day significantly
reduced sebum secretion in spayed female (Ebling, 1967) or castrated male
rats (Ebling and Skinner, 1967), whether or not exogenous androgen was given
at the same time. These findings confirm conclusions originally based on
measurements of gland size (Ebling, 1957 b).

It has been proposed that in man oestrogens may act systemically by
suppressing the secretion of endogenous androgens (Strauss and Pochi, 1963).
Most of the evidence from animal experiments supports the view that oestrogens act peripherally on the sebaceous gland, though not necessarily at the
same sites as testosterone. For example, if oestrogens act by suppressing
endogenous androgen, a significant reduction in mitotic activity in the sebaceous glands must occur, and this does not appear to be so (Ebling, 1957 b,
1967). Moreover, the effect of oestrogen can be demonstrated against that of
exogenous androgen; oestradiol suppresses sebum production to a greater extent in castrated male rats or spayed female rats treated with testosterone

than in rats without exogenous testosterone (Ebling, 1967; Ebling and Skinner, 1967). Finally the action of oestrogens has been shown in adrenalectomized female rats which presumably are without sources of endogenous androgen (Ebling, 1970 a).

## Action of anti-androgenic steroids

Non-oestrogenic steroids which antagonise the action of androgens might be expected to inhibit the sebaceous glands. A-norprogesterone (Lerner, Bianchi and Borman, 1960), has been shown to reduce their size (Lerner, 1964), and so has cyproterone acetate (Neumann and Elger, 1966). The effect was shown in castrated mice treated with testosterone propionate, so it clearly involved an antagonism and not a suppression of endogenous androgen production. The view has already been put that as oestrogens do not significantly inhibit sebaceous mitosis they act neither by suppressing endogenous androgen production nor by directly antagonising the effect of testosterone on mitosis. It would appear to follow that the action of a non-oestrogenic anti-androgenic steroid should be distinguishable from that of an oestrogen. Both compounds ought to reduce sebum production, but whereas oestrogens would inhibit intracellular synthesis with little or no effect on cell division, anti-androgens ought to reduce cell division.

This hypothesis was first tested using 17α-methyl-B-nortestosterone (Ebling, 1967) (Fig. 1). It is clear that sebum production of spayed female rats treated with exogenous testosterone is significantly reduced by either oestradiol or 17α-methyl-B-nortestosterone. But whereas oestradiol had no significant effect on cell division, the anti-androgenic steroids reduced it in proportion to the inhibition of sebum production.

A 6-cyclopropyl substituent was found to increase the anti-androgenic activity of 17α-methyl-B-nortestosterone by about 2.6 times as assayed in rats by inhibition of seminal vesicle weight, or about 1.5 times when measured by inhibition of androgen stimulated chick comb growth (Saunders and Ebling, 1969). This compound, 6α-6β ethylene 17α-methyl-B-nortestosterone had a significant oestrogenic activity as measured by ability to cause vaginal cornification and uterine growth in ovariectomized female rats. When tested in rats it had a similar potency to 17α-methyl-B-nortestosterone in suppressing sebum production, but this appeared to be entirely due to suppression of mitosis. i.e. its anti-androgenicity as distinct from its oestrogenicity. Δ-1 chlormadinone acetate (Ebling, 1970 b) and cyproterone acetate have also been assayed by this procedure.

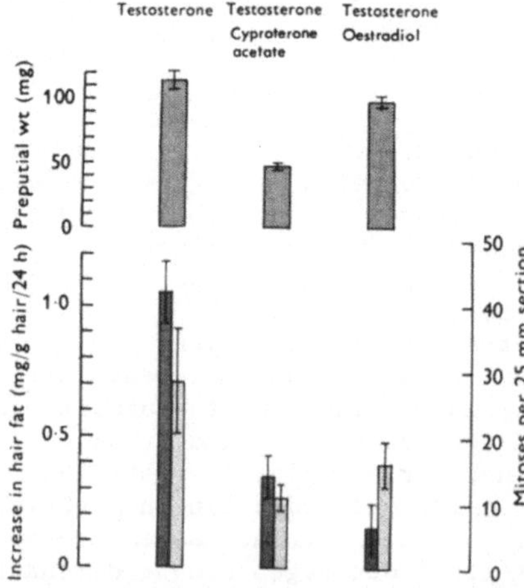


Fig. 2. Effects of cyproterone acetate and of oestradiol on preputial and sebaceous glands of testosterone treated female rats. Means and standard errors for 12 rats.

The effects of cyproterone acetate were compared with those of oestradiol
in an experiment which is illustrated in Fig. 2. Cyproterone acetate
(5 mg/24 h), given to rats treated with testosterone (0.2 mg/24 h) very significantly reduced sebum production, the incidence of sebaceous mitoses and
preputial gland weight. Oestradiol (2 - 4 μg/24 h), given with testosterone,
had an even greater effect on sebum production, but was without comparable
effects on mitosis or on preputial gland weight.


## Role of the pituitary

The demonstration that the sebaceous glands of the rat only respond to testosterone in the presence of the pituitary preceded any attempt to measure
sebaceous secretion (Ebling, 1957 a; Lasher, Lorincz and Rothman, 1955). The
existence of an anterior pituitary hormone with tropic activity for sebaceous, preputial and Harderian glands was claimed (Lorincz and Lancaster,
1957), although further purification of such a substance was not achieved
until much later (Woodbury, Lorincz and Ortega, 1965 a, b). The assay used
by these workers was, however, the response of the preputial gland to progesterone. Since the sebaceous glands do not appear to react to progesterone
irrespective of whether the pituitary is present, and since the preputial
glands show a considerable response to testosterone even in the absence of
the pituitary, it seems possible that these authors had embarked on a search
for a "preputiotrophic" hormone which might not necessarily have "sebotrophic" activity.

The invention of methods for estimating sebum production in the rat has
made possible further studies of the role of pituitary hormones. It has been
shown that hypophysectomy reduces the level of sebum production below that
of the castrated rat (Ebling, Ebling and Skinner, 1969 a). But whereas testosterone increases sebum production in castrated rats, it has an insignificant
effect in hypophysectomized-castrated animals. Even fairly large doses of
testosterone, of the order of 0.6 mg/day, have failed to produce significant
increases in sebaceous secretion (Ebling, Ebling, Skinner and White, 1970 b).
However, the view that the sebaceous glands of hypophysectomized rats are
unresponsive to testosterone has been questioned. Nikkari and Valavaara
(1969) treated hypophysectomized female rats with testosterone propionate
and estimated the skin lipids by washing the whole animal at 2 day intervals
with acetone. Although their conclusion that 0.2 mg/day of testosterone propionate had an effect on sebum secretion may be questionable, it is difficult not to accept their claim that a dose of 1 mg/day produced an increase
in skin surface lipid, and they have recently (Nikkari and Valavaara,
1970 a) again obtained a significant increase in sebum production when 1 mg
of testosterone propionate per 4 days was given to 12 week-old hypophysectomized female rats. Thody and Shuster (1970) using similar methods have also
claimed that the rat sebaceous gland remains sensitive to testosterone after
hypophysectomy. There is probably little point in protracted debate over
this issue. The important question is not whether the sebaceous glands of
hypophysectomized animals will react at all to testosterone, but whether
they show as great a response as in castrated animals with intact pituitaries. Neither Nikkari and Valavaara (1970 a) nor Thody and Shuster (1970)
provide evidence that the responses of hypophysectomized rats and rats with
intact pituitaries are identical.

If, therefore, we accept the view that the pituitary is necessary for the
sebaceous glands fully to respond to testosterone, the question arises of
what pituitary hormones are concerned. A sebotropic preparation made according to the method of Woodbury et al. (1965 b) proved completely incapable
of restoring the response of hypophysectomized male rats to testosterone
(Ebling et al. 1969 a). On the other hand it has been possible to restore
this response (Fig. 3) with a growth hormone preparation (Squibb Lot No.
GH3), a prolactin preparation  (Squibb ovineluteotrophin Lot 53273 - 002R),
which appeared to be free of somatotrophic activity (Ebling et al. 1969 a),
and a thyrotrophic hormone preparation (Ferring SF 1822), with a very slight

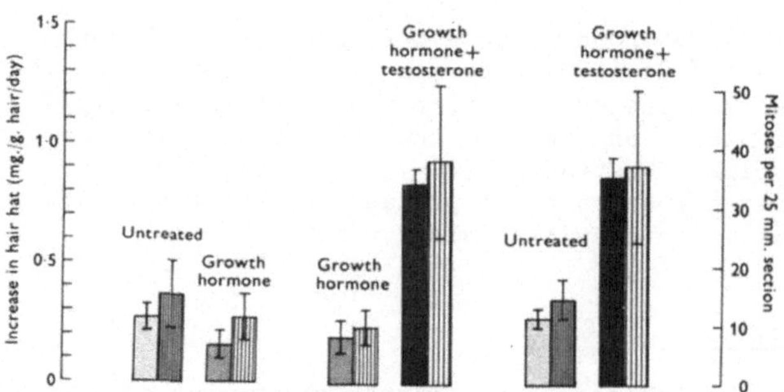

Fig. 3. Effect of growth hormone plus testosterone on secretion of sebum
(left hand columns) and mitosis in the sebaceous glands (right hand
columns) in hypophysectomized-castrated rats. Comparisons between
9, 13 and 12 pairs of litter-mates respectively. The vertical lines
indicate the S. E. From Ebling et al. (1969) by permission of the
Journal of Endocrinology.


growth hormone activity (Ebling, Ebling and Skinner, 1970 a). Since the
growth hormone preparation was incapable of restoring the response of the
preputial gland to progesterone (Ebling et al. 1969 b), its action appeared
to be distinct from the 'sebotrophic' activity as defined by Woodbury et al.
(1965 a, b). So far, we do not know whether the hormone preparations used by
us contain a true sebotrophic (as distinct from preputiotrophic) factor as
an impurity, or whether they contain sebotrophic amino acid sequences within
their molecules.


## Effect of adrenocorticotrophic hormone

It seemed appropriate to discover whether adrenocorticotrophic hormone (ACTH),
a peptide of known constitution, could facilitate the response of the sebaceous glands to testosterone, in common with somatotrophin, prolactin and
thyrotrophin. In fact, ACTH proved incapable of exercising such a permissive
action (Ebling et al. 1970 b). But the experiments produced the interesting
result that, given alone, ACTH was capable of increasing both sebum secretion and the incidence of mitoses in the sebaceous glands of hypophysectomized-castrated male rats (Fig. 4). However, no significant changes in the
weight of the preputial glands or the seminal vesicles were produced. It
seemed a reasonable assumption that the administration of ACTH to hypophysectomized-castrated rats might restore sebaceous activity to the level of
castrates by stimulating the production of adrenal steroids of low androgenicity to which the sebaceous gland might be sensitive. What made this hypothesis difficult to accept was the fact that the potent androgen, testosterone, was by itself incapable of doing this, unless a pituitary hormone
other than ACTH was present.
   The dilemma can only be solved in one way, namely by the assumption that
whereas the sebaceous gland of the hypophysectomized rat cannot fully uti-
lize and respond to testosterone, it can respond to other androgenic steroids even in the absence of the pituitary factor.


## Action of other androgenic steroids

To test the above hypothesis the action of various steroids on sebaceous
secretion in castrated and hypophysectomized-castrated rats has been studied.
   The results of an experiment on castrated rats are shown in Fig. 5. It
can be seen that 5α-dihydrotestosterone in a dose of 0.2 mg/day had an

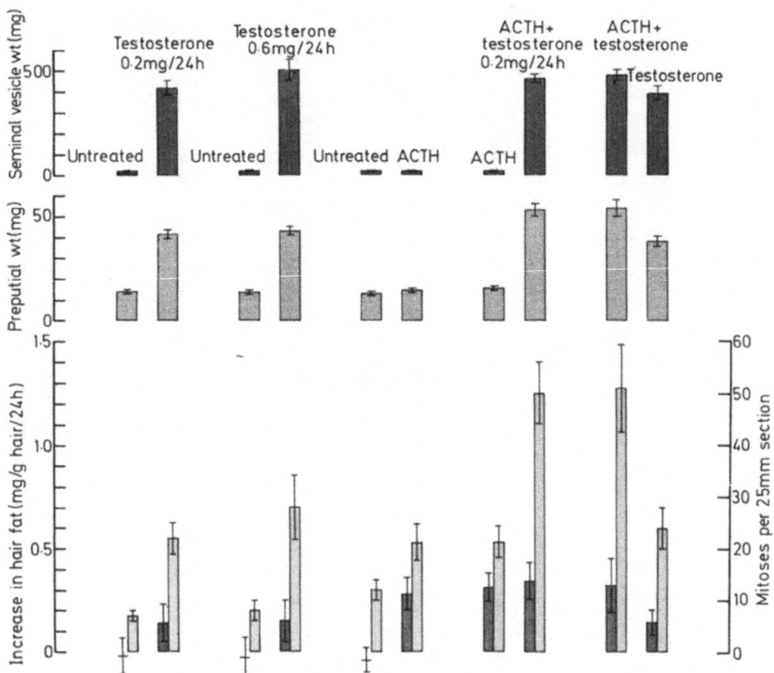


Fig. 4. Effects of testosterone and adrenocorticotrophic hormone (ACTH) on
secretion of sebum (lowest diagram, left-hand columns), mitosis in
the sebaceous glands (right-hand columns), weight of preputial
glands and weight of seminal vesicles in hypophysectomized-castrated
rats. Comparisons between 13, 10, 17, 20 and 14 pairs of litter-
mates respectively. The vertical lines indicate the S.E.
From Ebling et al. (1970) by permission of the Journal of Endocrinology.

effect on sebum production and on sebaceous mitosis equivalent to that of a
similar dose of testosterone. Androstenedione produced some effect on both
measures, but statistically significant effects only in the larger dose of
2.0 mg/day. Androsterone had no effect at the lower dosage; only three rats
were treated with 2.0 mg/day and the results suggest that at least the incidence of mitoses was increased.

In hypophysectomized-castrated rats (Fig. 6), testosterone - as in many
previous experiments - failed to produce any significant increase in sebaceous secretion, and the level in the treated rats remained far below that
of untreated rats. In contrast very significant responses in respect of both
sebum production and sebaceous mitosis were produced by a similar dose of
5α-dihydrotestosterone. Androstenedione, in a similar dose, produced a significant increase in sebum production, and in a dose of 2.0 mg/day was even
more effective.

Dehydroepiandrosterone and androsterone were also administered to hypophysectomized-castrated rats at dose levels of 0.2 and 2.0 mg/day. For both
compounds the interesting result was obtained that significant increases in
sebaceous mitoses were demonstrated (Fig. 7) whereas effects on sebum production were insignificant and almost negligible. Indeed, each of these steroids showed this effect on mitosis at a dose which was too small to produce
significant effects on preputial glands or seminal vesicles.

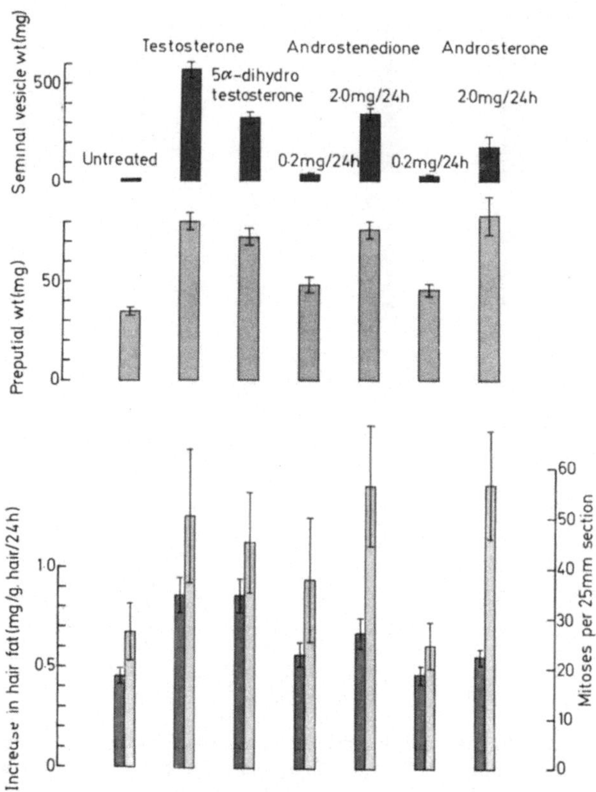


Fig. 5. Effects of testosterone, 5α-dihydrotestosterone, androstenedione and
androsterone on secretion of sebum (lowest diagram, left hand columns), mitosis in the sebaceous glands (right-hand columns), weight
of preputial glands and weight of seminal vesicles in castrated rats.
There were 10 rats in each group except the last (2.0 mg androsterone) which had 3. The vertical lines indicate the S.E.
From Ebling et al. 1971.

In summary, these results show two clear and interesting features. The
first is that the sebaceous glands of hypophysectomized rats will respond
by increased sebum production to 5α-dihydrotestosterone and androstenedione,
but not to testosterone. The second is that sebum production and mitosis do
not run in parallel. All androgenic steroids that have been studied, including testosterone, produce some increase in sebaceous mitosis even in the
absence of the pituitary, and dehydroepiandrosterone and androsterone, even
in small doses, produce significant increases in mitosis without any significant changes in sebaceous secretion. These results reinforce a previously
expressed view that steroid hormones act both on cell division and on intracellular synthesis and that these effects can be disengaged.
On the current evidence, therefore, it is reasonable to put the view that
the utilization of testosterone and the response to it by the sebaceous
glands are dependent upon a pituitary hormone of as yet unidentified nature.
It is well established that steroids are transformed in skin and this subject has already been discussed in a paper in this symposium. The evidence

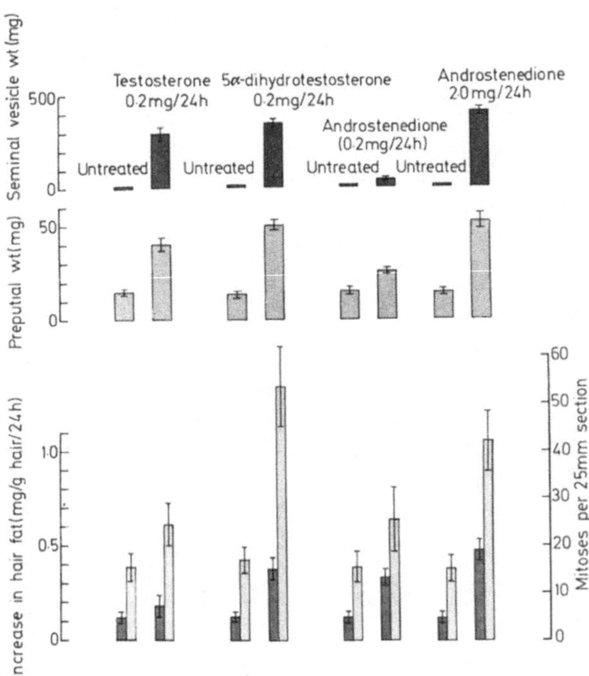


Fig. 6. Effects of testosterone, 5α-dihydrotestosterone and androstenedione
on secretion of sebum (lowest diagram, left-hand columns), mitosis
in the sebaceous glands (right-hand columns), weight of preputial
glands and weight of seminal vesicles in hypophysectomized-castrated
rats. Comparisons between 13, 15, 13 and 13 pairs of litter mates
respectively. The vertical lines indicate the S.E.
From Ebling et al. 1971.

for and against the view that the sebaceous glands are a site for these
transformations has been reviewed elsewhere (Ebling, 1970).

In the course of responding to testosterone, androgen-sensitive target
organs such as the preputial glands and seminal vesicles convert the steroid
to 5α-dihydrotestosterone, a potent androgen, and to the much less potent
androsterone. Since in hypophysectomized rats the sebaceous glands will respond both to androstenedione and to 5α-dihydrotestosterone, it seems more
likely that testosterone is converted via androstenedione, 5α-androstene-
dione, to 5α-dihydrotestosterone in that order (Fig. 8), than that androstenedione is converted by way of testosterone to 5α-dihydrotestosterone. It is
a reasonable hypothesis, therefore, that it is the conversion of testosterone to androstenedione which is influenced by the pituitary hormone. Whether
this influence is exerted on 17β-dehydrogenation, or on splitting testosterone from its conjugates, is not known, but it is a problem that we hope to
pursue.

## Apocrine glands

In conclusion it may be pointed out that the sebaceous glands are not the
only cutaneous glands to be influenced by steroid hormones. A number of spe-

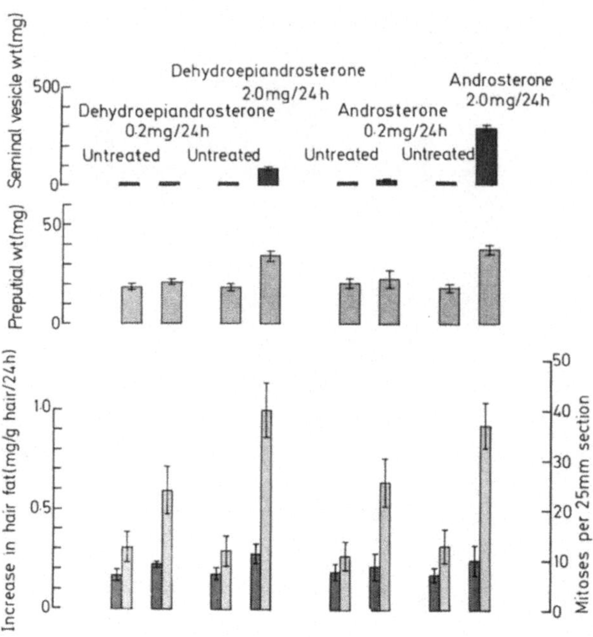


Fig. 7. Effects of dehydroepiandrosterone and androsterone on secretion of
sebum (lowest diagram, left-hand columns) mitosis in the sebaceous
glands (right-hand columns), weight of preputial glands and weight
of seminal vesicles in hypophysectomized-castrated rats. Comparisons
between 19, 18, 13 and 16 pairs of litter mates respectively. The
vertical lines indicate the S.E.
From Ebling et al. (1971).

cialized cutaneous glands are not of sebaceous origin but are apocrine
glands. The rabbit, for example, has three such glands: the anal, inguinal
and submandibular. They are all stimulated by androgens and greatly inhibited by oestrogens (Wales and Ebling, 1971; see also Strauss and Ebling,
1970). What is of particular interest, however, is that they produce metabolites which are concerned with external communication within the rabbit's
environment - so called pheromones. The chin and anal glands seem particularly to be concerned with marking territory (Mykytowycz, 1965, 1966 a, c;
Mykytowycz and Dudzinski, 1966), whereas the inguinal glands produce a sex
attractant (Mykytowycz, 1966 b). There is evidence that odoriferous components of identical composition are produced by the submandibular and anal
glands but that a different material, probably a steroid, is secreted by the
inguinal gland (Wales and Ebling, 1971).

The interest of this discovery is that it provides a biological reason
for the conversion of steroids by the skin glands. It seems quite possible
that the cutaneous glands respond to steroid hormones produced within the
body and convert them into pheromones which serve an external purpose.

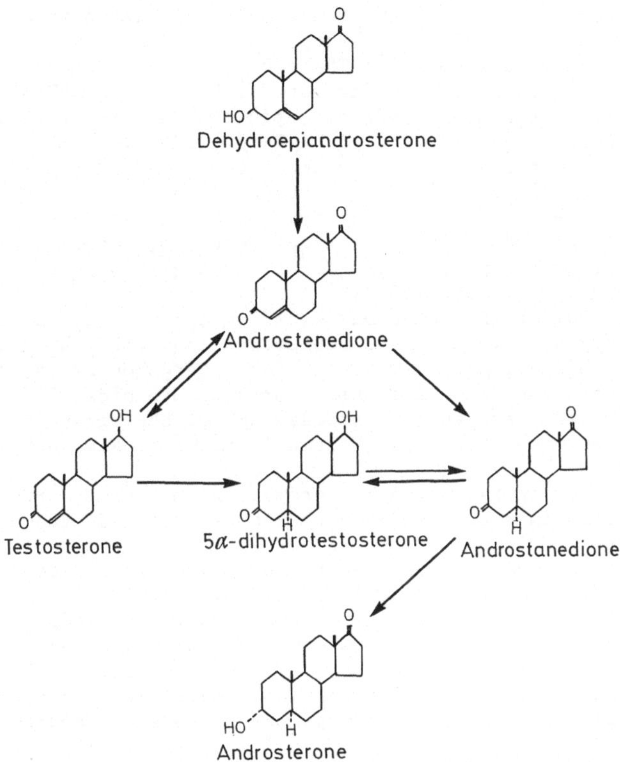


Fig. 8. Some possible pathways of androgenic steroid transformation.

## References


Allen, J.M.: Influence of hormones on cell division; II. Testosterone propionate. Anat. Rec. 128, 515 (1957).
-- The influence of hormones on cell division: II. Time response of ear,
seminal vesicle, coagulating gland and ventral prostate of castrate male
mice to a single injection of testosterone propionate. Exp. Cell. Res.
14, 142-148 (1958).
Archibald, A., Shuster, S.: Bioassay of androgen using the rat sebaceous
gland. J. Endocr. 37, xxii. (1967).
-- -- The measurement of sebum secretion in the rat. Brit. J. Derm. 82,
146-151 (1970).
Bischitz, P.G., Snell, R.S.: A study of the effect of ovariectomy, oestrogen
and progesterone on the melanocytes and melanin in the skin of the female
guinea-pig. J. Endocr. 20, 312-319 (1960).
Bullough, W.S.: Oestrogens, carbohydrate metabolism and mitosis. Ciba Foundation Colloquia on Endocrinology. 6, 278 (1953).
-- van Oordt, G.J.: The mitogenic actions of testosterone propionate and of
oestrone on the epidermis of the adult male mouse. Acta endocr. (Kbh.) 4,
291-305 (1950).
Carter, S.B.: The influence of oestrone on the division of cells. J. Endocr.
9, 19-29 (1953).

Dyson, M., Joseph, J.: The effect of androgens on tissue regeneration.
  J. anat. (Lond.) 103, 491–505 (1968).
Ebling, F.J.: Sebaceous glands. I. The effect of sex hormones on the seba-
  ceous glands of the female albino rat. J. Endocr. 5, 297–302 (1948).
-- Changes in the sebaceous glands and epidermis during the oestrous cycle
  of the albino rat. J. Endrocr. 10, 147–154 (1954).
-- The action of testosterone on the sebaceous glands and epidermis in ca-
  strated and hypophysectomized male rats. J. Endocr. 15, 297–306 (1957a).
-- The action of testosterone and oestradiol on the sebaceous glands and
  epidermis of the rat. J. Embryol. exp. Morph. 5, 74–82 (1957b).
-- Failure of progesterone to enlarge sebaceous glands in the female rat.
  Brit. J. Derm. 73, 65–68 (1961).
-- Hormonal control of the sebaceous gland in experimental animals. In Advan-
  ces in biology of skin; vol. 4. Sebaceous glands. Ed. by Montagna, W.,
  Ellis, R.A., Silver, A.F. 200–219. Oxford: Pergamon Press, 1963.
-- The action of steroids on the skin. Hormonal Steroids, Biochemistry,
  Pharmacology and Therapeutics. Proceedings of the First International
  Congress on Hormonal Steroids, Volume 1. 537–551. New York: Academic
  Press, 1964.
-- The action of an antiandrogenic steroid, 17α-methyl-B-nortestosterone, on
  sebum secretion in rats treated with testosterone. J. Endocr. 38,
  181–185 (1967).
-- Steroid hormones and sebaceous secretion. Advances in Steroids. Vol. 2.
  Ed. Briggs, M.H. London: Academic Press, p. 1. 1970a.
-- Factors influencing the response of the sebaceous glands of the rat to
  androgen. Brit. J. Derm. 82, Supplement 6, 10–14 (1970b).
-- Ebling, E., McCaffery, V., Skinner, J.: The response of the sebaceous
  glands of the hypophysectomized-castrated male rat to 5α-dihydrotesto-
  sterone, androstenedione, dehydroepiandrosterone and androsterone. J.
  Endocr. 51, 181–190 (1971).
-- -- Skinner, J.: The influence of pituitary hormones on the response of
  the sebaceous glands of the rat to testosterone. J. Endocr. 45, 245–256
  (1969a).
-- -- -- The influence of the pituitary on the response of the sebaceous and
  preputial glands of the rat to progesterone. J. Endocr. 45, 257–263
  (1969b).
-- -- -- The effects of thyrotrophic hormone and of thyroxine on the response
  of the sebaceous glands of the rat to testosterone. J. Endocr. 48, 83–90
  (1970a).
-- -- -- White, A.: The response of the sebaceous glands of hypophysectomized
  castrated male rats to adrenocorticotrophic hormone and to testosterone.
  J. Endocr. 48, 73–81 (1970b).
-- Johnson, E.: The control of hair growth. Symp. zool. Soc. Lond. 12,
  97–130 (1964a).
-- -- The action of hormones on spontaeous hair growth cycles in the rat.
  J. Endocr. 29, 192–201 (1964b).
-- Skinner, J.: The measurement of sebum production in rats treated with
  testosterone and oestradiol. Brit. J. Derm. 79, 386–393 (1967).
Gelfant, S.: A study of mitosis in mouse ear epidermis *in vitro*. V. Effects
  of hormones. Exp. Cell Res. 21, 603–615 (1960).
Glenn, E.M., Gray, J.: Effect of various hormones on the growth and histology
  of the gerbil (*Meriones unguiculatus*) abdominal sebaceous gland pad.
  Endocrinology 76, 115–123 (1965).
Groot, C.A. de, Lely, M.A.v.d., Kooij, R.: The effect of progesterone on the
  sebaceous glands of the rat. Brit. J. Derm. 77, 617–621 (1965).
Hamilton, J.B., Montagna, W.: The sebaceous glands of the hamster. I. Morpho-
  logical effects of androgens on integumentary structures. Amer. J. Anat.
  86, 191–234 (1950).
Haskin, D., Lasher, N., Rothman, S.: Some effects of ACTH, cortisone, proge-
  sterone and testosterone on sebaceous glands in the white rat. J. invest.
  Derm. 20, 207–211 (1953).

Huggins, C., Parson, F.M., Jensen, E.V.: Promotion of growth of preputial
    glands by steroids and the pituitary growth hormone. Endocrinology 57,
    25-32 (1955).
Jackson, D., Ebling, F.J.: The effect of oestradiol on moulting in the
    guinea-pig (*Cavia porcellus* L.) J. Endocr. 48, lv-lvi. (1970).
Jensen, E.: Studies on the hormonal regulation of the mouse hair cycle.
    Thesis: University of Copenhagen, 1958.
Johnson, E.: Quantitative studies of hair growth in the albino rat. II. The
    effect of sex hormones. J. Endocr. 16, 351-359 (1958).
Korenschevsky, V., Dennison, M.: The effect on male rats of the simultaneous
    administration of male and female sexual hormones and the relation to the
    assay of the hormones. Biochem. J. 28, 1486-1499 (1934).
-- -- The histological changes in the sex organs of spayed rats induced by
    testosterone and oestrone. J. Path. Bact. 43, 345-356 (1936).
Kuppermann, H.S.: Hormone control of dimorphic pigmentation in the golden
    hamster. Anat. Rec. 88, 442 (1944).
Lapière, C.: Modifications des glandes sebacées par des hormones sexuelles
    appliquées localement sur la peau de souris. C.R. Soc. Biol. (Paris) 147,
    1302 (1953).
Lasher, N., Lorincz, A.L., Rothman, S.: Hormonal effects on sebaceous glands
    in the white rat. II. The effect of the pituitary-adrenal axis. J. invest.
    Derm. 22, 25-29 (1954).
-- -- -- Hormonal effects on sebaceous glands in the white rat. III. Evidence
    for the presence of a pituitary sebaceous gland tropic factor. J. invest.
    Derm. 24, 499-505 (1955).
Lerner, L.J.: Hormone antagonists: inhibitors of specific activities of
    oestrogen and androgen. Recent Progr. Hormone Res. 20, 435-490 (1964).
-- Bianchi, A., Borman, A.: A-Norprogesterone, an androgen antagonist. Acta
    endocr. (Kbh.) Suppl. 51, 869 (1960).
Lorincz, A.L., Lancaster, G.: Anterior pituitary preparation with tropic acti-
    vity for sebaceous, preputial, and Harderian glands. Science 126, 124-125
    (1957).
Martan, J.: Effect of castration and androgen replacement on the supracaudal
    gland of the male guinea pig. J. Morph. 110, 285-298 (1962).
-- Price, D.: Comparative responsiveness of supracaudal and other sebaceous
    glands in male and female guinea pigs to hormones. J. Morph. 121, 209-222
    (1967).
Montagna, W., Kenyon, P.: Growth potentials and mitotic division in the seba-
    ceous glands of the rabbit. Anat. Rec. 103, 365-380 (1949).
Mykytowycz, R.: Further observations on the territorial function and histolo-
    gy of the submandibular cutaneous (chin) glands in the rabbit, *Oryctolagus
    cuniculus* (L.). Anim. Behav. 13, 400-412 (1965).
-- Observations on odoriferous and other glands in the Australian wild
    rabbit, *Oryctolagus cuniculus* (L.), and the hare, *Lepus europaeus* P.
    I. The anal gland. CSIRO Wild. Res. 11, 11-29 (1966a).
-- Observations on the odiferous and other glands in the Australian wild
    rabbit, *Oryctolagus cuniculus* (L.), and the hare, *Lepus europaeus* P.
    II. The inguinal glands. CSIRO Wild. Res. 11, 49-64 (1966b).
-- Observations on odoriferous and other glands in the Australian wild
    rabbit, *Oryctolagus cuniculus* (L.), and the hare, *Lepus europaeus* P.
    III. Harder's, lachrymal, & submandibular glands. CSIRO Wildl. Res. 11,
    65-90 (1966c).
-- Dudzinski, M.L.: A study of the weight of odoriferous and other glands in
    relation to social status and degree of sexual activity in the wild
    rabbit, *Oryctolagus cuniculus* (L.) CSIRO Wildl. Res. 11, 31-47 (1966).
Neumann, F., Elger, W.: Effect of a new antiandrogenic steroid 6-chloro-17
    hydroxy-1α-2-methylene-pregna-4, 6-diene-3, 20-dione acetate (cyproterone
    acetate) on the sebaceous glands of mice. J. invest. Derm. 46, 561-572 (1966).
Nikkari, T., Valavaara, M.: The production of sebum in young rats: effects
    of age, sex, hypophysectomy and treatment with somatotrophic hormone and
    sex hormones. J. Endocr. 43, 113-118 (1969).

-- -- The influence of age, sex, hypophysectomy and various hormones on the
composition of the skin surface lipids of the rat. Brit. J. Derm. 83,
459-472 (1970a).

-- -- Effects of androgen and prolactin on the production rate and composition of sebum in hypophysectomized female rats. J. Endocr. 48, 373-378
(1970b).

Saunders, H.L., Ebling, F.J.: The antiandrogenic and sebaceous gland inhibi-
tory activity of 6α, 6β ethylene 17α-methyl B-nortestosterone. J. invest.
Derm. 52, 163-168 (1969).

Snell, R.S., Bischitz, P.G.: A study of the effect of orchidectomy on the
melanocytes and melanin in the skin of the guinea-pig. Z. Zellforsch. 50,
825-834 (1959).

-- -- The effect of large doses of estrogen and estrogen and progesterone
on melanin pigmentation. J. invest. Derm. 35, 73-82 (1960).

Strauss, J.S., Ebling, F.J.: Control and function of skin glands in mammals.
Mem. Soc. Endocr. 18, 341-371 (1970).

-- Pochi, P.E.: The human sebaceous gland: its regulation by steroidal hormones and its use as an end organ for assaying androgenicity *in vivo*.
Recent Prog. Hormone Res. 19, 385-444 (1963).

Thody, A.J., Shuster, S.: The effects of hypophysectomy and testosterone on
the activity of sebaceous glands of castrated rats. J. Endocr. 47,
219-224 (1970).

Wales, N.A.M., Ebling, F.J.: The control of apocrine glands of the rabbit
by steroid hormones. J. Endocr. In press. 1971.

Wells, L.J.: Pigmentation of the scrotum as a sensitive indicator for androgen. Anat. Rec. 91, 305-306 (1945).

Wilde, P.F., Ebling, F.J.: Preliminary observations on the composition of
skin surface fat from rats treated with testosterone and estradiol.
J. invest. Derm. 52, 362-365 (1969).

Woodbury, L.P., Lorincz, A.L., Ortega, P.: Studies on pituitary sebotropic
activity. I. A new sensitive assay method for sebotropic activity based
on beta-gluruconidase content of preputial glands. J. invest. Derm. 45,
362-363 (1965a).

-- -- -- Studies on pituitary sebotropic activity. II. Further purification
of a pituitary preparation with sebotropic activity. J. invest. Derm. 45,
364-367 (1965b).

Symp. Dtsch. Ges. Endokrin. 17, 33–42 (1971)
© by Springer-Verlag

# Klinische Aspekte der Steroidwirkung an der Haut
## Clinical Aspects on the Effects of Steroids on the Skin

KURT WINKLER

Dermatol. Abt. des Städt. Krankenhauses Berlin-Britz
Mit 3 Abbildungen

## Summary

Androgens are of some importance in the treatment of xanthomatosis and se-
nile pruritus. They stimulate the secretion of sebum. This can be inhibited
by anti-androgens, oestrogens and oral contraceptives which, therefore, can
be employed in the treatment of seborrhoea and acne.

Oestrogens cause proliferation of the epithelium and increase the turgidi-
ty of the old skin and appear to intensify the anti-inflammatory action of
corticoids. Undesirable effects of oestrogens resp. of oral contraceptives
are: Precipitation of porphyria cutanea tarda, pigmentation of the facial
skin, development of spider naevi and candida vaginitis.

Progestogens are employed in the treatment of progressive scleroderma.
As regards topical corticoids undesirable effects such as atrophy of the
skin, striae, purpura, steroid-acne are discussed. The case of an 11 year-
old boy who developed Cushing's syndrome after topical corticoid application
is described.

Finally, allergic reactions following steroid therapy are discussed.

Die Wirkung der Steroide ist nicht auf ein bestimmtes Organ oder Organ-
system beschränkt. Sie spielen nicht nur eine wichtige Rolle bei der Regu-
lierung des Wachstums, der Fortpflanzung und des Stoffwechsels, sondern sie
sind z. T. auch an unspezifischen Anpassungsreaktionen beteiligt. Es ist
ganz natürlich, daß Substanzen mit so vielfältigen Eigenschaften auch am
Integument angreifen. Wir sind in der günstigen Lage, ihren Effekt hier meistens ohne besondere Untersuchungsmethoden zu erkennen.

Beginnen wir mit der Bedeutung der *Androgene* für die Haut. Diese Steroide
haben einen gewissen Wert für die Unterstützung der Behandlung von Xanthomatosen. Androgene führen zu einer Verminderung der Alphalipoproteine sowie
der Lipoproteine sehr niedriger Dichte. Besonders bei solchen Hypertrigly-
ceridämien, bei denen diese Lipoproteinfraktion stark vermehrt ist, kann es
nach Behandlung mit Androgenen zu einer drastischen und anhaltenden Verminderung der Triglyceridspiegel im Serum kommen (Schlierf). Husmann u. Mitarb.
haben einen Abfall der Serumlipide unter der Gabe von 150 mg Mesterolon
(Proviron) täglich festgestellt. Die Autoren halten die Anwendung dieses
Mittels bei Krankheitsbildern, die mit einer Hyperlipämie einhergehen, für
eine wertvolle Bereicherung der therapeutischen Möglichkeiten. So kann man
Mesterolon in geeigneten Fällen bei Männern zur Unterstützung der *Behandlung
von Xanthomatosen* empfehlen.

Androgene haben ferner einen Einfluß auf die Alterungsvorgänge der Haut.
So sahen Wright u. Mitarb. bei Männern im Alter von 60 - 79 Jahren nach
Testosteronönanthatgaben meist eine Verbreiterung der Epidermis und eine Zunahme der Grundsubstanz der Kutis. Goldzieher u. Mitarb. beobachteten schon
1952 nach äußerlicher Testosteronanwendung eine Epidermisproliferation.

Die Schweiß- und Talgsekretion läßt im Alter nach, die Haut wird trocken.
Diese Fettverarmung wird als wesentliche Ursache für den generalisierten
Pruritus senilis angenommen, der 25% aller Menschen nach dem 70. Lebensjahr
befällt (Zakon). Wir pflegen bei Altersjuckreiz Androgene, und zwar Proviron,
zu geben. Die günstige Wirkung der Androgene beim Altersjuckreiz ist vielleicht damit zu erklären, daß sie die Talgsekretion anregen.

Die stimulierende Wirkung der Androgene auf die Talgdrüsen ist durch
zahlreiche Untersuchungen, u. a. auch von Ebling, sichergestellt. Selbst
ganz schwach wirksame Androgene wie anabole Steroide fördern bei Frauen und
Kindern die Talgsekretion (Krüskemper). Eine solche Stimulierung der Talgdrüsen dürfte nur beim Pruritus senilis erwünscht sein, sonst ist es unser
therapeutisches Ziel, die gesteigerte Talgsekretion, die Seborrhoe, eher zu
hemmen.

Da Androgene die Talgproduktion fördern, lag die Annahme nahe, daß *Steroide mit antiandrogener Wirkung* die Talgproduktion hemmen. Wir prüften die
Wirkung der Testosteronantagonisten Cyproteron und Cyproteronacetat zunächst
am mongolischen Gerbil. Dieser hat an der Bauchhaut eine sogenannte Abdomi-
naldrüse, deren Größe androgenabhängig ist. Die Abbildung 1 zeigt die durchschnittlichen Gewichte der Abdominaldrüsen von weiblichen Tieren, die Abbildung 2 die von männlichen. Die Zahlen in den Säulen geben die untersuchte
Tierzahl an. Die 1. Säule der Abb. 1 zeigt die Drüsengewichte von intakten
weiblichen Tieren, die 6. Säule die von kastrierten weiblichen Tieren. Man
sieht, daß das Drüsengewicht bei kastrierten weiblichen Tieren niedriger ist
als bei intakten. Die 2. Säule zeigt die Gewichte von kastrierten Tieren,
die 12 Tage tgl. 0,1 mg Testosteronpropionat erhalten haben: das Drüsengewicht hat mächtig zugenommen. Die 3. Säule zeigt, daß die Wirkung von 0,1 mg
Testosteronpropionat durch 1,5 mg Cyproteronacetat vollständig aufgehoben
wird. Auch bei männlichen Tieren (Abb. 2) wird durch 1,5 mg Cyproteronacetat
tgl. über 12 Tage lang der Effekt von 0,1 mg Testosteronpropionat weitgehend
gehemmt (Säule 3). Eine völlige Aufhebung, wie bei weiblichen Tieren, tritt
aber bei dieser Dosis noch nicht auf.

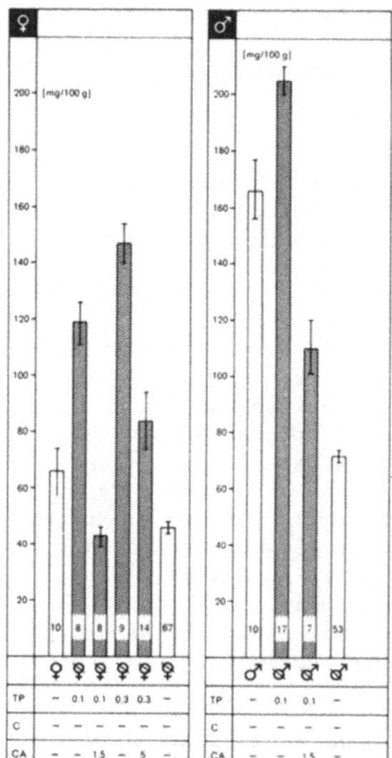

Abb. 1. Drüsengewicht weiblicher intakter (Säule 1) und weiblicher
kastrierter Tiere (Säule 6).

Bei weiblichen kastrierten Tieren ist die Wirkung von 0,1 mg
Testosteronpropionat (TP) auf
das Drüsengewicht (Säule 2)
durch 1,5 mg Cyproteronacetat
(CA) aufgehoben (Säule 3). Die
Wirkung von 0,3 mg Testosteronpropionat auf das Drüsengewicht
(Säule 4) ist durch 5 mg Cyproteronacetat stark gehemmt
(Säule 5).

Abb. 2. Drüsengewicht männlicher intakter (Säule 1) und männlicher
kastrierter Tiere (Säule 5).
Bei männlichen kastrierten Tieren ist die Wirkung von 0,1 mg
Testosteronpropionat auf das
Drüsengewicht (Säule 2) durch
Cyproteronacetat weitgehend gehemmt (Säule 3).

Die hemmende Wirkung der Testosteronantagonisten auf die Talgdrüsenfunk-
tion, die wir beim Tier feststellten, konnten wir durch Untersuchungen beim
Menschen bestätigen. Wir behandelten über 200 Patienten, die an Seborrhoe

und Akne litten, täglich mit 200 mg Cyproteron innerlich. In der Mehrzahl
der Fälle bestätigten die Patienten, daß der Talgfluß 2 bis 3 Wochen nach
Beginn der Behandlung aufhörte, d. h., daß die Gesichts- und Kopfhaut trok-
ken wurde. Einzelne Behandelte klagten sogar darüber, daß die Haut infolge
Trockenhaut spannte, so daß sie die Gesichtshaut einfetten mußten. Ein bis
drei Wochen nach Beendigung der Cyproteroneinnahme kam es wieder zu einem
stärkeren Fettfilm. Unsere klinischen Beobachtungen konnten wir durch die
gravimetrische Methode der Bestimmung der Sebumproduktion nach Strauss u.
Mitarb. objektivieren.

Die Wirkung des Cyproterons ist bei der Seborrhoe wesentlich besser als
bei der Akne. Das ist auch zu erwarten, da bei der Akne neben der Seborrhoe
noch andere Faktoren ursächlich eine Rolle spielen.

Wirksamer als Cyproteron ist Cyproteronacetat, das zusammen mit Äthinylöstradiol von Hammerstein u. Mitarb. bei aknekranken Frauen mit gutem Erfolg
gegeben wurde.

Auch durch massive *Östrogengaben* läßt sich die Talgproduktion einschrän-
ken. Östrogene hemmen die Gonadotropinsekretion und vermindern dadurch die
Androgensekretion aus den Testes oder den Ovarien. Deshalb werden Östrogene
seit vielen Jahren zur Behandlung der Akne verwandt. Schreus hat 1956
Progynon-Depot 10 mg empfohlen, das er bei Frauen alle 4 Wochen unmittelbar
nach der Regel gab. Er ging später zur oralen Östrogentherapie über und verabreichte täglich 0,02 mg Äthinylöstradiol, bei Männern fortlaufend, bei
Frauen nur vom 10. bis 24. Tag ab Mensesbeginn. Diese hohen Dosen können bei
Männern zur Gynäkomastie, bei Frauen zu Zyklusstörungen führen. Deshalb werden heute zur Behandlung der Akne bei Frauen sog. Ovulationshemmer (OH) in
der üblichen Dosierung empfohlen (Kümmel, Pochi u. Mitarb., Hopf).

Auch das im OH enthaltene Gestagen kann einen Einfluß auf die Talgproduktion und damit auf die Akne haben. Günstig erschien wegen seiner antiandro-
genen Wirkung das Gestagen "Chlormadinonacetat", das aber zur Zeit nicht im
Handel ist.

Man muß ferner bedenken, daß solche Gestagene, die eine androgene Nebenwirkung haben wie Norethisteronacetat, Norgestrel, Medroxyprogesteronacetat
bei einer besonderen Empfindlichkeit der Talgdrüsen einen ungünstigen Einfluß auf die Akne haben können.

Östrogene sind ferner zur Behandlung der *Altershaut* herangezogen worden.
Sie führen zu einer Proliferation des Epithels (Hohlweg), zu einer Wasser-
retention der Kutis, zu einer Erhöhung des Hauttugors (Blank), zu einer Erweiterung der Gefäße, ja sogar zu einer echten Gefäßneubildung (Nikolowski).
Sie machen die Haut elastischer und glatter und bewirken eine Zunahme des
Fettgewebes (Napp). Eller u. Eller sahen nach Östrogensalben einen deutlichen proliferativen Effekt am Epithel, der abhängig vom Lebensalter war. An
atrophierter Haut wurden Größenzunahme der Epidermiszellen und lebhafte Tei-
lungen beobachtet. Pockrandt beobachtete nach Einreibung einer Dien-oestrol-
diacetat-Salbe an der weiblichen und männlichen Greisenhaut histologische
Veränderungen, die für eine Verjüngung der Haut charakteristisch waren.

Goldzieher u. Mitarb. fanden bei 27 älteren Frauen nach lokaler Östrogen-
anwendung eine Proliferation der Epidermis, eine Verlängerung der Retelei-
sten und Papillen, eine Vermehrung der Keratohylingranula, eine Neubildung
von elastischen Fasern und eine Zunahme der Vaskularisation. Bei Männern,
die wegen eines Prostatakarzinoms Cyren-A erhielten, sah Bosse eine feine
weiche Haut ohne Faltenbildung, Schlaffheit und ohne Alterspigmentierungen.

Nach innerlicher Gabe von Östriolsuccinat (2 mg tgl. oral, 6 Monate lang)
wird die atrophische Epidermis kastrierter Frauen breiter (Rauramo u.
Mitarb.).

Man darf allerdings nicht außer acht lassen, daß nach längerer Östrogen-
anwendung Nebenwirkungen wie Uterusblutungen auftreten können. Das gilt
gleichfalls für östrogenhaltige Externa. Diese unterliegen auch der Rezept-
pflicht und werden deshalb von der kosmetischen Industrie nicht angeboten.

Eine ähnliche Wirkung auf die Altershaut wie die Östrogene hat das Pregnenolon. Diese Steroidverbindung führt nach wiederholter lokaler Anwendung
an der Meerschweinchenhaut zur Anregung des Zellwachstums und zu einer Zell-

regeneration in der Epidermis mit Vermehrung der Zellen und Erhöhung der die
Epidermis aufbauenden Zellagen (Schaaf u. Mitarb.).

Neben einer Epidermisverdickung soll auch eine bessere Hydratation der
Epidermis nach äußerlicher Anwendung von Pregnenolon nachweisbar sein
(Tronnier). Die Verwendung von Hautpflegemitteln auf der Basis dieses Ste-
roids bedeutet einen Fortschritt in der Pflege alternder Haut (Schaaf,
Soltermann). Pregnenolon ist z.B. in der Sana-Creme enthalten, von der wir
einen guten Eindruck haben.

Mit dem Mangel an Östrogenen in Zusammenhang gebracht werden gewisse
Palmar- und Plantarkeratosen. Haxthausen bezeichnete diese als *Keratoma
postklimactericum*. Hierbei entstehen bei Frauen kurz vor oder nach Beginn
des Klimakteriums an Handtellern und Fußsohlen papulöse, verhornte Efflores-
zenzen. Für die endokrine Auslösung dieser Veränderungen sprechen die guten
Erfahrungen, die verschiedene Autoren mit der Östrogenbehandlung machten.
Allerdings dürften außer dem Absinken des Steroidhormonspiegels noch weitere
Faktoren eine Rolle spielen (Greither), da das Krankheitsbild nur ganz selten auftritt.

Ferner *senken Östrogene* ähnlich wie Androgene den *Lipidspiegel* im Blut
und können deshalb zusätzlich bei der Xanthomatose bei Frauen gegeben werden.

Schließlich haben Östrogene, wie experimentelle Untersuchungen gezeigt
haben, *entzündungshemmende Wirkungen* (Spangler u. Mitarb.). Die Zugabe von
Östrogenen (45 - 300 mg Dihydrodiäthylstilböstrol oder 576 - 1440 mg Chloro-
trianisen) erlaubt nach Spangler u. Mitarb. bei schweren und chronischen
Hautaffektionen eine signifikante Reduktion der Kortikoiddosis.

Auch unerwünschte Erscheinungen an der Haut können nach der Verabreichung
von Östrogenen auftreten. So lösen sie in seltenen Fällen einen Schub einer
*Porphyria cutanea* tarda aus oder führen zur Manifestation der Erkrankung
(Thivolet u. Mitarb., Vail). Hierbei kommt es an lichtexponierten Stellen,
besonders im Gesicht und an den Handrücken, zur Blasenbilden nach Sonnenbe-
strahlung und Alltagstraumen. Hyperpigmentierungen begleiten die Veränderungen. Eine Hypertrichose besonders in der Jochbogengegend ist recht charakteristisch.

Roenigk u. Mitarb. stellten 28 Porphyriefälle zusammen, die nach Östrogenbehandlung aufgetreten waren. Hierunter befanden sich 20 Männer, die wegen eines Prostatakarzinoms Östrogene (meist Diäthylstilböstrol) erhalten
hatten. Im Hinblick auf die große Zahl der Kranken, die wegen eines Prostatakarzinoms mit Östrogenen behandelt wurden, ist das Auftreten einer Porphyrie allerdings sehr selten.

Eine Porphyrie können auch Ovulationshemmer (OH) auslösen (Roenigk u.
Mitarb.). Degos beobachtete eine Frau, bei der es danach zu einem Schub
einer Porphyrie kam. Nach Absetzen des OH und nach einigen Aderlässen bildeten sich die Erscheinungen zurück, und die Porphyrinwerte normalisierten
sich.

Wenn die Verabreichung von Östrogenen eine Porphyrie auslösen oder ver-
schlechtern kann, so ist dasselbe während der Gravidität zu erwarten, da ja
während dieser Zeit der Östrogenspiegel erheblich ansteigt. Vine u. Mitarb.
prüften 27 Schwangerschaften bei 16 Frauen mit Porphyrie: 21mal trat eine
Verschlimmerung des Leidens auf. Zimmermann u. Mitarb. berichteten über eine
Porphyrie, die zum ersten Mal während der Gravidität manifest wurde.

Die Pathogenese der durch Östrogene induzierten Porphyria cutanea tarda
ist nicht völlig geklärt. Vielleicht verändern Östrogene unter bestimmten
Bedingungen ein oder mehrere Enzyme in der Leber.

Eine weitere unerwünschte Nebenwirkung der Östrogene auf die Haut ist
eine *Pigmentzunahme*. Diese wird auch durch die Gestagene gefördert. Während
der Schwangerschaft sind beide Hormone vermehrt. So sehen wir während dieser
Zeit bei 2/3 der Frauen ein Chloasma uterinum, eine sogenannte masque gravi-
dique auftreten. Sie bildet sich nach der Entbindung gewöhnlich zurück.

Eine Pigmentzunahme wird oft beobachtet bei der Verabreichung von OH und
zwar sowohl von Kombinations- als auch von Sequentialpräparaten. Die Pigmentierungen treten gewöhnlich symmetrisch auf und bestehen aus unregelmäßig
gestalteten braunen, mehr oder weniger scharf begrenzten Flecken. Die große

Zahl der Melanozyten in der Gesichtshaut und die Lichtexposition erklären
die Lokalisation. Ein vorangegangenes Chloasma gravidarum weist auf eine besondere Disposition zur Pigmentvermehrung hin. Auch wenn während der Schwangerschaft kein Chloasma aufgetreten war, so können nach OH trotzdem Pigmentierungen entstehen.

Die Häufigkeit des Vorkommens von Pigmentierungen nach OH wird verschieden angegeben. Nach Jelinek tritt sie bei 5 bis 8% der Frauen, nach Resnik
bei 29% auf. Sie nimmt entsprechend der Hormonmenge, der Behandlungsdauer
und der Sonnenexposition zu. Nach Vankos tritt das Chloasma nach einjähriger
Einnahme bei 4%, nach fünfjähriger Einnahme bei 37% auf. Tronnier u. Mitarb.
gaben 25 Versuchspersonen OH und stellten fest, daß es während fünfmonatiger
Verabreichung bei 20% der Fälle zu einer vermehrten Pigmentierung kam.

Um dieser kosmetischen Störung vorzubeugen, sollte man Präparate mit
einem möglichst geringen Östrogen-Gestagengehalt geben. Außerdem ist es
wahrscheinlich günstig, die Pille abends zu geben, damit die Menge der zirkulierenden Hormone am Tage während der Sonneneinwirkung möglichst gering
ist.

OH können ferner eine *Fotosensibilisierung* hervorrufen. So berichten
Erickson u. Mitarb. über eine polymorphe Lichtdermatitis, die bei einer
31jährigen Negerin während der Einnahme von Enovid E auftrat. Nach Absetzen
des OH bildeten sich die Hauterscheinungen zurück, kamen aber nach Einnahme
von anderen Ovulationshemmern sowie von Äthinylöstradiol erneut zum Aus-
bruch. Deshalb scheint das Östrogen zur Lichtsensibilisierung beigetragen zu
haben. Das Auftreten der Lichtdermatitis war besonders überraschend wegen
der dunklen Hautfarbe der Patientin, bei der man gerade durch den Pigment-
reichtum einen Lichtschutz erwartet hatte. Eine Porphyrie wurde in diesem
Falle ausgeschlossen.

Bei Frauen, die OH einnehmen ebenso wie während der Schwangerschaft entwickeln sich oft *Teleangiektasien* und Spidernaevi (Jelinek). Sie finden sich
besonders im Gesicht, im Nacken und an den oberen Brustkorbpartien.

Die Entstehung der Teleangiektasien kann darauf zurückgeführt werden, daß
der Tonus der glatten Muskulatur unter OH und während der Gravidität abnimmt
(Ludwig). Es kommt zu einer ödematösen Auflockerung der Gefäßwandstruktur.
Gestagene begünstigen eine Gefäßweitstellung (Pfeifer u. Mitarb.).

OH fördern ebenso wie die Gravidität das Auftreten einer *Candida-Vulvo-
vaginitis* (Porter u. Mitarb.). Diese entsteht meist erst nach neunmonatiger
Verabreichung (Jelinek). OH führen offensichtlich zu einem vermehrten Glyko-
gengehalt der Scheide, der den Milchsäuregehalt steigert. Milchsäure fördert
das Wachstum von Candida.

*Gestagene* haben ferner eine Bedeutung bei der Behandlung der *progressiven
Sklerodermie* (Korting u. Mitarb.). Progesteron führt zu einer Hemmung der
Kollagenfaserbildung und damit zu einer Auflockerung und Aufweichung des
Bindegewebes. Diese Wirkung, die nicht auf das Bindegewebe des Uterus be-
schränkt ist, wird auch durch synthetische Gestagene vom Typ des Äthinyl-
nortesteronacetats hervorgerufen.

Der Sklerodermie liegt offenbar eine gesteigerte Kollagenneubildung und
ein verlangsamter Abbau des Kollagens zugrunde. Ein Maß für den Abbau des
Kollagens ist die Hydroxyprolinausscheidung im Harn. Bei Sklerodermiepatien-
ten liegt der Spiegel in bzw. unter der Norm (Holzmann u. Mitarb.). Durch
Progesteron läßt sich ein Ansteigen der Hydroxyprolinausscheidung bei Sklerodermiekranken auf mehr als das Doppelte erreichen. Nicht nur diese bio-
chemischen Untersuchungsergebnisse sprechen für eine günstige Beeinflussung
der Bindegewebsmechanismen bei Sklerodermiekranken, sondern auch die klinischen Befunde. So konnten Holzmann u. Mitarb. bei 30 Kranken mit Gestagenen
bessere Durchblutungsverhältnisse der Akren, eine auffällige Verringerung
der Raynaudsymptomatik und eine erhebliche Erweichung der Indurationen mit
besserer Beweglichkeit der Gelenke erreichen. Eigene Beobachtungen zeigten
bei einer allerdings geringen Fallzahl von Sklerodermiekranken keine wesentliche Besserung des Leidens nach Gestagenen.

Von allen Steroiden haben die *Kortikoide* wegen ihrer antiphlogistischen,
antiproliferativen, antiakantholytischen und juckreizstillenden Eigenschaf-

ten für die Behandlung der Hautkrankheiten die größte Bedeutung. So ist seit
der Einführung dieser Steroide die Prognose verschiedener lebensbedrohender
Dermatosen wesentlich günstiger geworden. Als Beispiele seien genannt: der
Pemphigus, das Lyell-Syndrom, das Erythema exsudativum multiforme majus, das
Quinckesche Ödem.

Zu berücksichtigen ist, daß unter einer innerlichen oder äußerlichen Kor-
tikoidbehandlung fast alle Hautkrankheiten sich so verändern können, daß
eine exakte klinische wie histologische Diagnose erheblich erschwert, wenn
nicht unmöglich wird (Schimpf). Es sei betont, daß Kortikoide nur symptoma-
tisch oder morbostatisch wirken.

Zur Beeinflussung gewisser umschriebener Dermatosen werden Kortikoid-
Kristallsuspensionen (KKS) intra- oder subläsional injiziert. Als Beispiele
seien genannt: der Lichen sclerosus et atrophicus vulvae, der identisch ist
mit der Kraurosis vulvae (Grimmer), der Lichen simplex chronicus Vidal, das
Keloid. Bei diesem führt die Abnahme der Fibroblastenzahl, die Hemmung der
Fibrillogenese, die Rarefizierung des Kollagens vielfach zu einer deutlichen
Abflachung.

Eine unerwünschte Wirkung der Injektion von KKS ist eine umschriebene,
rosa-bläulich verfärbte, leicht eingesunkene *Hautatrophie*, die 4 - 6 Wochen
nach der Einspritzung entstehen kann und sich im Verlauf von 1 - 3 Jahren
zurückbildet (Nasemann).

Am häufigsten werden die Kortikoide äußerlich angewandt. Hierfür gebraucht man (Abb. 3)

1. halogenfreie Kortikoid-Externa, die meist Hydrokortison oder Prednisolon
   enthalten.
2. halogenierte Kortikoid-Externa, die ein oder zwei Fluoratome oder ein
   Fluor- und ein Chloratom enthalten.

Kortikoid-Externa lassen sich heute aus der Ekzem-Therapie nicht mehr weg-
denken.

Es ist hervorzuheben, daß etwa 80% aller Dermatitiden auf alle Kortikoide,
also auch auf eine 1%ige Hydrokortisonsalbe, in den handelsüblichen Konzentrationen und Präparaten gleich gut und gleich schnell ansprechen (Schieferstein, Tronnier). Dagegen lassen sich z. B. die Lichenkrankheiten, die
Psoriasis und der Erythematodes nur durch halogenierte Kortikoid-Externa gut
beeinflussen.

Eine Ausnahme macht hier vielleicht Hydrokortison-17α-butyrat, das in
0,1%iger Konzentration offenbar gleich stark bei der Psoriasis wirkt wie
0,1%iges Triamcinolon-acetonid (Poland u. Mitarb.). Das ist praktisch wichtig, da Hydrokortison im Vergleich zu den halogenierten Kortikoiden in der
gleichen Dosis weniger systemische und kaum lokale Nebenwirkungen verursacht.
Hydrokortisonbutyrat ist nicht im Handel.

Die vasokonstriktorische Wirkung der halogenierten Kortikoid-Externa ist
so stark, daß sie beim Ulcus cruris zu Nekrosen und sogar zu Blutungen führen können (Bjornberg u. Mitarb., Szysymar u. Mitarb.). Man beseitigt mit
diesen Präparaten unter Okklusivverbänden überschießende Granulationen und
ersetzt hierdurch die Ätzung mit dem Lapisstift. Bei mangelhaften Granulationen sind sie kontraindiziert.

Besonders die halogenierten Kortikoid-Externa vermindern die Mitoserate
in der Epidermis, hemmen die Basalzellproliferation und beeinflussen deshalb
die Psoriasis günstig. Der antimitotische Effekt der halogenierten Kortikoid-
Externa kann bei längerem Gebrauch zu einer atrophischen, schlaffen Haut mit
runzeliger, zigarettenpapierdünner Epidermis (*Kortikoidhaut*) führen. Hierzu
kann es auch nach langzeitiger innerlicher Kortikoidverabreichung kommen.

Halogenierte Kortikoide bremsen offenbar sogar das Wachstum des Nagel-
häutchens. Deshalb werden solche Präparate zuweilen zur Maniküre gebraucht
(Steigleder).

Durch die Atrophie der Epidermis und den Kollagenschwund werden die Haut-
gefäße sichtbarer. Es treten *Teleangiektasien* auf, oder die vorhandenen ver-
mehren sich (Götz u. Mitarb., Bandmann). Das hat praktische Bedeutung bei
der Rosacea, die durch langfristigen Gebrauch halogenierter Kortikoid-
Externa verschlimmert werden kann (Sneddon).

| Chemische Kurz-<br>bezeichnung der<br>unveresterten<br>Kortikoide | Kortikoid-<br>Prozentgehalt<br>nebenstehender<br>Präparate | Beispiele für<br>Handelspräparate |
|---|---|---|
| **Halogenfreie Kortikoid-Externa** | | |
| Hydrokortison | 0,33, 1,0 | Sanatison |
| Prednisolon | 0,33 | Sanapredon |
| 6-Methylprednisolon | 0,25 | Medrate |
| **Halogenierte Kortikoid-Externa** | | |
| Triamcinolon | 0,1 | Volon A |
| Fluorometholon | 0,025 | Delmeson |
| Dexamethason | 0,04 | Sanamethason |
| Betamethason | 0,1 | Betnesol-V |
| Fluorandrenolon | 0,05 | Sermaka |
| Fluocinolon | 0,01-0,2 | Jellin |
| Flumethason | 0,02 | Locacorten |
| Fluprednyliden | 0,15-0,075 | Etacortin |
| Fluocortolon | 0,2-0,5 | Ultracur, Ultralan |
| Clocortolon | 0,2 | Kaban |

Abb. 3. Äußerlich angewandte Kortikoide

Infolge der Hautatrophie treten nicht nur die Gefäße deutlicher hervor, sondern auch die Talgdrüsen. Sie werden vor allem am Hals als kleine weiße Papeln in der atrophischen Haut erkennbar: *Cutis linearis punctata colli* (stippled skin).

In der atrophischen Kortikoid-Haut oder unabhängig von ihr können *Striae* auftreten. Sie entstehen nach innerlicher oder seltener nach äußerlicher Kortikoid-Anwendung infolge Schädigung oder Zerstörung der elastischen Fasern. Sie treten fast nur bei jüngeren Menschen auf (Thiers u. Mitarb.) und kommen nie im Gesicht, am Hals und nie an Händen und Füßen vor. Sie bilden sich in der Regel nach Anwendung von Kortikoid-Externa unter Okklusivverbänden bes. an den Extremitäten (Chernosky u. Mitarb., Carr u. Mitarb.). Sie werden aber auch ohne Okklusivverbände beobachtet und zwar im Leisten-Oberschenkel-Bereich. Hier schaffen Dünnheit und höhere Temperatur des Integumentes die gleichen Bedingungen wie ein Okklusivverband. Wir sahen einen 28jährigen Mann, der 8 Monate lang verschiedene fluorierte Kortikoidsalben im Genitofemoralbereich eingerieben hatte. Es waren hochrote breite Striae am Skrotum und den gegenüberliegenden Oberschenkelflächen entstanden.

Durch *Zugabe eines Anabolikums* kann man die Mitosehemmung eines Kortikoid-Externums aufheben (Bullough u. Mitarb.). Hiermit soll einer Hautatrophie entgegengewirkt werden. Auf diesem Prinzip beruht das neue Präparat Dexatopic.

Zuweilen sieht man eine mehr oder weniger ausgedehnte *Kortikoidpurpura*, die auf eine vermehrte Gefäßbrüchigkeit, wahrscheinlich infolge der katabolen Wirkung der Kortikoide zurückgeführt werden kann. Die Gerinnungsfaktoren sind hierbei normal.

Sehr selten kommt es nach äußerlicher Anwendung halogenierter Kortikoide, etwas häufiger nach innerlicher Gabe von Kortikoiden, zu einer Steroid-Akne. Hierbei handelt es sich nicht um die klassische Form der Akne vulgaris mit Seborrhoe und Komedonenbildung, sondern um akneforme Knötchen auf trockener Haut. Diese Knötchen finden sich bes. im Gesicht, an der Brust und am Rücken. Vielfach liegt der Steroid-Akne eine Hefebesiedlung zugrunde (Orfanos).

Es muß betont werden, daß wir nach örtlicher Anwendung von Hydrokortison und Prednisolon die genannten Nebenerscheinungen (Atrophie, Striae, Teleangiektasien, Akne, Pigmentverschiebungen) nicht beobachteten. Deshalb empfehlen wir, an frei getragenen Hautstellen bei langzeitiger Anwendung halogenierte Externa nur in schwacher Konzentration zu gebrauchen oder vielleicht besser halogenfreie Präparate zu bevorzugen.

Kortikoide hemmen die lokale sowie die humorale Infektabwehr des Organismus. Vor allem unter Okklusivverbänden treten Follikulitiden und Soormykosen auf. Es empfiehlt sich deshalb, bei dieser Anwendungsweise nur Externa mit entsprechenden Zusätzen zu gebrauchen.

Besonderes Interesse gilt der Frage, ob Nebenwirkungen systemischer Art nach Kortikoid-Externa auftreten können. Werden halogenierte Präparate in größerer Menge *unter Okklusivverbänden* angewandt, so kann es zu einem *Absinken des Plasmakortisolspiegels*, der 17-Hydroxykortikoide und der 17-Ketosteroide im Harn kommen (March u. Mitarb., Scoggins u. Mitarb., Gill u. Mitarb., Kirketerp). Wir stellten eine deutliche Abnahme der Kortikosteroidwerte im Blut und Harn nach Anwendung von 40 g Fluocortolon-Salbe unter Okklusivverbänden fest.

Bei Erwachsenen genügen die resorbierten Kortikoidmengen offenbar nicht, um systemische Nebenwirkungen hervorzurufen. Bei Kindern kann es hierzu ausnahmsweise kommen (Feinblatt u. Mitarb.), wie folgender Fall zeigt:

Ein 11jähriger Junge war wegen einer Erythrokeratodermia figurata variabilis Mendes da Costa seit frühester Jugend wiederholt längere Zeit mit Kortikoid-Externa behandelt worden. Wir wandten am Stamm und an den Extremitäten verschiedene halogenierte Kortikoidsalben an, auf dem Kopf benutzten wir im Wechsel eine Triamcinolon-Tinktur und halogenierte Salben unter Okklusivverbänden. Nach dreiwöchiger Behandlung entwickelte sich ein *Cushing-Syndrom* mit Striae cutis, Mondgesicht, Büffelnacken, Stammfettsucht. Innerhalb weniger Wochen kam es zu einer Gewichtszunahme von 8 kg. Der Blutdruck war mit 125/85 Hg leicht erhöht. Es fanden sich keine Veränderungen an der Sella turcica. Eine in der Universitätskinderklinik durchgeführte Funktionsdiagnostik ließ einen NNR-Tumor weitgehend ausschließen (Weber). Als wesentliches Argument gegen einen Tumor ist der erhaltene Rhythmus im Plasmakortisol-Tagesprofil anzusehen. Nach Absetzen der Kortikoidexterna nahm das Gewicht wieder ab, und das Mondgesicht bildete sich zurück.

Eine seltene Nebenerscheinung, die fast alle Steroide verursachen können, sind *allergische Reaktionen*. So berichteten Savel u. Mitarb. über zwei Frauen, die während der Einnahme eines OH ein Erythema exsudativum multiforme und ein Erythema nodosum bekamen. Die Autoren konnten in vitro eine Überempfindlichkeit der Lymphozyten gegenüber dem im OH enthaltenen Östrogen, dem Mestranol, nachweisen. Dagegen führen Baden u. Mitarb. ein Erythema nodosum, das während der Einnahme eines oralen Kontrazeptivums auftrat, auf das darin enthaltene Gestagen zurück.

Gegenüber Progesteron sind von mehreren Autoren (Shelley u. Mitarb., Hoischen u. Mitarb., Mosonyi u. Mitarb., Jones u. Mitarb.) Autosensibilisierungen beschrieben worden.

Vielleicht ist auch der *Herpes gestationis* hier zu nennen. Dieses Leiden, das meist für eine Sonderform des Morbus Duhring gehalten wird, beginnt gewöhnlich in der 2. Schwangerschaftshälfte und ist gekennzeichnet durch juckende, papulo-vesikulöse oder bullöse Veränderungen. Es bildet sich in der Regel innerhalb eines Monats nach der Entbindung zurück. Lynch u. Mitarb. beobachteten eine Frau, bei der durch Gestagengaben ein Rezidiv ausgelöst wurde. Ranneberg u. Mitarb. berichteten über einen Fall von Herpes gestationis, der sich nach der Entbindung wesentlich besserte, aber nicht völlig abheilte. Nach Einnahme weniger Tabletten eines Gestagens trat eine Verschlimmerung der Hauterscheinungen in einem nie vorher gesehenen Maße mit quälendem Juckreiz auf. Morgan beobachtete eine Frau, bei der ein Herpes gestationis 6 Monate über die Geburt hinaus durch einen OH unterhalten wurde, der eine Schwangerschaft verhindern sollte.

Paradox erscheint, daß Glukokortikoide, wenn auch nur ganz selten, allergische Reaktionen hervorrufen können. So beobachteten Bonner u. Mitarb. nach innerlicher Verabreichung von Prednison ein makulopapulöses z. T. bullöses Exanthem, das nach Hydrokortisongaben abheilte und nach erneuter Prednisongabe wieder ausbrach. Comaish sah ein Exanthem bei einer Frau einmal nach intraartikulärer sowie zweimal nach oraler Prednisongabe. Testungen mit hochgereinigtem Prednison, Prednisolon und Hydrokortison zeigten stark positive Reaktionen.

Nach äußerlicher Kortikoidanwendung wurden ebenfalls allergische Hautreaktionen beschrieben. So konnte Wiegel 37 Kontaktallergien gegenüber kortikoidhaltigen Externa aus der Literatur zusammenstellen und 4 eigene Beobachtungen hinzufügen. Bei diesen Kontaktallergien handelte es sich um folgende Kortikoide:

Hydrokortison-Azetat, Triamcinolon-Acetonid, Triamcinolon, Dexamethason, 6-alpha-Chlor-Prednison-Azetat, 6-alpha-Methylprednisolon, Natrium-Hydrokortison-Succinat. In einem Teil der Fälle wurden die Kontaktallergien durch Epikutantestungen sichergestellt. Praktisch kann man derartige Allergien im Hinblick auf ihre große Seltenheit vernachlässigen.

Ich habe versucht, die wichtigsten Wirkungen der Steroide an der Haut zu zeigen. Auf einige Probleme bin ich bei der Kürze der Zeit nicht eingegangen. Ich nenne hier die Bedeutung des Androgenmetabolismus in der Haut für die Entstehung der Akne und die Bedeutung des verminderten Dehydroepiandrosteron-Gehaltes im Plasma für die Pathogenese der Psoriasis (Holzmann u. Mitarb.). Weitere Forschungen bes. über den Metabolismus der Steroide in der Haut werden sicher die Pathogenese einiger Dermatosen klären und neue therapeutische Möglichkeiten zeigen.


## Literatur


Baden, H.P., Holcomb, F.D.: Arch. Derm. 98, 634 (1968).

Bandmann, H.-J.: in: Fortschr. der prakt. Derm. u. Venerol. Bd. VI, Berlin-Heidelberg-New York: Springer 1970.

Bjornberg, A., Hellgren, L.: Arch. Derm. 92, 52 (1965).

Blank, I.H.: J. Amer. med. Ass. 164, 412 (1957).

Bonner, C.D., Homburger, F.: New Engl. J. Med. 256, 131 (1957).

Bullough, W.S., Homan, J.D., Laurence, E.B., Overbeek, G.A.: J. Endocr. 41, 453 (1968).

Carr, R.D., Belcher, R.W.: Acta derm.-venereol. (Stockh.) 49, 508 (1969).

-- Hamilton, J.F.: Arch. Derm. 99, 26 (1969).

Chernosky, M.E., Knox, J.M.: Arch. Derm. 90, 15 (1964).

Comaish, St.: Brit. J. Derm. 81, 919 (1969).

Degos, R. et al.: Ann. Derm. Syph. (Paris) 96, 5 (1969).

Ebling, F.J.: J. Embryol. exp. Morph. Vol. 5, Part 1, pp. 74-82 (1957).

Eller, J.J., Eller, W.D.: ref. in Derm. Wschr. 122, 1064 (1950).

Erickson, C.L.R., Peterka, E.S.: J. Amer. med. Ass. 203, 980 (1968).

Feinblatt, B.I. et al.: Amer. J. Dis. Child. 112, 218 (1966).

Gill, K.A., Baxter, D.L.: Arch. Derm. 89, 734 (1964).

Goldzieher, J.W., Roberts, I.S., Rawls, W.B., Goldzieher, M.A.: Arch. Derm. 66, 304 (1952).

Götz, H., Reichenberger, M.: Münch. med. Wschr. 110, 1913 (1968).

Greither, A.: Hdb. d. Haut u. Geschlechtskrankheiten. Berlin-Heidelberg-New York: Springer 1969, III,2 (155).

Grimmer, H.: Z. Haut- u. Geschl.-Kr. 45, 535 (1970).

Hammerstein, J., Cupceancu, B.: Dtsch. med. Wschr. 94, 829 (1969).

Hohlweg, W.: Zbl. Gynäk. 76, 1890 (1954).

Hoischen, W., Steigleder, G.K.: Dtsch. med. Wschr. 91, 398 (1966).

Holzmann, H., Korting, G.W.: Dtsch. med. Wschr. 93, 1721 (1968).

- Morsches, B., Gebhardt, R., Hoffmann, G., Menzel, P., Oertel, G.W.: Z. Haut- u. Geschl.-Kr. 45, 579 (1970).

Hopf, G.: Therapiewoche 18, 687 (1968).

Husmann, F., Suchan, P., Gross, W.: Verh. dtsch. Ges. inn. Med. Bd. 74, München: J.F. Bergmann 1968.

Jelinek, J.E.: Arch. Derm. 101, 181 (1970).

Jones, W.N., Gordon, V.H.: Arch. Derm. 99, 57 (1969).

Kirketerp, M.: Ugeskr. Laeg. 127, 405 (1965).

Kortin, G.W., Holzmann, H.: Die Sklerodermie und ihr nahestehende Bindegewebsprobleme, Stuttgart: Thieme 1967.

Krüskemper, H.L.: Anabole Steroide. Stuttgart: Thieme 1963.
Kümmel, J.: Med. Welt (Stuttg.) 17, 138 (1966).
Ludwig, H.: in: H. Teller, Ergebnisse der Angiologie. Stuttgart: Schattauer
    1971.
Lynch, F.W., Albrecht: R.J.: Arch. Derm. 93, 446 (1966).
March, C., Kerbel, G.: J. Amer. med. Ass. 187, 676 (1964).
Morgan, J.K.: Brit. J. Derm. 80, 456 (1968).
Mosonyi, L., Halmy, L., Csiky, Th.: Hautarzt 18, 412 (1967).
Napp, J.H.: zit. bei K. Winkler.
Nasemann, Th.: Hautarzt 21, 46 (1970).
Nikolowski, W.: cosmetologica 19, 193 (1970).
Orfanos, C.: Fortschr. Med. 86, 106 (1968).
Pochi, P.E., Strauss, J.S.: Arch. Derm. 95, 47 (1967).
Pockrandt, H.: Zbl. Gynäk. 76, 1889 (1954).
Polano, M.K., Suurmond, D., Lely, M.A. v.d., Warnaar, P.: Brit. J. Derm. 83,
    93 (1970).
Porter, P.S., Lyle, J.S.: Arch. Derm. 93, 402 (1966).
Pfeifer, G.W., Schreiner, K.: in: H. Teller, Ergebnisse der Angiologie.
    Stuttgart: Schattauer, 1971.
Ranneberg, K.M., Holzmann, H.: Med. Welt (Stuttg.) 21, 1727 (1970).
Rauramo, L., Punnonen, R.: Z. Haut- u. Geschl.- Kr. 44, 463 (1969).
Resnik, S.: J. Amer. med. Ass. 199, 95 (1967).
Roenigk, H.H., Gottlob, M.E.: Arch. Derm. 102, 260 (1970).
Savel, H., Madison, J.F., Meeker, C.I.: Arch. Derm. 101, 187 (1970).
Schaaf, F.: Probleme dermatologischer Grundlagenforschung. Heidelberg:
    A. Hüthig 1969.
- Gross, F.: Arch. klin. exp. Derm. 205, 312 (1957).
Schieferstein, G.: Z. Haut-, u. Geschl.-Kr. 41, 354 (1966).
Schimpf, A.: Fortschr. Med. 87, 910 (1969).
Schlierf, G.: Dtsch. med. Wschr. 95, 1858 (1970).
Schreus, H.Th.: Münch. med. Wschr. 98, 728 (1956).
Scoggins, R.B., Kliman, B.: J. invest. Derm. 45, 347 (1965).
Shelley, W.B., Preucel, R.W., Spoont, S.S.: J. Amer. med. Ass. 190, 35
    (1964).
Sneddon, J.: Brit. med. J. 5645, 671 (1969).
Soltermann, W.: Praxis 49, 18 (1960).
Spangler, A.S., Antoniades, H.N., Inderbitzin, T.M., Sotman, S.L.: Fortschr.
    Med. 87, 473 (1969).
- - - - J. clin. Endocr. 29, 650 (1969).
Steigleder, G.K.: Klin. Mbl. Augenheilk. 148, 800 (1966).
Strauss, J.S., Pochi, P.E.: J. invest. Derm. 36, 293 (1961).
Szyszymar, B., Malaszynska-Olkiewicz, E.: ref. Zbl. Haut- u. Geschl.-Kr.
    128, 9 (1971).
Thiers, J., Perrot, H., Arcano, F.: Dermatologica (Basel) 135, 455 (1967).
Tronnier, H.: Aesthet. Med. 15, 381 (1966).
- Landarzt 45, 985 (1969).
- Hoppe-Seyler, G.: Therapiewoche 19, 1024 (1969).
Vail, J.T.: J. Amer. med. Ass. 201, 671 (1967).
Vankos, J.: Ther. hung. 18, 36 (1970).
Vine, S. u. Mitarb.: zit. bei Vail.
Weber, B.: persönliche Mitteilung.
Wiegel, O.: Med. Welt (Stuttg.) 19, 828 (1968).
Winkler, K.: Hormonbehandlung in der Dermatologie 2. Auflage. Berlin:
    Walter de Gruyter 1969.
Wright, E.T., McGillis, T.J., Sobel, H.: Dermatologica (Basel) 140, 124
    (1970).
Zakon: zit. bei Jänner Aesthet. Med. 15, 378 (1966).
Zimmermann, T.S. u. Mitarb.: zit. bei Vail.

Symp. Dtsch. Ges. Endokrin. 17, 43–47 (1971)
© by Springer-Verlag

# Hormone und Behaarung

## Hormones and Hair Growth

E. Ludwig

Universitäts-Hautklinik Hamburg
Mit 7 Abbildungen[+]

## Summary

Contrary to what is generally believed, human scalp and body hair is affected by only a limited number of hormones. Major and therefore clinically
important changes in the type and distribution of the hair are induced only
by testosterone and other biologically active androgens. They are responsible for the growth and maintenance not only of male sexual but also for the
so-called ambisexual hair. Male amounts of androgens are also required for
the development of pattern alopecia (common baldness) in genetically predis-
posed subjects of both sexes. If the production of androgens is lacking or
deficient, androgen-dependent types of hair fail to grow, or if previously
present are lost.

Increased levels of circulating androgens are of consequence only in the
female, where they give rise either to hirsutism or to androgenetic alopecia
or both. The clinical picture of the latter condition is described. Emphasis
is laid on the fact that androgenetic alopecia is as indicative of increased androgen production (above what is considered female level) as is
hirsutism.

Instances of increased or reduced responsiveness of hair follicles to
androgenic stimulation, which should never be assumed unless any overproduc-
tion of androgens has carefully been excluded, are a healthy young man without terminal hair on the trunk, a "hairless woman", and a heavily bearded
woman nursing her baby. Future investigations will have to clarify the biochemical basis of such pronounced differences in the end organ-responsiveness.


Haare sind im Grunde genommen, ähnlich wie der Talg, ein holokrines Sekret
der Haut. Während die Zellen der Talgdrüsen verfetten und den Talg bilden,
der sich auf der Hautoberfläche ausbreitet, verhornen die Zellen der Haar-
matrix und werden zu Haaren, die nach einer unterschiedlich langen Wachstums-
phase ausfallen, man könnte auch sagen, ausgeschieden werden. Da bei vielen
Endokrinopathien Hautveränderungen auftreten – man denke an die Pigmentierung beim M. Addison, die Striae beim Cushing-Syndrom, die Necrobiosis lipoi-
dica beim Diabetes – dürfte man bei endokrinen Störungen auch ein häufiges
Vorkommen von Haarveränderungen erwarten. Tatsächlich führen aber nur wenige
Hormone zu klinisch manifesten Veränderungen am Haarkleid. Die eindrucksvoll-
sten Haarveränderungen, nämlich die Alopecia areata universalis, bei der jede Art von Behaarung verloren geht, und das unter der Bezeichnung Hypertri-
chosis lanuginosa acquisita bekannte paraneoplastische Syndrom, bei dem ein
völlig ungehemmtes Wachstum aller Arten von Haaren stattfindet, haben nach
bisherigen Untersuchungen kein endokrines Substrat.

---

[+] Halbtonbilder s. Anhang S. 185

## Haarwachstum und Hormone

Mit Ausnahme weniger Hautareale wie Handteller, Fußsohlen, Lippenrot entstehen zwischen dem zweiten und fünften Foetalmonat am ganzen Körper einmalig Haaranlagen. Mit Ausnahme des Kopfes ist die Dichte der gebildeten Haarfollikel überall etwa die gleiche. Die beim Erwachsenen feststellbaren Unterschiede in der Dichte der Körperbehaarung beruhen auf einer "Verdünnung" als Folge unterschiedlichen Wachstums der Haut im Bereich einzelner Areale. Im Laufe des Lebens sind es genetische und hormonelle Faktoren sowie das Alter, die über den Typus des einzelnen Haares, über das Verteilungsmuster und den Ablauf des Haarcyclus, worunter man die Sequenz von Wachstum, Ausfall und Ersatz des Haares versteht, entscheiden. Die genetische Konstellation steckt den Rahmen ab, innerhalb dessen sich die fördernden oder hemmenden Einflüsse der Hormone in Abhängigkeit auch von der Ansprechbarkeit des Erfolgsorgans manifestieren können, und bestimmt, was "genetically permissible" ist.

Unsere derzeitigen Kenntnisse über die Wirkung der hypophysären Hormone, der Hormone der Schilddrüse, der Epithelkörperchen, des Insulins und selbst der Hormone der Nebennierenrinde - sofern es sich nicht um Androgene handelt - auf die Behaarung des Menschen sind recht lückenhaft. Das liegt einerseits daran, daß die im Tierversuch, vornehmlich an Nagern, gewonnenen Erkenntnisse nur bedingt auf den Menschen übertragen werden können, und andererseits an der sich erst in der letzten Zeit anbahnenden Zusammenarbeit zwischen Endokrinologen und am Haar besonders interessierter Dermatologen.

Da die durch Ausfall oder Überproduktion der genannten Hormone bedingten Haarveränderungen nur Randerscheinungen klinischer Syndrome darstellen und außerdem bekannt sind, sollen hier nur die Steroidhormone besprochen werden.

## Cortisol und Corticosteroide

Erfolge bei der Behandlung der Alopecia areata mit hohen Dosen von Corticosteroiden haben zu der auch unter Dermatologen weit verbreiteten Ansicht geführt, daß diese besondere Alopecieform mit einem Cortisolmangel zusammenhänge oder zumindest hormonell bedingt sei. Diese Annahme ist aber unbegründet. Bei der Alopecia areata liegt kein Cortisolmangel vor und außerdem hat Cortisol in physiologischen Mengen keinen nachweisbaren Einfluß auf das Haarwachstum des Menschen. Bei der Nebennierenrindeninsuffizienz gibt es keine Veränderungen der Behaarung, die sich nicht durch den Ausfall der adrenalen Androgene erklären ließen. Umgekehrt dürfte der bei Fällen von M. Cushing beobachtete Hirsutismus auf eine gleichzeitig gesteigerte adrenale Androgenproduktion zurückzuführen sein (5). Die Wirkung der Corticosteroide bei der Alopecia areata beruht im wesentlichen auf der Auflösung eines entzündlichen peribulbären Infiltrates, das eine normale Haarbildung verhindert (2). Es handelt sich somit um einen antiphlogistischen Corticosteroideffekt.

## Weibliche Sexualhormone

Ähnlich wie dem Cortisol werden auch den Oestrogenen und dem Progesteron vielfältige Wirkungen auf die Behaarung der Frau zugeschrieben. Die Gründe hierfür sind einerseits die bekannte Tatsache, daß das Kopfhaar der Frauen länger wächst als das der Männer, und andererseits die in der Schwangerschaft und in der Menopause vorkommenden Veränderungen des weiblichen Haarkleides.

Solange genaue Untersuchungen ausstehen, erscheint es nicht statthaft, für Erscheinungen wie z. B. die in der Menopause oft zunehmende Behaarung des Gesichtes und die fortschreitende Abnahme der Achselbehaarung einfach einen Mangel an weiblichen Sexualhormonen verantwortlich zu machen. Gesichert ist die wohl durch den erhöhten Oestrogen- und Progesteronspiegel bedingte Verlangsamung des Haarcyclus in der Schwangerschaft, die eine Zunahme

der Haarfülle zur Folge hat, und die Häufigkeit eines Effluviums 3 Monate
nach der Entbindung, für dessen Zustandekommen sicher auch das Geburtstrauma
und neuerdings auch die Thromboseprophylaxe mit Anticoagulantien verantwortlich gemacht werden kann. Ebenso sicher ist aber auch, daß die Ovariektomie
keine nennenswerten Veränderungen des Haarkleides zur Folge hat (10), und
daß bei ovarieller Agenesie trotz cyclusgerechter Oestrogen-Gestagenbehand-
lung die Bildung der Scham- und Achselhaare ausbleibt (3).


## Androgene

Die für die Behaarung mit Abstand wichtigsten Steroidhormone sind das Testosteron und - in Abhängigkeit von ihrer biologischen Aktivität - die übrigen
Androgene. Mangel und Überproduktion an Androgenen führen zu augenfälligen,
diagnostisch bedeutsamen Veränderungen am Haarkleid. Auf eine knappe Formel
gebracht, bewirkt Testosteron das Wachstum all jener Körperhaare, die man
als männlich sexuell und als ambisexuell bezeichnet, und löst bei Vorliegen
einer entsprechenden genetischen Veranlagung jene retrograde Metamorphose
der Kopfhaare des Scheitels aus, deren Endresultat die Glatze ist, und zwar
sowohl beim Mann als auch bei der Frau. Wenig bekannt ist, daß Testosteron
auch ohne Vorliegen einer Anlage zur Glatzenbildung bei fast allen jungen
Männern und bei vielen Frauen zum Verlust eines schmalen Streifens von Kopf-
haaren entlang der Stirnhaargrenze und im Bereich der Stirnschläfenwinkel
führt (4). Die Haare dieser Regionen nehmen also eine Zwischenstellung zwischen echten männlich sexuellen und ambisexuellen Haaren ein. Den Beweis dafür, daß auch für das Wachstum der sog. ambisexuellen Behaarung der Frau
Testosteron erforderlich ist, liefern Frauen mit Gonadendysgenesie. Achsel-
und Schambehaarung treten bei ihnen trotz cyclusgerechter Behandlung mit
Oestrogen und Gestagen erst nach Verabreichung von Testosteron auf, um wieder auszufallen, sobald die Androgenzufuhr abgesetzt wird (3). Aufgrund dieser Befunde wäre es sinnvoll, anstelle von männlich sexuellen und ambisexuellen von androgen-abhängigen Haaren zu sprechen.
    Mangel an Testosteron hat, je nachdem ob er prae- oder postpuberal auf-
tritt, entweder das Ausbleiben der Bildung der genannten Haararten oder deren Verlust zur Folge. Daher das Fehlen aller androgen-abhängigen Haare bei
Eunuchen, Eunuchoiden, bestimmten Formen des Hypogonadismus, bei Gonadendysgenesie, dem Sheehan-Syndrom und dem M. Addison bei der Frau. Bei letzterem
reicht die ovarielle Androgenproduktion allein nicht zur Aufrechterhaltung
der Sexualbehaarung aus.
    Die Haarfollikel des Mannes werden durch die zirkulierenden physiologi-
schen Androgenmengen bereits maximal stimuliert. Daher kann weder eine gesteigerte endogene Produktion noch eine exogene Zufuhr das Wachstum der Körperbehaarung steigern oder eine gegebenenfalls bestehende Glatzenbildung be-
schleunigen. "Weibliche" Mengen an zirkulierenden Androgenen reichen normalerweise nicht aus, um die den Körperfollikeln der Frau innewohnenden Poten-
zen zu realisieren. Steigt die endogene Androgenproduktion aber an oder werden Androgene von außen zugeführt, so können auch die Körperhaarfollikel der
Frau Terminalhaare hervorbringen. Ebenso können bei Vorliegen einer entsprechenden Anlage auch die Kopfhaare der Frau jene Involution erfahren, die bei
der Glatzenbildung des Mannes stattfindet. Der klinische Ausdruck einer vermehrten androgenen Stimulierung der Körperhaarfollikel bei der Frau ist der
*Hirsutismus*, der der Kopfhaarfollikel die *andro-genetische Alopecie*. Daß
beide Symptome häufig gleichzeitig auftreten, ist einleuchtend. Da der Hirsutismus in den meisten Fällen auf ein Plus an Androgenen hinweist, das irgend-
wo im Organismus auf irgendeine Weise entsteht, sollte man eine gesteigerte
Androgenempfindlichkeit der Haarfollikel erst dann in Betracht ziehen, wenn
eine Mehrproduktion an Androgenen mit Hilfe aller zur Verfügung stehenden
Untersuchungsmethoden ausgeschlossen worden ist (8). Das Gleiche gilt auch
für die seltene praemature Pubarche (12).

<u>Die androgenetische Alopecia</u>

Neueren Datums und daher nicht so allgemein bekannt wie der Hirsutismus ist
die androgenetische Alopecie (6). Sie ist klinisch durch eine mehr oder weniger ausgeprägte, sich im Laufe von Jahren entwickelnde Lichtung des Haares
am Scheitel und durch das Erhaltenbleiben eines Haarsaumes entlang der
Stirnhaargrenze gekennzeichnet (Abb. 1 u. 2). Dieser frontale Haarsaum ist
auch noch in weit fortgeschrittenen Fällen zu erkennen (Abb. 3). Hinzu kommt
als fast konstantes Symptom eine ausgeprägte Seborrhoea oleosa. Die urinäre
Testosteronausscheidung ist fast in allen Fällen signifikant erhöht (1, 9).
Die gegenüber der Norm um rund 100% erhöhten Mittelwerte entsprechen fast
der Erhöhung, die man beim sog. idiopathischen Hirsutismus findet (1). Daraus ergibt sich, daß der androgenetischen Alopecie die gleiche diagnostische
Bedeutung zukommt wie dem Hirsutismus. Die klassische Form der androgenetischen Alopecie scheint mit mäßig erhöhten Androgenwerten korreliert zu
sein (7). Erreicht die Androgenproduktion männliche Werte, so entsteht eine
Glatze, die sich nicht von der männlichen unterscheidet (Abb. 4). In solchen
Fällen ist fast immer auch ein Hirsutismus vorhanden.


<u>Ansprechbarkeit der Haarfollikel</u>

Die schon erwähnte unterschiedliche Ansprechbarkeit der Haarfollikel auf
androgene Stimulierung erklärt eine Reihe klinischer Befunde, deren Deutung
aufgrund der gemessenen Androgenwerte nicht möglich ist. Zu erwähnen ist
zunächst die sehr unterschiedliche Körperbehaarung bei gleichaltrigen gesunden Männern mit einer urinären Testosteronausscheidung (bestimmt nach der
Methode von Voigt u. Mitarb., 11) im altersentsprechenden Normbereich
(Abb. 5 u. 6). Bei der testikulären Feminisierung dürfte es die generelle
Androgenresistenz der Erfolgsorgane und damit auch der Körperhaarfollikel
sein, die das Fehlen der Sexualbehaarung bedingt. Bei der berühmten Barbuda
des Malers Ribera (Abb. 7) spricht trotz des imposanten Bartes die durch
Säugling und Laktation belegte Mutterschaft gegen eine stark erhöhte endogene Androgenproduktion. Es dürfte sich um einen Fall von sog. idiopathischen Hirsutismus handeln, bei dem in der Regel eine leicht bis mittelmäßig
erhöhte Androgenproduktion mit einer gesteigerten Ansprechbarkeit der Körperhaarfollikel zusammentrifft.
     Die Klärung der biochemischen Grundlagen so unterschiedlicher Endorgan-
ansprechbarkeit wäre vielleicht eine ebenso lohnende wie reizvolle Aufgabe
für die Endokrinologie in den 70er Jahren.


<u>Literatur</u>

  1. Apostolakis, M., Ludwig, E., Voigt, K.D.: Testosteron-Oestrogen- und
     Gonadotropinausscheidung bei diffuser weiblicher Alopecie. Klin. Wschr.
     <u>43</u>, 9-15 (1965).
  2. Braun-Falco, O., Rassner, B.: Klinik, Pathogenese und Therapie der Alo-
     pecia areata. In: Fortschritte der praktischen Dermatologie und Venereo-
     logie hrsg. von Marchionini, A. Berlin-Heidelberg-New York: Springer,
     1965.
  3. Greenblatt, R.B.: Factors influencing the growth of sexual hair. In:
     Greenblatt, R.B.: The hirsute female. Springfield, Ill.: Thomas, 1963.
  4. Hamilton, J.B.: Patterned loss of hair in man: types and incidence.
     Ann. N. Y. Acad. Sci. <u>53</u>, 708-728 (1951).
  5. Liddle, G.W.: Cushing's syndrome and problems of hirsutism. In: Green-
     blatt, R.B.: The hirsute female. Springfield, Ill.: Thomas, 1963.
  6. Ludwig, E.: Die androgenetische Alopecie bei der Frau. Arch. klin. exp.
     Derm. <u>219</u>, 558-564 (1964).

7. -  The role of sexual hormones in pattern alopecia. In: Biopathology of
   pattern alopecia. Hrsg. von Baccaredda - Boy, A., Moretti, G., Frey, J.R.
   Basel/New York: Karger, 1968.
8. Mansuwan, K., Kalant, N.: Urinary excretion of testosterone in idiopathic
   hirsutism. Proc. Soc. exp. Biol. (N.Y.) $\underline{119}$, 911-914 (1965).
9. Piérard, J., Kint, A., Backer, J.v.: Sur le role de l'hormone male dans
   l'álopécie féminine. Arch. belges Derm. $\underline{24}$, 409-414 (1968).
10. Summers, V.K.: The role of the adrenal cortex and gonads in the control
    of sexual hair distribution. Acta med. scand. $\underline{136}$, 105-111 (1949).
11. Voigt, K.D., Volkwein, U., Tamm, J.: Eine Methode zur Bestimmung der
    Testosteron-Ausscheidung im Urin. Klin. Wschr. $\underline{42}$, 642-644 (1964).
12. Zurbrügg, R.P., Gardner, L.I.: Urinary $C_{19}$ steroids in two girls with
    precocious sexual hair. J. clin. Endocrin. $\underline{23}$, 704-708 (1964).

Symp. Dtsch. Ges. Endokrin. _17_, 49–56 (1971)
© by Springer-Verlag

# Über die cutane Penetration von Sexualsteroid-hormonen

## Cutaneous Penetration of Sexual Steroid Hormones

K. H. KOLB UND P. E. SCHULZE

Forschungslaboratorium der Schering AG, Dept. Biodynamik, Berlin
Mit 4 Abbildungen

## Summary

Systematic pharmacokinetic tests were conducted in the skin of guinea pigs
and hairless mice by topical application of radioactively labeled ethinyl
estradiol, testosterone, progesterone, norethisterone acetate, d-norgestrel,
and 17α-hydroxyprogesterone capronate. The animal tests were complemented
by investigations in human skin.

Animal testing was carried out in vivo and in isolated skin. Testing in
humans was done in clinically normal skin, "stripped" skin, and skin biopsies.

Quantitative determinations in the individual skin layers or media were
complemented by qualitative microautoradiographs. The active ingredients had
been incorporated each time in a water/oil emulsion, an oil/water emulsion,
and a fatty base, particle size and concentration being the same. For each
steroid hormone investigated, the penetration characteristics were demonstrated as a function of time. In some cases, the route of penetration was
shown and the extent of penetration described as a function of formulation.

Active ingredient release was vehicle-dependent in both species and the
sequence identical. About ten times as much ethinyl estradiol and testosterone penetrated the skin from the fatty base as the oil/water emulsion. The
rate of penetration of the progestational steroid hormones was highest from
oil/water emulsions. Distribution in the skin, which is also vehicle-dependent, is stated in addition to the penetrated portion.

In humans, penetration of clinically normal skin and skin from which the
stratum corneum had been removed was compared and described. It was shown
that with a change of skin condition the active ingredients, having their
own time-related penetration characteristics, penetrate quickly from the
contact surface, but that transport from the deeper layers of the dermis does
not take place at the same time.

1929 kam Zondek (15) nach der Einreibung alkoholischer Tinkturen von
oestrogenen Hormonen zu der Schlußfolgerung, daß diese, topisch angewandt,
klinisch fast so wirksam seien wie die Injektion der Hormone, aber andererseits die Sexualhormone, als ölige Lösungen appliziert, weniger wirksam seien.
Diesem schon biopharmazeutische Gesichtspunkte beinhaltenden Zitat folgten
zahlreiche weitere Bestätigungen der percutanen Resorbierbarkeit von Sexualsteroidhormonen (1, 2, 3, 5, 8, 9, 10, 11, 14).

Die Ergebnisse der verschiedenen Experimentatoren widersprechen sich zu
einem großen Teil. Die Gründe hierfür sind neben methodischer Unvergleich-
barkeit vor allem in der Komplexität des Penetrationsvorganges selbst zu sehen. Denn neben den morphologisch faßbaren Widerständen, die durch physiologische und anatomische Eigenschaften des Applikationsorganes Haut – einem
der inhomogensten Organe des Körpers – bedingt sind, gibt es mehrere Faktoren, bedingt durch die physiko-chemischen Eigenschaften, deren Einfluß auf

die cutane Penetration zwar im einzelnen theoretisch und experimentell berechenbar, aber in ihrer Summe nicht vorbestimmbar sind.

Neben den physiko-chemischen Eigenschaften der Wirkstoffe wird die Diffusion und Penetration auch von den Eigenschaften der verwendeten Grundlage, dem Vehikel, bestimmt.

Bei der externen Anwendung von Sexualsteroidhormonen interessiert die Frage nach Geschwindigkeit und quantitativem Verlauf der Diffusion des inkorporierten Pharmakons aus dem Vehikel, dessen Abgabe an die obere Hautschicht, Weg und Tiefe des Eindringens in die Haut, das Durchdringen der Haut und der Übergang in Transportmedien.

Im folgenden wird versucht, vom Standpunkt der Pharmakokinetik aus für jeden dieser Schritte die Beziehung zur Quantität der Wirkstoffe zu beschreiben als Grundlage für die nachzufolgende Erkenntnis der funktionellen Aktivität relativiert zur Quantität.

In unseren Versuchen wurde neben Humanversuchen zur Ermöglichung systematischer Untersuchungen tierische Haut in in vivo und in vitro Modellversuchen verwandt. Neben sogenannten haarlosen Mäusen verwandten wir überwiegend Meerschweinchen, weil bei dieser Spezies die Kenntnis über das Verhalten der Haut – wie Zellproliferation und Mitoseaktivität – gegen äußere Faktoren am besten bekannt ist. Für die Bewertung der Tierversuche gilt jedoch der Vorbehalt, daß eine Kongruenz des Verhaltens zwischen Tier- und Menschenhaut a priori nicht bestehen muß und selbst eine partielle Übertragung der am Tier gewonnenen Erkenntnisse nur dann möglich ist, wenn ein gleiches Verhalten an menschlicher Haut gefunden wird.

Wir haben aus der Gruppe der Oestrogene die Penetration von Aethinyloestradiol-$^3$H untersucht. Wie auch bei den folgenden Untersuchungen wurde der Wirkstoff in gleicher Teilchengröße, in gleicher Konzentration jeweils in eine Öl-Wasser-Emulsion, in eine Wasser-Öl-Emulsion und in eine Fettsalbe inkorporiert. Von jeder galenischen Zubereitung wurde die gleiche Menge auf eine stets gleich große depilierte Hautfläche aufgetragen.

Im ersten Modellversuch (Abb. 1) wurde die zeitabhängige Penetrationsrate

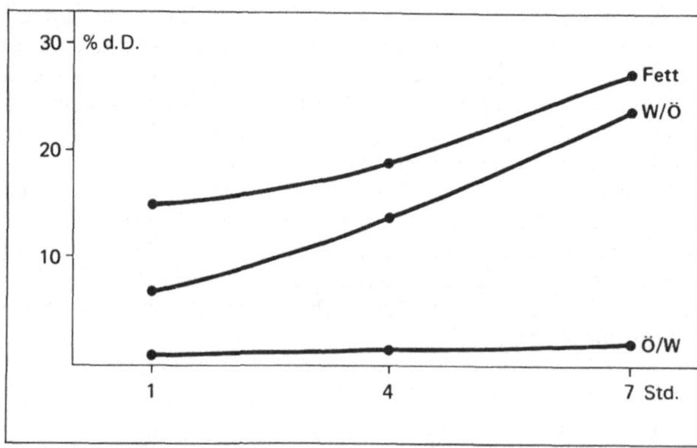


Abb. 1. Penetration von Aethinylöstradiol 6-7$^3$H (0,1%ig in Ö/W-W/Ö-Fettgrundlage) in vitro an Meerschweinchen
Wiederbestimmung in der Haut in % der appl. Dosis

an Meerschweinchen untersucht. Der in der Haut befindliche Anteil wird hier in Prozent der applizierten Dosis wiedergegeben. Dargestellt ist wie bei allen nachfolgenden Untersuchungen der Mittelwert aus fünf Einzeluntersuchungen.

Die Graphik zeigt die Abhängigkeit des Penetrationsausmaßes von der galenischen Formulierung. So ist die Freigabe und Penetration von Aethinyl-

oestradiol aus einer Wasser-Öl-Emulsion und aus einer Fettsalbe zu jedem Zeitpunkt um ein vielfaches größer als aus einer Öl-Wasser-Emulsion. Aus dieser
war nach 7-stündiger Kontaktzeit noch 98% unresorbiert.

Die Begründung für die Untersuchung der Wirkstoffe in den drei klassischen galenischen Formulierungen - Creme, Salbe und Fettsalbe - waren prag-
matische, anwendungstechnische Überlegungen. Daneben sollte durch die Anwendung dieser differenten Zubereitungsformen die Overton-Meyersche Lipoidlös-
lichkeitstheorie und die Theorie, nach der für die Penetration der oberfläch-
lichen Epidermisbarriere einem Wasser-Fett-Verteilungskoeffizienten, der nahe
beim Wert 1 liegt, Bedeutung zukäme, eine praxisnahe Bestätigung finden.
Vorweggenommen sei, daß beide Größen im in vivo Experiment durch andere Faktoren - wie etwa Molekulargröße, Partikelgröße, Viskosität und Hydratation -
überlagert werden.

Da vielfach Sexualsteroidhormone in alkoholischer Lösung angewandt wurden, haben wir das quantitative Penetrationsverhalten einer aethanolischen
0,1%igen Aethinyloestradiollösung untersucht. In dieser Verabreichungsform
entsprach die Penetrationsrate in Größenordnung der nach Anwendung einer
Öl-Wasser-Emulsion.

Wohl infolge einer raschen Dissoziierung wurde zu jeder Untersuchungszeit
zwar eine maximale Wirkstoffbeladung des Str. disjunctum und conjunctum im
Stratum corneum gefunden, von dort jedoch erfolgte nur ein geringfügiger Abtransport in die tiefer liegenden Schichten.

In Tab. 1 wird versucht, nach 7-stündiger Kontaktzeit die $^3$H-Verteilung
in Abhängigkeit der galenischen Formulierung in Zuordnung zu den einzelnen
Hautschichten in Prozent der applizierten Dosis wiederzugeben. Diese selbst-

| 7 Std. | Ö/W | W/Ö | Fett |
|---|---|---|---|
| Cornea | 0.905 | 12.491 | 15.415 |
| Epidermis | 0.341 | 4.321 | 5.321 |
| Str.Papill. | 0.010 | 2.124 | 1.945 |
| Str. Reticul. | 0.003 | 2.321 | 1.832 |
| Subcutis | 0.012 | 2.953 | 3.053 |

verständlich grob schematische Übersicht veranschaulicht, daß in jedem Fall
der Hauptanteil der penetrierten Menge im Stratum corneum nachweisbar war,
wobei nach Zerlegung der Cornea wiederum die größte Menge im Str. disjunctum
und conjunctum nachgewiesen wurde.

Neben den absoluten quantitativen Unterschiedsmerkmalen im Radioaktivi-
tätsgehalt nach Anwendung der Öl-Wasser-Emulsion gegenüber denen unter Applikation der Wasser-Öl-Emulsion und Fettsalbe wurde in den beiden letzteren
Fällen eine vergleichbare und nur geringe Aktivitätsverdünnung in Corium und
Subcutis gemessen.

Um Kenntnis über die Passagezeit durch Epidermis und Corium und damit zum
Übergang in gefäßführende Schichten zu erhalten, wurde ein in vitro Durch-
strömungsmodell benutzt. Die Versuchsanordnung ist nach dem Prinzip aufge-
baut, daß auf eine Kammer mit einem Volumen von 5 ml, auf deren Boden sich
eine Zufluß- und Abflußöffnung befindet, frisch exzidierte, von der Subcutis
getrennte Haut so befestigt wird, daß die vorgegebene Penetrationsfläche
(238 qmm) sich zentral über dem Kammervolumen befindet. Unter Temperaturkon-
stanz (37° C) wird das nunmehr durch die Haut geschlossene System mit einer
Nährlösung (Krebs-Ringer-Bicarbonat-Puffer) gefüllt und durch Fotozellen
gesteuert in kurzzeitigen Abständen entleert und wieder aufgefüllt.

Abb. 2 gibt anhand der Wiederbestimmung im Durchströmungsmedium Einblick
über die Passagezeit und deren Abhängigkeit von der galenischen Zubereitungs-
form. Bereits 30 Minuten p.a. konnte unter Anwendung der Wasser-Öl-Emulsion
und der Fettsalbe eine definierbare Radioaktivitätsmenge im Durchströmungsmedium nachgewiesen werden. Bei Versuchsende war nach Applikation von Aethinyloestradiol in der Salbe bzw. Fettsalbe die $^3$H-Menge gegenüber der Freiga-

be aus der Öl-Wasser-Emulsion um etwa das 10fache größer. Zwischen 4 und 6
Stunden erfolgte, bezogen auf % der Eingabe/30 Minuten, das Maximum der
Freigabe.

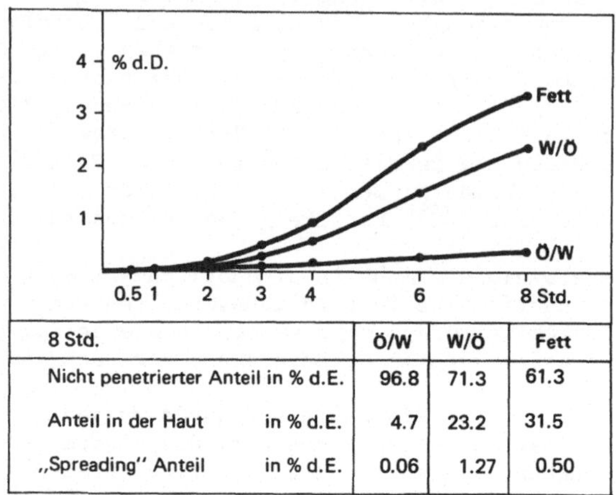



| 8 Std. | Ö/W | W/Ö | Fett |
|---|---|---|---|
| Nicht penetrierter Anteil in % d.E. | 96.8 | 71.3 | 61.3 |
| Anteil in der Haut        in % d.E. | 4.7 | 23.2 | 31.5 |
| „Spreading" Anteil        in % d.E. | 0.06 | 1.27 | 0.50 |

Abb. 2. Penetration von Aethinylöstradiol 6-7[3]H in vitro an der Haut von
Meerschweinchen
Wiederbestimmung im Durchströmungsmedium in % der appl. Dosis

In der unteren Hälfte der Tabelle ist neben dem nicht penetrierten Anteil
die in der Haut wiederbestimmte Menge angegeben. Auch in diesem Modell war
die Penetration aus der Fettsalbe am größten und gegenüber dem penetrierten
Anteil aus der Wasser-Öl-Emulsion um ein vielfaches größer.
   Mit derselben Versuchsanordnung wurde die Penetration an der exzidierten
Haut haarloser Mäuse untersucht (Abb. 3). Begründung für diese Versuchsan-

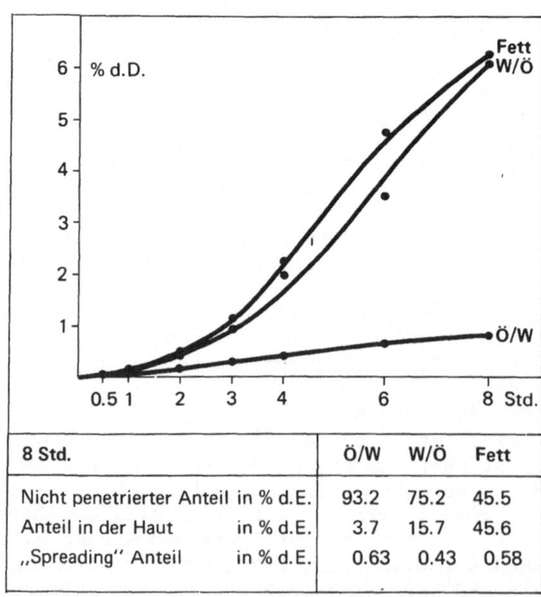



| 8 Std. | Ö/W | W/Ö | Fett |
|---|---|---|---|
| Nicht penetrierter Anteil in % d.E. | 93.2 | 75.2 | 45.5 |
| Anteil in der Haut        in % d.E. | 3.7 | 15.7 | 45.6 |
| „Spreading" Anteil        in % d.E. | 0.63 | 0.43 | 0.58 |

Abb. 3. Penetration von Aethinylöstradiol 6-7[3]H in vitro an der Haut von
haarlosen Mäusen
Wiederbestimmung im Durchströmungsmedium in % der appl. Dosis

ordnung war zum einen die Besonderheit dieser Tierart, bei der die Penetration über die nur rudimentär angelegten Haarfollikel stark reduziert sein mußte und zum anderen die generelle Bestätigung der bisher vorliegenden Kenntnisse und Abhängigkeiten an einem anderen Hautmodell.

Der Penetrationsweg von Aethinylöstradiol bei den haarlosen Mäusen – nachgewiesen durch Mikroautoradiographien – verlief überwiegend transfollikulär. Bezogen auf die Gesamtpenetration betrug der transepidermale-interzelluläre Abtransport nur zwischen 10 und 20%. Beim Meerschweinchen zum Vergleich konnte qualitativ nur ein follikulärer Penetrationsweg nachgewiesen werden. Der überwiegende Anteil wurde in den Haarfollikeln und, entsprechend der Beladung absteigend aufgeführt, in Talg- und Schweißdrüsen nachgewiesen.

Die in der Abbildung kumulativ angegebene Wiederfindung im Durchströmungsmedium zeigt, daß Wirkstoffreigabe und Penetration in identischer Reihenfolge mit den an der Meerschweinchenhaut erhobenen Befunden vehikelabhängig war. Nach 8 Stunden wurde aus der Wasser-Öl-Emulsion und Fettsalbe gegenüber der Öl-Wasser-Emulsion auch hier etwa das 10fache abtransportiert. Unterschiedlich war gegenüber der Meerschweinchenhaut die durchweg etwa doppelt so große Penetrationsquantität und auffallend war ferner, daß nahezu die Hälfte des in der Fettsalbe inkorporierten Wirkstoffes in der Haut nachweisbar war. Der maximale Abtransport, bezogen in % der Eingabe/30 Minuten, fand zwischen der 3. und 6. Stunde statt.

Aus der Gruppe der C 19-Steroide untersuchten wir das Penetrationsverhalten von radioaktiv markiertem Testosteron. In Abb. 4 ist an Meerschweinchen

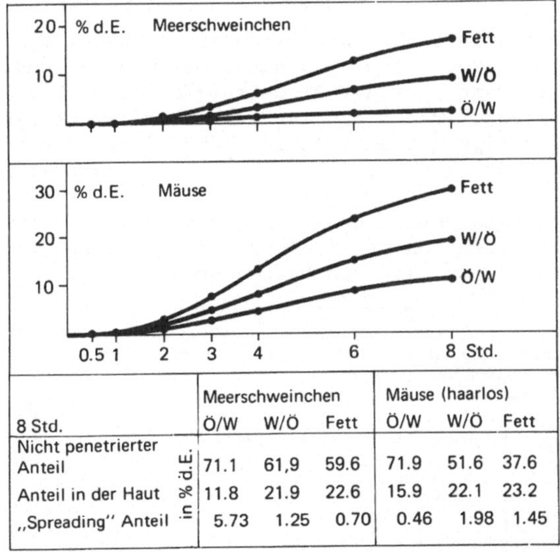


| 8 Std. | | Meerschweinchen | | | Mäuse (haarlos) | | |
|---|---|---|---|---|---|---|---|
| | | Ö/W | W/Ö | Fett | Ö/W | W/Ö | Fett |
| Nicht penetrierter Anteil | | 71.1 | 61,9 | 59.6 | 71.9 | 51.6 | 37.6 |
| Anteil in der Haut | in % d.E. | 11.8 | 21.9 | 22.6 | 15.9 | 22.1 | 23.2 |
| „Spreading" Anteil | | 5.73 | 1.25 | 0.70 | 0.46 | 1.98 | 1.45 |

Abb. 4. Penetration von Testosteron 6-7$^3$H (0.1%ig in Ö/W-W/Ö-Fettgrundlage in vitro an der Haut von Meerschweinchen und haarlosen Mäusen Wiederbestimmung im Durchströmungsmedium in % der appl. Dosis

und haarloser Maus neben der Wiederbestimmung im Durchströmungsmedium der nicht resorbierte Anteil und die in der Haut nach 8-stündiger Kontaktzeit wiedergefundene Menge in % der applizierten Dosis wiedergegeben.

Bei beiden Tierarten ist das Penetrationsausmaß abhängig vom angewandten Vehikel und verläuft im Prinzip gleichsinnig. Lediglich bei der haarlosen Maus sind penetrierte Menge und damit korreliert Resorptionsausmaß und die in der Haut bestimmbare Menge größer. Analog wie unter der Anwendung von Aethinyloestradiol ist die Freigabe und Penetration von Testosteron aus einer Fettgrundlage am besten und aus einer Öl-Wasser-Emulsion am geringsten. Bei beiden Tierarten liegt die Penetrationsrate aus der Wasser-Öl-Emulsion in der Mitte zwischen den beiden Extremwerten.

Da jeweils die Versuche nach 8 Stunden - also zu einem Zeitpunkt, zu dem die quantitative Resorption noch nicht abgeschlossen war - beendigt wurden, interessierte die Frage nach dem Penetrationsausmaß nach 24-stündiger Kontaktzeit. In keinem tierexperimentellen Modell konnte zwischen 8 und 24 Stunden unabhängig von der galenischen Formulierung eine nennenswerte Steigerung der Penetration gefunden werden. Die Ursache hierfür liegt überwiegend im limitierten Penetrationsweg.

Von den Sexualsteroidhormonen ist bislang die quantitative Penetration von gestagen wirksamen Hormonen am wenigsten untersucht (4, 6, 12, 13). In der nachfolgenden Tabelle (Tab. 2) wird die über 8 Stunden erfolgte Penetration von Progesteron, Norethisteron-acetat, d-Norgestrel und 17α-Hydroxyprogesteron-capronat in Abhängigkeit der galenischen Formulierungen vergleichend dargestellt.

| Penetrationsrate 0 - 8 Std. in % d. appl. Dosis | Ö/W | W/Ö | Fett |
|---|---|---|---|
| Progesteron-$^3$H    0.1 % | 21.3 | 17.2 | 14.5 |
| Norethisteronacetat-$^3$H    0.1 % | 10.4 | 3.6 | 3.5 |
| d-Norgestrel-$^3$H    0.1 % | 14.3 | 11.7 | 9.5 |
| 17α-Oxyprogesteroncapronat-$^3$H    0.1 % | 9.6 | 8.1 | 4.9 |

In jedem Fall penetriert von den Gestagenen - obwohl z.B. die Lipoidlöslichkeit von Norethisteron-acetat gegenüber der von Aethinyloestradiol besser ist - die größte Menge aus der Öl-Wasser-Emulsion. Die aus den Öl-Wasser-Emulsionen penetrierte Menge ist gegenüber der Freigabe aus den Fettsalben - unberücksichtigt der unterschiedlichen individuellen quantitativen Penetrationscharakteristik - bei jedem der gestagenen Steroide annähernd doppelt so groß.

Nach den systematisch durchführbaren Tierexperimenten nun abschließend zum Penetrationsverhalten von Aethinyloestradiol-$^3$H, Testosteron-$^3$H und Norethisteron-acetat-$^3$H an klinisch hautgesunden Probanden (Tab. 3).

| Penetration von radioaktiv markierten Sexual-Steroidhormonen nach topischer Applikation an klinisch gesunder Haut beim Menschen in % der appl. Dosis | | | |
|---|---|---|---|
| Zeit in Std. p.a. | Testosteron-$^3$H | Aethinyloestradiol-$^3$H | Norethisteron-acetat $^3$H |
| 1 | 6.9 | 10.6 | 2.9 |
| 4 | 7.1 | 14.2 | 3.7 |
| 8 | 8.5 | 18.2 | 4.8 |
| 24 | 10.5 | 20.4 | 6.3 |

Die Steroid-Hormone wurden in 0,1%iger Zubereitung in gleicher Menge (50 mg) auf gleicher Fläche (4 cm$^2$) jeweils in derselben Rückenhautregion appliziert. In der Tabelle sind Freigabe und Penetration aus einer einheitlichen Formulierung - einer Wasser-Öl-Emulsion - wiedergegeben. Die Angabe in % der topisch applizierten Dosis entspricht dem Medianwert aus jeweils fünf Einzeluntersuchungen.

Eine Wertung nach der penetrierten Quantität zeigt, daß ab der vierten Stunde von Aethinyloestradiol gegenüber Testosteron annähernd das doppelte und gegenüber Norethisteron-acetat das vierfache penetrierte. Rund die Hälfte der über 24 Stunden insgesamt resorbierten Menge penetrierte in jedem Fall bereits nach 1stündiger Kontaktzeit. Von der 8. zur 24. Stunde vergrößerte sich die Penetrationsrate nurmehr um rund 10 - 20%.

Zu diesen Versuchen sei noch angemerkt, daß nach Anwendung von 10 bzw.
100fach erhöhten Wirkstoffmengen daraus kein linearer Anstieg der penetrierten Menge korrelierte. So penetrierte gegenüber einer 0,1%igen Zubereitung
unter der Anwendung einer 1%igen Testosteron-Wasser-Öl-Emulsion um rund 5%
mehr und nach Anwendung einer 10%igen Zubereitung nur um rund 10% mehr.
    Experimentelle Arbeiten (7) zeigen, daß das Penetrationsverhalten an kli-
nisch gesunder Haut gegenüber dem von Haut, an der das Stratum corneum entfernt war, unterschiedlich ist.
    Tab. 4 zeigt, daß nach Entfernung des Str. corneum durch die Stripping
Technik für diese Stoffe die Hauptbarriere für die Penetration entfernt ist.
Nach wenigen Stunden entsprach die penetrierte Menge nahezu der applizierten. Dieses Modell veranschaulicht, daß das Ausmaß der Penetration wesentlich von dem Hautzustand abhängt.

| Penetration von radioaktiv markierten Sexualsteroidhormonen nach topischer Applikation an „gestrippter" Haut beim Menschen in % der appl. Dosis | | | |
| --- | --- | --- | --- |
| Zeit in Std. p.a. | Testosteron-$^3$H | Aethinyloestradiol-$^3$H | Norethisteron-acetat-$^3$H |
| 1 | 93.1 | 74.3 | 53.4 |
| 4 | 97.5 | 96.5 | 80.1 |
| 8 | 98.2 | 98.5 | 90.2 |

    Durch Biopsien ließ sich jedoch nachweisen, daß an "gestrippter" Haut
zwar die Wirkstoffe von der Kontaktfläche aus sehr rasch und nahezu quantitativ penetrieren, daß jedoch der Abtransport aus den tieferen Schichten der
Dermis nicht zeitgleich erfolgte.


<u>Zusammenfassung</u>

An der Haut von Meerschweinchen und haarlosen Mäusen wurden systematische
pharmakokinetische Untersuchungen nach topischer Applikation von radioaktiv
markiertem Aethinyloestradiol, Testosteron, Progesteron, Norethisteron-
acetat, d-Norgestrel und 17$a$-Hydroxyprogesteron-capronat durchgeführt.
    Die Tierversuche wurden durch Untersuchungen an menschlicher Haut ergänzt.
    Die Experimente wurden am Tier in vivo sowie an isolierter Haut und am
Menschen an klinisch gesunder und "gestrippter" Haut sowie an Hautbiopsien
durchgeführt.
    Die quantitativen Bestimmungen in den einzelnen Hautschichten bzw. Medien
wurden durch qualitative Mikroautoradiographien ergänzt.
    Die Wirkstoffe waren bei gleicher Teilchengröße und gleicher Konzentra-
tion jeweils in einer Wasser-Öl-Emulsion, einer Öl-Wasser-Emulsion bzw. in
einer Fettgrundlage inkorporiert.
    Für jedes der untersuchten Steroid-Hormone wurde zeitabhängig die Penetra-
tionscharakteristik nachgewiesen, z. T. der Penetrationsweg aufgezeigt und
die Abhängigkeit des Penetrationsausmaßes von der galenischen Formulierung
dargestellt.
    Wirkstofffreigabe erfolgte bei beiden Tierarten vehikelabhängig und in
identischer Reihenfolge.
    Von Aethinyloestradiol und Testosteron penetrierte aus der Fettgrundlage
gegenüber der Öl-Wasser-Emulsion etwa das 10fache. Von den gestagen wirksamen Steroid-Hormonen war die Penetrationsrate aus den Öl-Wasser-Emulsionen
am größten.
    Neben dem penetrierten Anteil wird die Verteilung in der Haut, die gleich-
falls vehikelabhängig erfolgt, angegeben.
    An menschlicher Haut wurde die Penetration an klinisch gesunder Haut mit
der nach Entfernung des Stratum corneum verglichen und dargestellt, daß bei

verändertem Hautzustand zwar die Wirkstoffe mit eigener zeitabhängiger Penetrationscharakteristik von der Kontaktfläche aus rasch penetrieren, daß jedoch der Abtransport aus den tieferen Schichten der Dermis nicht zeitgleich
erfolgt.


Literatur

1. Feldmann, R.J., Howard, I.: J. invest. Derm. 52, 89 (1969).
2. Frost, P.: J. invest. Derm. 46, 584 (1966).
3. Goldzieher, J.W.: Arch. Derm. 66, 304 (1952).
4. - Baker, R.E.: J. invest. Derm. 35, 215 (1960).
5. Jadassohn, W., Uehlinger, E., Zuercher, W.: Helvet. med. Acta 4, 149
   (1937).
6. Malkinson, F.D.: J. invest. Derm. 31, 1 (1958).
7. - J. invest. Derm. 31, 19 (1958).
8. Miescher, K., Gasche, P.: Helvet. med. Acta 7, Suppl. VI, 93 (1940/41).
9. Moore, C.R.: IAMA 111, 11 (1938).
10. Reiss, F.: Amer. J. Med. Sci. 252, 588 (1967).
11. Schaaf, F., Gross, F.: Arch. Klin. exp. Derm. 205, 312 (1957).
12. Shipley, N.: Steroids 5, 75 (1965).
13. Sternberg, T.H.: Curr. ther. Res. 3, 469 (1961).
14. Zondek, B.: Lancet 234, 1107 (1938).
15. - Klin. Wschr. 2229 (1929).

Symp. Dtsch. Ges. Endokrin. 17, 57–83 (1971)
© by Springer-Verlag

# Physiologie und Pathophysiologie der Corpus Luteum-Funktion bei der Frau

## Physiology and Pathophysiology of Corpus Luteum Function in Women

JOSEF ZANDER UND BENNO RUNNEBAUM

Universitäts-Frauenklinik München
Mit 15 Abbildungen[+]

Summary

Recent advances in the development of highly sensitive and more specific
assays for plasma FSH, LH and HCG and of rapid protein displacement and radio-
immunological methods for the determination of steroids permit further
knowledge regarding the corpus luteum function during the menstrual cycle and
during pregnancy.
    In regard to the physiology and pathophysiology of human corpus luteum
function the most significant data are summarized:
- Steroid production -
(1)  The corpus luteum forms progesterone, 20α-dihydroprogesterone, 17α-
     hydroxyprogesterone, oestradiol-17β and oestrone.
(2)  From the quantitative aspect the corpus luteum produces primarily pro-
     gesterone, 17α-hydroxyprogesterone and 20α-dihydroprogesterone.
(3)  During pregnancy the corpus luteum secretes progesterone until term. At
     term, however, the progesterone secretion is several times smaller than
     in the luteal phase of the menstrual cycle.
(4)  Androstenedione is formed in ovarian tissue and secreted into the blood.
     A higher secretion of androstenedione, however, by the ovary bearing a
     corpus luteum has not yet been proven.
(5)  Probably dehydroepiandrosterone is also formed by the ovary, but it is
     not established whether this substance is specifically formed by the
     corpus luteum.
(6)  It is suggested that the direct secretion of testosterone by the ovary
     is quite small. At the present time there is no evidence for a specific
     formation of testosterone by the corpus luteum.
(7)  The formation of steroid sulfates in the ovary has not been shown until
     now.
- Steroid production during various stages of the menstrual cycle and of
  pregnancy -
(8)  The majority of recent plasma progesterone data indicates that progeste-
     rone is already formed by the Graafian follicle prior to ovulation.
(9)  Progesterone production increases until the middle of the luteal phase
     of the cycle. If implantation of the fertilized egg does not occur, the
     production of progesterone steadly decreases until menstruation. In the
     case of implantation of the egg, there is an increased production of
     progesterone by the corpus luteum shortly after this process.
(10) The corpus luteum graviditatis forms more progesterone, at least in the
     first weeks of pregnancy, than the corpus luteum mentruationis. Relia-
     ble information regarding the luteal part in total progesterone produc-
     tion in early pregnancy is not available at the present time.

---

[+] Halbtonbilder s. Anhang S. 189

(11) The relationship between progesterone and 20α-dihydroprogesterone is
     changed in pregnancy in comparison to the menstrual cycle.
(12) The formation of 17α-hydroxyprogesterone appears to be increased at the
     time of ovulation and in the luteal phase of the cycle. Furthermore the
     production of this substance is probably augmented by the early corpus
     luteum graviditatis.
(13) Oestrogens are predominantly formed during the phase of ovulation and
     in the middle of the luteal phase of the cycle. A reliable differentia-
     tion between the luteal and placental part of oestrogen production dur-
     ing early pregnancy is not possible at this time.
(14) Available data concerning the question of changes in androgen produc-
     tion during the menstrual cycle and during pregnancy are not sufficient.
- Circadian variations in plasma steroids during the luteal phase of the
  cycle and during pregnancy -
(15) There is no significant change in peripheral plasma progesterone con-
     centration during the luteal phase of the cycle and during early preg-
     nancy. In the last trimester of pregnancy, however, a significant
     increase in plasma progesterone concentration is observed between 4 p.m.
     and 8 p.m.
(16) In regard to diurnal variations of plasma oestrogens during the luteal
     phase of the cycle and in early pregnancy, no data are available. In
     late pregnancy a higher plasma oestriol concentration is observed in
     the morning.
(17) Significant changes in diurnal plasma testosterone concentration do not
     seem to occur during the luteal phase of the menstrual cycle.
- In vitro steroid biosynthesis -
(18) Under in vitro conditions, the synthesis of several steroids including
     progesterone, androstenedione, oestradiol-17β and oestrone from acetate
     has been demonstrated in human corpus luteum menstruationis and gravi-
     ditatis.
(19) In vitro steroid biosynthesis appeared to be increased in corpora lutea
     graviditatis as compared to corpora lutea menstruationis.
(20) Under the influence of HCG the in vitro incorporation of acetate into
     steroids is significantly increased in corpora lutea menstruationis and
     graviditatis.
(21) As shown in many in vitro experiments, the main pathway of steroid bio-
     synthesis in the corpus luteum via progesterone - 17α-hyproxyprogester-
     one - androstenedione - oestradiol-17β is well established.
(22) In addition a second pathway of steroid synthesis in the corpus luteum
     via 17α-hydroxypregnenolone - dehydroepiandrosterone is discussed.
- Regulation of the corpus luteum function under physiological conditions -
(23) Probably there exists a close correlation between the LH peak in the
     ovulatory phase and the transformation of the Graafian follicle into a
     corpus luteum.
(24) At present it is unknown which factors determine the life span of the
     corpus luteum during the menstrual cycle.
(25) The development of the corpus luteum and its maintenance in early preg-
     nancy appear to be dependent upon the formation of HCG by the tropho-
     blast.
(26) It is most likely that HCG stimulates progesterone synthesis in the
     corpus luteum, at least in the earliest stages of pregnancy.
- Corpus luteum insufficiency -
(27) The present knowledge about malformation and functional insufficiency
     of the human corpus luteum is poor. Some recent results indicate that
     disturbances in the regulative mechanisms as well as in the production
     of steroids are associated with corpus luteum insufficiency. The plasma
     concentration of progesterone and 17α-hydroxyprogesterone is signifi-
     cantly lower in women with short luteal phases.
(28) The aim of this presentation was to demonstrate some significant aspects
     of the physiology and pathophysiology of human corpus luteum function.

Am 23. April des vergangenen Jahres vor genau 100 Jahren wurde in Leob-
schütz, einer kleinen Stadt in Oberschlesien Ludwig Fraenkel geboren
(Abb. 1). Er erkannte als erster die Bedeutung der Corpus-luteum-Funktion
für den menstruellen Zyklus, für die Implantation und für die Erhaltung des
befruchteten Eies.

Schon im Verlauf seiner ersten klinischen Tätigkeit in einer privaten
Frauenklinik in Breslau begann Fraenkel mit seinen Arbeiten über das Corpus
luteum. Er richtete sich hierzu aus eigenen Mitteln ein Laboratorium ein.
Die Anregung verdankte er seinem Lehrer und Freund, dem Breslauer Embryolo-
gen Gustav Born. Dieser ist letztlich auch der Vater der Hypothese, daß das
Corpus luteum graviditatis nach seinem anatomischen Aufbau und seiner Entwicklung eine Drüse mit innerer Sekretion sein müsse, welche die Ansiedlung
und Entwicklung des befruchteten Eies im Uterus veranlasse. Schwer erkrankt
ersuchte Born den jungen Fraenkel, diese Idee experimentell zu verfolgen.

Schon im Alter von 33 Jahren legte dann Fraenkel (1903) anhand von zahlreichen Experimenten am Kaninchen sowie sorgfältigen Beobachtungen bei der
Frau hinreichende Beweise für die Richtigkeit der Annahme von Gustav Born
vor. In seiner klassischen Arbeit im Archiv für Gynäkologie kam er zu folgender Schlußfolgerung: "Die Corpora lutea haben die Funktion, die Insertion
der Eier zu ermöglichen und ihre Weiterentwicklung zu sichern."

Dieses - wie Fraenkel es bezeichnete - definitive Gesetz war für die weitere Erforschung des Corpus luteum und seiner Sekretionsprodukte von funda-
mentaler Bedeutung.

Im Alter von 67 Jahren wurde Fraenkel, der 1923 die Leitung der Universitäts-Frauenklinik in Breslau übernommen hatte, von den Nationalsozialisten
vertrieben. Er fand eine neue Heimat in Montevideo in Uruguay, wo er seine
Lehr- und Forschungstätigkeit fortsetzen konnte. Unter anderem beschäftigte
er sich weiter intensiv mit Untersuchungen über die Funktion des Corpus
luteum bei Schlangen, Schafen und zuletzt bei Gürteltieren. Mehr als 300
Veröffentlichungen stammen aus seiner Hand. Er starb vor 20 Jahren im Alter
von 81 Jahren auf einer Europareise.

Seine letzte Arbeit ging wenige Tage nach seinem Tode bei der Redaktion
des Archivs für Gynäkologie ein. Sie trug den Titel "Zur Histo-Physiologie
des Corpus luteum" und behandelte das Corpus luteum eines in Südamerika und
in Texas vorkommenden Gürteltieres, welches mehrere Embryonen, aber nur ein
einziges Corpus luteum aufweist (Fraenkel, 1952).

Die These von Ludwig Fraenkel, nach der das Corpus luteum die Insertion
des Eies ermöglicht und seine Weiterentwicklung sichert, besteht heute noch
zu Recht. Es hat sich lediglich gezeigt, daß in Bezug auf die Bedeutung der
Corpora lutea für die weitere Erhaltung der Gravidität zahlreiche artspezi-
fische Unterschiede bestehen.

Für die wesentlichen physiologischen Funktionen des Corpus luteum bei der
Frau ergibt sich heute folgendes Bild:
1. Das Corpus luteum entsteht aus einem reifen Graaf'schen Follikel unmittelbar nach der Ovulation.
2. Die Lebensdauer des Corpus luteum von etwa 14 Tagen und damit die Länge
   der zweiten Zyklusphase ist gegenüber der ersten Zyklusphase relativ kon-
   stant.
3. Unter der Einwirkung der vom Corpus luteum spezifisch gebildeten Gestagene wird die Uterusschleimhaut sekretorisch transformiert.
4. Kommt es nicht zur Insertion eines befruchteten Eies, so bildet sich das
   Corpus luteum zurück. Als Folge wird die transformierte Schleimhaut mit
   der menstruellen Blutung abgestoßen.
5. Die Gegenwart eines Corpus luteum ist Voraussetzung für die Implantation
   eines befruchteten Eies in der Uterusschleimhaut. Kommt diese zustande,
   so bleibt das Corpus luteum auch über den Zeitpunkt der an sich erwarte-
   ten menstruellen Blutung hinaus erhalten.
6. Die ungestörte Entwicklung des befruchteten Eies wird in den ersten
   Schwangerschaftswochen durch das Corpus luteum graviditatis garantiert.
   Die weitere Entwicklung der Schwangerschaft bis zur Geburt ist ohne Gegenwart eines Corpus luteum möglich.

Im folgenden soll versucht werden, den gegenwärtigen Stand unseres Wissens über die Bildung und Sekretion von Steroiden in Corpora lutea des Zyklus und der Schwangerschaft, über die Regulation der Corpus luteum-Funktion sowie über die Störungen seiner Funktion in den Grundzügen darzustellen.

## Konzentration von Steroiden in Corpora lutea

Folgende Steroide wurden bisher in menschlichen Corpora lutea nachgewiesen: Progesteron, 20α-Dihydroprogesteron, Androstendion, Östradiol-17β und Östron (Tab. 1). In quantitativer Hinsicht steht Progesteron an der Spitze.

Tab. 1. <u>Steroide in menschlichen Corpora lutea</u>

| Substanz | µg/g Feuchtgewebe | | Autoren |
|---|---|---|---|
| | C.L.-Phase | Gravidität | |
| Progesteron | 8,3–32,0 | 6,9–42,7 | Zander, 1957, 1958; Zander u. Mitarb., 1958; Maeyama u. Mitarb., 1970 |
| 20α-OH-Progesteron | 5,6 | 1,5 | Zander, 1958; Zander u. Mitarb., 1958 |
| 17α-OH-Progesteron | 1,7 | | Zander, 1958 |
| 16α-OH-Progesteron | | 0,29 | Zander u. Mitarb., 1962 |
| Androstendion | 0,6 | | Zander, 1958; Simmer u. Voss, 1960 |
| DHEA (?) | 0,04 | | Simmer u. Voss, 1960 |
| Östradiol-17β | | 0,48 | Zander u. Mitarb., 1959 |
| Östron | | 0,53 | Zander u. Mitarb., 1959 |

C.L. - Corpus luteum

Es folgen 20α-Dihydroprogesteron und 17α-Hydroxyprogesteron. Androgene und Östrogene wurden nur in Mengen unter 1 µg/g Feuchtgewebe nachgewiesen.

Detaillierte Untersuchungen über Steroide im Corpus luteum im Verlauf des Zyklus und der Schwangerschaft liegen nur für Progesteron vor. Abb. 2 zeigt

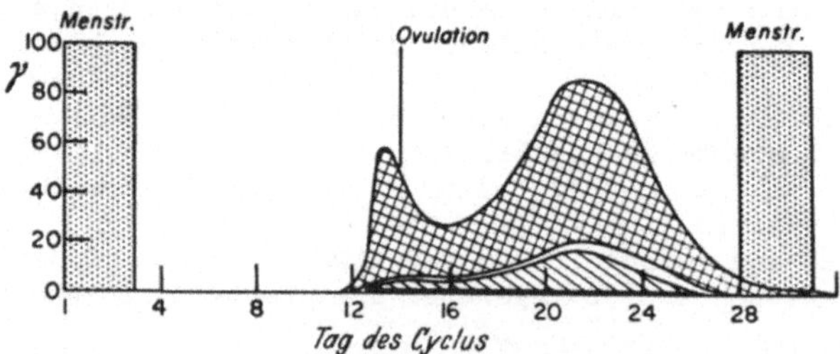


Abb. 2. Gesamtmenge von Progesteron (oberes Feld) sowie von 20α-Dihydroprogesteron und 20β-Dihydroprogesteron (unteres Feld) in 7 Follikeln und 32 Corpora lutea im Verlauf des Zyklus (nach Zander u. Mitarb., 1958).

die Gesamtmengen von Progesteron in Corpora lutea im Verlauf des Zyklus. Es findet sich ein eindeutiger Gipfel in der Mitte der Corpus luteum-Phase. Zweifellos ist Progesteron auch schon vor der Ovulation im sprungreifen Follikel nachweisbar. Daten über die Konzentration von Progesteron in µg/g Feuchtgewebe sind in Tab. 2 zusammengestellt. Sowohl die eigenen (Zander u. Mitarb., 1958) als auch die kürzlich veröffentlichten Ergebnisse von Maeyama u. Mitarb. (1970) zeigen keine signifikante Veränderung der Progesteronkon-

Tab. 2. Progesteron in menschlichen Corpora lutea

| Status | µg/g Feuchtgewebe | | | |
| --- | --- | --- | --- | --- |
| | Zander u. Mitarb.(1958) | | Maeyama u. Mitarb. (1970) | |
| C.L.Phase 1.Drittel | 12,1± 6,3 | (9) | 32,0± 7,7 | (6) |
| C.L.Phase 2. Drittel | 22,0±14,7 | (11) | 28,6± 8,1 | (9) |
| C.L.Phase 3. Drittel | 8,3± 4,1 | (9) | 26,7± 5,0 | (7) |
| Gravidität Mens II/III | 17,5±21,6 | (4) | 42,0±11,7 | (12) |
| Gravidität Mens III/IV | 9,9± 6,4 | (6) | | |
| Gravidität Mens IX/X | 6,9± 5,0 | (5) | | |

C.L. – Corpus luteum
( ) – Anzahl der Bestimmungen

zentration im Verlauf des Zyklus. Im Corpus luteum graviditatis ist bis zum
Ende der Schwangerschaft Progesteron nachweisbar. Die Zunahme der gesamten
Progesteronmenge im Corpus luteum bis zur Mitte der Corpus luteum-Phase und
ihre spätere Abnahme dürfte in erster Linie auf die Zu- bzw. Abnahme der
Gewebemasse oder in anderen Worten auf das Wachstum und die Rückbildung des
Corpus luteum zurückzuführen sein.


## Sekretion und Produktion von Steroiden

1963 haben wir gemeinsam mit Mikhail u. Allen die Sekretion von Progesteron,
$20\alpha$-Dihydroprogesteron, $17\alpha$-Hydroxyprogesteron und Androstendion durch direkten Nachweis dieser Substanzen im Venenblut von Ovarien mit Corpora
lutea nachgewiesen (Mikhail u. Mitarb., 1963). Ihre Konzentration war im
Ovarialvenenblut wesentlich höher als im peripheren venösen Blut. Inzwischen
liegen weitere Daten vor (Tab. 3). Neben den genannten Steroiden wurden

Tab. 3. Steroide im Plasma der Vene von Ovarien mit Corpus luteum (C.L.)

| Substanz | µg/100 ml | | Autoren |
| --- | --- | --- | --- |
| | C.L.Phase | Gravidität | |
| Progesteron | 32,0 | 11,7-68,0 | Mikhail u. Mitarb., 1963; Mikhail, 1967, 1970; Le Maire u. Mitarb., 1970; |
| $20\alpha$-OH-Progesteron | 1,2 | 1,4 | Mikhail, 1967, 1970. |
| $20\beta$-OH-Progesteron | 0,01 | | Mikhail, 1967, 1970. |
| $17\alpha$-OH-Progesteron | 6,6 | 7,1 | Mikhail, 1967, 1970. |
| Androstendion | 0,8-3,4 | | Mikhail u. Mitarb., 1963; Horton u. Mitarb., 1966; Mikhail, 1967, 1970; Aakvaag u. Fylling, 1968. |
| Testosteron | 0,07-0,6 | | Horton u. Mitarb., 1966; Mikhail, 1967, 1970; Aakvaag u. Fylling,1968. |
| DHEA | 2,5 | | Mikhail, 1967, 1970. |
| DHEA-Sulfat | 47,5-74,2 | | Aakvaag u. Fylling,1968; Kalliala u. Mitarb., 1970. |
| Östradiol-17$\beta$ | 0,47 | | Mikhail, 1967, 1970. |
| Östron | 0,09 | | Horton, 1965; Mikhail 1967, 1970. |

Testosteron (Horton u. Mitarb., 1966; Mikhail, 1967, 1970; Aakvaag u.
Fylling, 1968), Dehydroepiandrosteron (Mikhail, 1967, 1970) sowie sein Sulfat (Aakvaag u. Fylling, 1968; Kalliala u. Mitarb., 1970), Östradiol-17β und
Östron (Mikhail, 1967, 1970) nachgewiesen.

Der Nachweis von 20β-Dihydroprogesteron ist bisher nicht eindeutig (Mikhail, 1970). In quantitativer Hinsicht steht von den freien Steroiden Progesteron wiederum an der Spitze. Es folgen 17α-Hydroxyprogesteron und 20α-Dihydroprogesteron. Insgesamt besteht im Ovarial-Venenblut eine ähnliche quantitative Verteilung der Steroidkonzentrationen wie im Corpus luteum-Gewebe.

In Tab. 4 sind aufgrund neuerer Daten die Quotienten Steroidkonzentration im Ovarialvenenplasma:Steroidkonzentration im Armvenenplasma dargestellt.

Tab. 4. <u>Verhältnis der Steroidkonzentration im Plasma der Vene von Ovarien mit Corpus luteum zur Steroidkonzentration im Armvenenplasma während der Lutealphase des menstruellen Zyklus.</u>

| Substanz | Ovarialvenenplasma<br>Armvenenplasma | Autoren |
|---|---|---|
| Progesteron | 30 | Runnebaum u. Mitarb., 1965; van der Molen u. Groen, 1965; Mikhail, 1967, 1970; |
| 20α-OH-Progesteron | 5 | Runnebaum u. Mitarb.,1965; van der Molen u. Groen,1965; Mikhail, 1967, 1970; Saxena u. Mitarb., 1968. |
| 20β-OH-Progesteron | 0,2 | Runnebaum u. Mitarb.,1965; Mikhail, 1967,1970; Saxena u. Mitarb.,1968. |
| 17α-OH-Progesteron | 33 | Runnebaum u. Mitarb.,1965; Mikhail, 1967, 1970; Strott u. Mitarb., 1969. |
| Androstendion | 15 | Horton,1965; Mikhail, 1967, 1970; Aakvaag, Fylling, 1968. |
| Testosteron | 1,2-5 | Aakvaag, Fylling, 1968. |
| DHEA | 3 | Mikhail, 1967,1970; Kalliala u. Mitarb., 1970. |
| DHEA-Sulfat | 1 | Migeon u. Mitarb.,1957; Aakvaag, Fylling, 1968; Kalliala u. Mitarb., 1970. |
| Östradiol-17β | 28 | Mikhail, 1967, 1970; Baird, 1968; Koreman u. Mitarb., 1969. |
| Östron | 8 | Mikhail, 1967, 1970; Baird, 1968. |

Der Quotient für Progesteron, 20α-Dihydroprogesteron, 17α-Hydroxyprogesteron, Androstendion, Dehydroepiandrosteron, Östradiol-17β und Östron spricht für Produktion dieser Substanzen in den Ovarien mit Corpora lutea. Dabei ist offen, ob die Substanzen im Corpus luteum selbst oder im übrigen Ovarialgewebe gebildet werden. Aus dem Quotienten für Testosteron und Dehydroepiandrosteronsulfat läßt sich keine Bildung in den Ovarien mit Corpora lutea ableiten.

In Tab. 5 sind neuere Daten für Steroidsulfate im Plasma der Vene von Ovarien mit Corpus luteum und im Armvenenplasma bei derselben Patientin zusammengestellt (Kalliala u. Mitarb., 1970). Sie sprechen ebenfalls gegen eine Bildung von Steroidsulfaten in Ovarien mit Corpus luteum.

Der eindeutige Beweis für die Sekretion von Progesteron, 20α-Dihydroprogesteron, 17α-Hydroxyprogesteron und Östradiol-17β unmittelbar im Corpus luteum wurde durch die quantitative Messung der Konzentrationen dieser Steroide im Plasma der Venen von Ovarien mit und ohne Corpus luteum erbracht (Tab. 6). Diese Daten zeigen weiter, daß Androstendion nicht im Corpus luteum-Gewebe sondern im übrigen Ovarialgewebe gebildet wird. Die Konzentra-

Tab. 5. Steroidsulfate im Plasma der Vene von Ovarien mit Corpus luteum
(C.L.) und im Armvenenplasma während der Lutealphase des Zyklus bei
derselben Patientin[1]

| Substanz | $\mu g/100\ ml^2$ | |
| --- | --- | --- |
| | Ovar mit C.L. | Armvene |
| Androsteron-Sulfat | 20,0 | 19,8 |
| DHEA-Sulfat | 74,2 | 78,0 |
| Epiandrosteron-Sulfat | 5,1 | 5,1 |
| 5-Androsten-3β-17β-diol-Sulfat | 4,2 | 4,8 |
| Pregnenolon-Sulfat | 9,6 | 8,9 |
| 5-Pregnen-3β,20α-diol-Sulfat | 16,0 | 16,6 |
| 5-Pregnen-3β,17α,20α-triol-Sulfat | 2,4 | 2,4 |

[1] nach Kalliala u. Mitarb., 1970
[2] Mittelwerte von jeweils 3 Frauen


Tab. 6. Steroide im Plasma der Vene von Ovarien mit und ohne Corpus
luteum (C.L.)

| Status | Substanz | $\mu g/100\ ml$ | | Autoren |
| --- | --- | --- | --- | --- |
| | | mit C.L. | ohne C.L. | |
| | Progesteron | 39,5 (4) | 1,7 (4) | Mikhail, 1967, 1970; |
| | 20α-OH-Progesteron | 1,6 (4) | 0,2 (3) | " |
| 3.-10.Tag | 17α-OH-Progesteron | 7,7 (4) | 2,0 (4) | " |
| nach der | Androstendion | 1,7 (4) | 1,7 (3) | " |
| Ovulation | Östradiol-17β | 0,5 (3) | 0,1 (3) | " |
| Ende der Gravidität | Progesteron | 24,3 (10) | 12,4 (10) | Le Maire u. Mitarb.,1970. |

( ) - Anzahl der Bestimmungen


tion dieses Androgens ist nämlich im Plasma der Vene von Ovarien mit Corpus
luteum die gleiche wie im Plasma der Vene des Ovariums ohne Corpus luteum.
Weiterhin ist nach diesen Daten anzunehmen, daß die Sekretion von Progesteron am Ende der Gravidität gegenüber der Sekretion in der Corpus luteum-
Phase erheblich vermindert ist (Le Maire u. Mitarb., 1970).
In Tab. 7 sind eine Reihe von Daten über die täglichen Bildungsraten verschiedener Steroide in der Lutealphase des menstruellen Zyklus zusammengestellt. Es ist dabei zu berücksichtigen, daß die Ergebnisse zum Teil mit
sehr verschiedenen Methoden ermittelt wurden. Ein vorsichtiger Vergleich erscheint jedoch erlaubt. Auch hier zeigt sich, daß die Progesteronbildung an
der Spitze steht. Seit den Untersuchungen von Ober u. Mitarb. (1954) ist
allgemein anerkannt, daß seine tägliche Produktionsrate etwa bei 20 mg
liegt. Es folgt 17α-Hydroxyprogesteron. Die tägliche Produktion von Östra-
diol-17β und Östron dürfte hingegen unter 0,5 mg liegen.
Zusammengefaßt ergibt sich aus den vorliegenden Untersuchungen unter in
vivo-Bedingungen für die Steroidbildung in Corpora lutea zur Zeit folgendes
Bild:
1. Das Corpus luteum bildet Progesteron, 20α-Dihydroprogesteron, 17α-Hydroxy-
   progesteron, Östradiol-17β und Östron.
2. Das Corpus luteum bildet in quantitativer Hinsicht überwiegend Progesteron, 17α-Hydroxyprogesteron und 20α-Dihydroprogesteron.
3. Das Corpus luteum graviditatis bildet Progesteron bis zum Ende der
   Schwangerschaft. Die Progesteronsekretion ist jedoch in dieser Zeit um
   ein vielfaches niedriger als in der Corpus luteum-Phase.

Tab. 7. <u>Produktionsraten in der Lutealphase des menstruellen Zyklus</u>

| Steroid | mg/Tag | Autoren |
|---|---|---|
| Progesteron | 22,1 | Ober u. Mitarb., 1954; |
| | | van der Molen u. Aakvaag, 1967; |
| 20α-OH-Progesteron | 1,0 | "     " |
| 17α-OH-Progesteron | 4,0 | Strott u. Mitarb., 1969; |
| Androstendion | 3,2 | Rivarola u. Mitarb., 1966; |
| Testosteron | 0,3 | "     " |
| Östradiol-17β | 0,09 | Eren u. Mitarb., 1967; |
| Östradiol/Östron | 0,3 | Brown, 1957; Zander u. Mitarb., 1959. |

4. Androstendion wird im Ovarialgewebe gebildet und an die Blutbahn abgege-
ben. Eine gegenüber dem übrigen Ovarialgewebe vermehrte Androstendionbil-
dung im Corpus luteum ist bisher nicht erwiesen.
5. Dehydroepiandrosteron wird sehr wahrscheinlich ebenfalls im Ovarium gebildet. Es ist noch unklar, ob im Corpus luteum eine spezifisch vermehrte
Dehydroepiandrosteronbildung erfolgt.
6. Testosteron wird, wenn überhaupt, so nur in geringen Mengen im Ovarium
gebildet. Für eine spezifische Testosteronbildung im Corpus luteum liegt
bisher kein eindeutiger Beweis vor.
7. Die Bildung von Steroidsulfaten im Ovarium ist bisher nicht erwiesen.


<u>Verlauf der Steroidproduktion im Zyklus und in der Schwangerschaft</u>

Die ersten indirekten Informationen über die Steroidproduktion von Progesteron im Verlauf des Zyklus und der Schwangerschaft wurden durch die Bestim-
mung der Pregnandiolausscheidung erhalten. Auf die Wiedergabe der bekannten
Kurven wird verzichtet.
Serienbestimmungen von Progesteron im peripheren Blut von Frauen im Verlauf des menstruellen Zyklus sind erst seit einigen Jahren mit Hilfe von
Doppelisotopenmethoden (Woolever, 1963), gaschromatografischen - (van der
Molen u. Groen, 1965), sowie Protein-Verdrängungsmethoden (Neill u. Mitarb.,
1967) möglich. Dabei findet sich ähnlich wie für Pregnandiol im Harn ein
Gipfel der Progesteronkonzentration in der Mitte der Corpus luteum-Phase. Er
liegt bis zu 10fach höher als die Progesteronkonzentration in der präovulatorischen Phase.
Keine Übereinstimmung besteht bisher in der Frage, ob schon unmittelbar
vor dem Follikelsprung ein Anstieg der Progesteronkonzentration im peripheren Blut erfolgt. In eigenen Untersuchungen beobachteten wir in einem
Plasma-Pool, welcher 4 Tage vor dem Anstieg der Basaltemperaturen gesammelt
wurde, eine 3fach höhere Progesteronkonzentration als in einem Plasma-Pool,
welcher 9 Tage vor dem Basaltemperaturanstieg gesammelt wurde (Runnebaum u.
Zander, 1967). Johansson u. Wide (1969) fanden erhöhte Plasmaprogesteronwer-
te zur Zeit des LH-Anstieges in der Zyklusmitte. Kürzlich zeigten Yussmann
u. Taymor (1970) in Bestimmungen mit 8-Stunden-Intervallen einen recht eindeutigen präovulatorischen Anstieg des Plasmaprogesterons bei 5 Patientinnen (Abb. 3). Die Ovulationsphase wurde in diesen Fällen durch eine Laparotomie gesichert. Schließlich beobachtete Mikhail (1970) kürzlich ebenfalls
im Plasma der Vene eines Ovariums mit reifem Follikel eine 3-4fach höhere
Progesteronkonzentration als im Plasma der Vene eines Ovariums ohne reifen
Follikel. Auf der anderen Seite konnten Neill u. Mitarb. (1967) sowie
Yoshimi u. Lipsett (1968) keinen präovulatorischen Anstieg der Progesteronkonzentration im peripheren Plasma beobachten.
An dieser Stelle ist auf die Schwierigkeit hinzuweisen, den genauen Zeit-
punkt des Eisprunges festzulegen. Sie spielt bei allen Versuchen, endokrine
Vorgänge mit dem Ereignis der Ovulation zu korrelieren, eine wesentliche

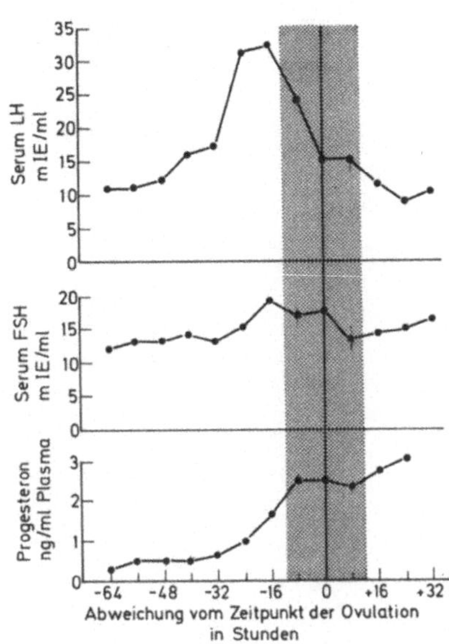


Abb. 3. Durchschnittliche, 8-stündliche Konzentrationen von LH, FSH und Progesteron bei 5 Frauen vor, während und nach der Ovulation.
(Nach Yussman u. Taymor, 1970)

Rolle. Leider steht uns bis heute keine Methode zur Verfügung, welche es
unter routinemäßigen Bedingungen erlaubt, den Zeitpunkt des Eisprunges zuverlässig zu bestimmen.

20α-Dihydroprogesteron wurde von unserer Gruppe im peripheren Blut der
Lutealphase identifiziert (Runnebaum u. Mitarb., 1965). Inzwischen liegen
auch für diese Substanz Serienbestimmungen im Verlauf des menstruellen Zyklus vor (van der Molen u. Groen, 1967; Saxena u. Mitarb., 1968). Abb. 4
zeigt den Verlauf der Plasmakonzentration von 20α-Dihydroprogesteron im Vergleich zu den Progesteronkonzentrationen sowie zum LH- und FSH-Gipfel.

Von Interesse sind Serienbestimmungen von 17α-Hydroxyprogesteron im Verlaufe des Zyklus (Strott u. Mitarb., 1969). Fotherby hatte schon 1960 gezeigt, daß der Gipfel der Pregnantriolausscheidung zeitlich vor dem Gipfel
der Pregnandiolausscheidung liegt.

Abb. 5 zeigt den Verlauf des 17α-Hydroxyprogesteronplasmaspiegels im Vergleich zum Progesteron sowie zum LH-Gipfel in der Zyklusmitte. 17α-Hydroxy-
progesteron zeigt gleichzeitig mit LH einen eindeutigen Gipfel. Auch vorher
ist schon ein Anstieg des 17α-Hydroxyprogesterons im Plasma zu beobachten.
Schließlich besteht ein weiterer Gipfel in der Mitte der Corpus luteum-Phase.
Über die physiologische Bedeutung dieses recht auffallenden Verhaltens des
17α-Hydroxyprogesterons ist nichts Sicheres bekannt.

Für die Plasmakonzentration der Androgene, Androstendion (Horton, 1965)
und Testosteron (Lobotsky u. Mitarb., 1964) wurden bisher im Zyklus keine
regelmäßig zu beobachtenden signifikanten Veränderungen nachgewiesen.

Abb. 6 zeigt Plasmatestosteronbestimmungen in drei menstruellen Zyklen,
welche wir gemeinsam mit van der Molen ermittelt haben (van der Molen, 1968).
Auch hier sind keine signifikanten Veränderungen der Testosteronkonzentration nachweisbar.

Anhaltspunkte für die ovarielle Sekretion von Östrogenen im Verlauf des
menstruellen Zyklus ergeben sich aus den bekannten doppelgipfligen Kurven
der Östrogenausscheidung mit dem Harn. Kürzliche Untersuchungen von Baird u.
Guevara (1969) sowie von Mikhail u. Mitarb. (1970), mit einer Doppelisoto-
penmethode bzw. mit einer radioimmunologischen Methode (Abb. 7) über die

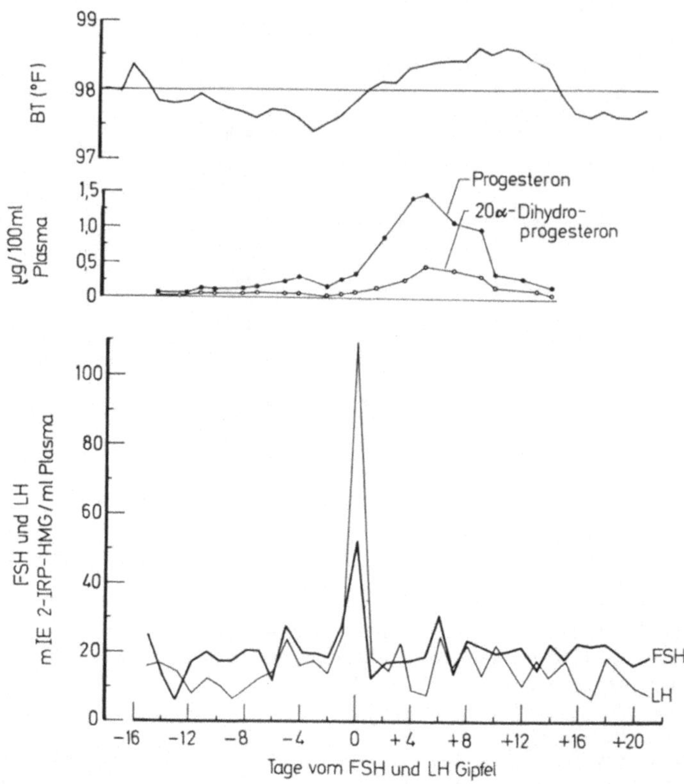


Abb. 4. Zusammenhang zwischen den durchschnittlichen Konzentrationen von
FSH, LH, Progesteron und 20α-Dihydroprogesteron sowie der Basaltem-
peratur bei Frauen während 17 menstrueller Zyklen.
(Nach Saxena u. Mitarb., 1968)

Östrogenkonzentration im peripheren Blut zeigen ebenfalls den sehr charakte-
ristischen doppelgipfligen Verlauf der Östrogenkonzentration.

Für die Beurteilung der Steroidbildung im Verlauf der Schwangerschaft besteht die Schwierigkeit der Abgrenzung der Produktion von Seiten des Corpus
luteum von der plazentaren Steroidproduktion. Der bekannte Anstieg der Pregnandiolausscheidung im Harn sowie der Progesteronkonzentration im Plasma
bis zur Geburt läßt naturgemäß eine solche Unterscheidung nicht zu. Serienbestimmungen der Plasmasteroide bei ein und derselben Frau in der Schwangerschaft liegen nur vereinzelt vor. Abb. 8 von Johansson (1969 a) zeigt den
Verlauf des Progesteronplasmaspiegels von der Konzeption bis zum 40. Tag
nach der Empfängnis. Der Verlauf dieser Kurve spricht für eine vermehrte
Progesteronsekretion von Seiten des Corpus luteum nach der Implantation des
Eies.

Von Interesse sind in diesem Zusammenhang Untersuchungen von Yoshimi u.
Mitarb. (1969) über die Plasmakonzentration von Progesteron und 17α-Hydroxy-
progesteron im Verlauf von Schwangerschaften, welche nach medikamentösen
Ovulationsauslösungen eintraten. Die Ergebnisse sind in Abb. 9 dargestellt.

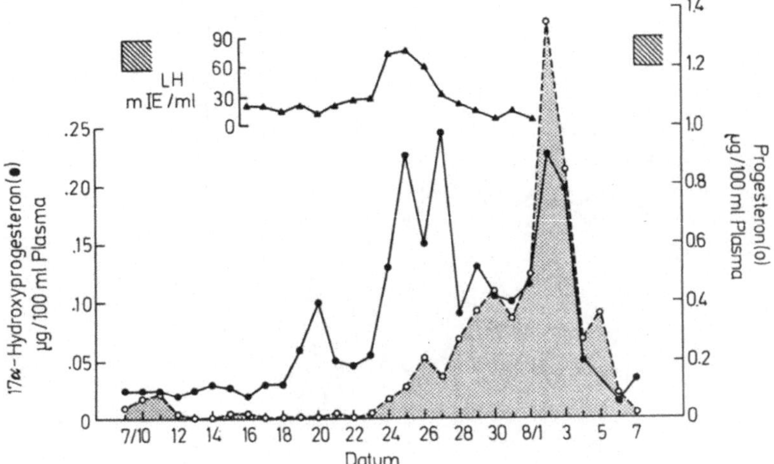



Abb. 5. Plasmaspiegel von LH, 17α-Hydroxyprogesteron und Progesteron im Verlaufe eines menstruellen Zyklus. Der Beginn der Menstruation ist jeweils schraffiert angegeben.
(Nach Strott u. Mitarb., 1969)

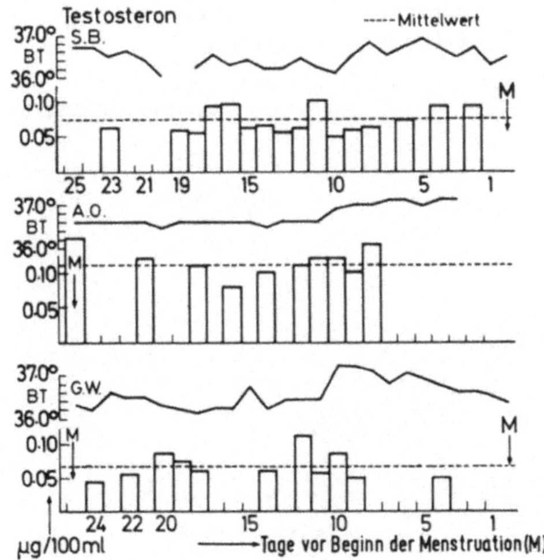



Abb. 6. Basaltemperaturen und Testosteronkonzentrationen im Plasma bei 3
Frauen während biphasischer Zyklen. (Nach van der Molen, Runnebaum
u. Zander; aus van der Molen, 1968)

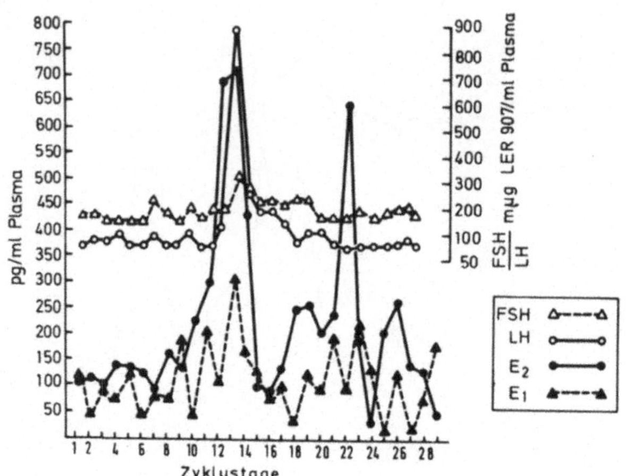


Abb. 7. Konzentrationen von FSH, LH, Östradiol-17β und Östron während eines
normalen menstruellen Zyklus.
(Nach Mikhail u. Mitarb., 1970)

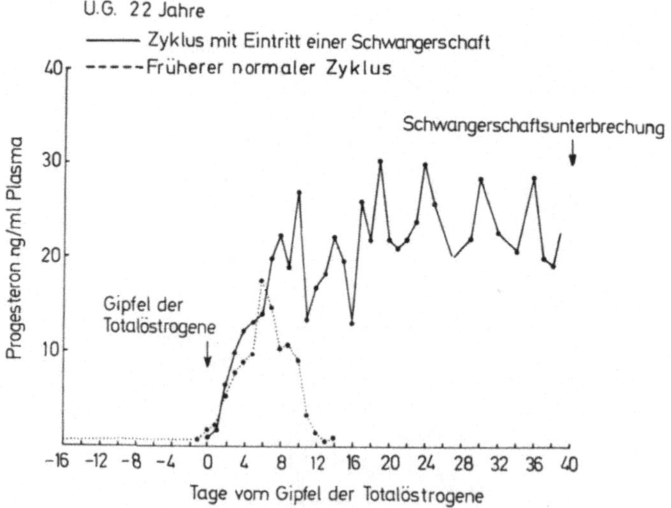


Abb. 8. Progesteronspiegel im Plasma bei einer Frau von der Ovulation bis
zur 8. Schwangerschaftswoche.
(Nach Johansson, 1969 a)

3 - 4 Wochen nach der Konzeption findet sich nach einem sehr deutlichen Anstieg ein Gipfel beider Substanzen. Progesteron steigt dann etwa ab der
8. Schwangerschaftswoche erneut an.

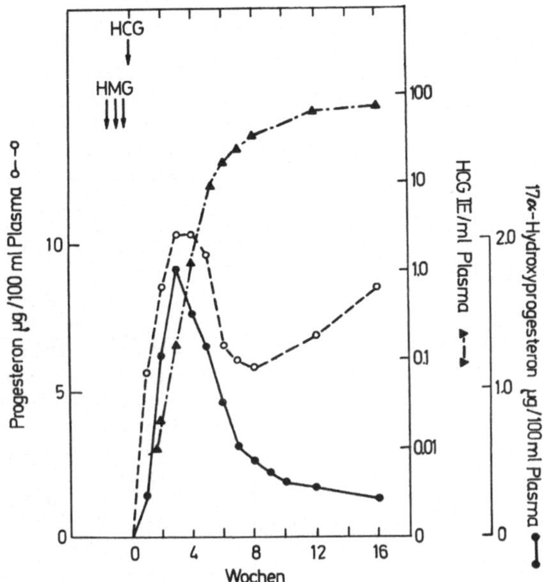


Abb. 9. Durchschnittliche Konzentrationen von Progesteron, 17α-Hydroxypro-
        gesteron und HCG im Plasma bei 9 Frauen nach medikamentöser Auslö-
        sung der Ovulation mit HMG-HCG.
        (Nach Yoshimi u. Mitarb., 1969)


Hingegen fällt 17α-Hydroxyprogesteron kontinuierlich ab. Nimmt man an,
daß 17α-Hydroxyprogesteron vorwiegend vom Corpus luteum gebildet wird, so
sprechen diese Befunde dafür, daß die Fähigkeit des Corpus luteum zur Bildung zumindest von 17α-Hydroxyprogesteron bis zur 10. Schwangerschaftswoche
nachläßt. Es bleibt offen, wie groß der Anteil des Progesterons ist, welcher
schon vor der 10. Schwangerschaftswoche von der Plazenta gebildet wird.
    Bemerkenswert ist auch eine kürzliche Untersuchung von Rubin u. Mitarb.
(1970), aus der hervorgeht, daß die Konzentration von 20α-Dihydroprogesteron
im Plasma bis zur 25. Schwangerschaftswoche, im Gegensatz zu dem sehr deutlichen Anstieg der Progesteronkonzentration, keine signifikanten Veränderungen zeigt.
    Nach Untersuchungen von Mizuno u. Mitarb. (1968) liegt die Plasmakonzen-
tration von Testosteron und Androstendion von der 8. Schwangerschaftswoche
an etwa doppelt so hoch wie in der Lutealphase des Zyklus. Mikhail u. Allen
(1967) fanden im Venenblut eines Ovariums mit Corpus luteum am Ende der
Schwangerschaft eine etwa 10fach höhere Androstendionkonzentration als in
der frühen Schwangerschaft. Hingegen war die Progesteronkonzentration etwa
6-7fach niedriger als in der frühen Schwangerschaft.
    Einige Daten über durchschnittliche Steroidkonzentrationen im peripheren
Venenplasma während der Lutealphase des Zyklus und in der ersten Hälfte der
Gravidität sind in Tab. 8 zusammengestellt.

Tab. 8. <u>Steroide im peripheren Armvenenplasma während der Lutealphase des</u> <u>Zyklus und in der 1. Hälfte der Gravidität</u>

| Substanz | μg/100 ml | | Autoren |
|---|---|---|---|
| | C.L.Phase | 1.Grav.-Hälfte | |
| Pregnenolon | 0,21 | | Bermudez u. Mitarb., 1970; |
| Progesteron | 0,7 -1,5 | 2,2 - 5,2 | van der Molen u. Groen, 1965; Runnebaum u. Mitarb., 1965; Johansson, 1969 b; |
| 20α-OH-Progesteron | 0,1 -0,4 | 0,5 - 1,0 | van der Molen u. Groen, 1965; Runnebaum u. Mitarb.,1965; Saxena u. Mitarb.,1968; Rubin u. Mitarb., 1970; |
| 20β-OH-Progesteron | 0,02-0,1 | | Runnebaum u. Mitarb.,1965; Saxena u. Mitarb., 1968; |
| 17α-OH-Progesteron | 0,05-0,2 | 0,3 | Runnebaum u. Mitarb.,1965; Strott u. Mitarb.,1969; Yoshimi u. Mitarb., 1969; |
| Androstendion | 0,15 | 0,33 | Horton, 1965; Mizuno u. Mitarb.,1968; Lobotsky u. Mitarb.,1964; Mizuno u. Mitarb., 1968; |
| Testosteron | 0,06 | 0,10 | |
| DHEA | 0,5 -1,5 | | van der Molen, 1968; |
| DHEA-Sulfat | 37,0 | | Mignon u. Mitarb., 1957; |
| Östradiol-17β | 0,02 | 0,05-0,09 | Roy u. Mitarb.,1965;Baird,1968; |
| Östron | 0,01 | 0,14-0,39 | "     "              " |

Die bisher vorliegenden Ergebnisse lassen für den Verlauf der Steroidproduktion im Corpus luteum des Zyklus und der Schwangerschaft folgende Annahme zu:

1. Die Mehrzahl der bisher vorliegenden Befunde spricht dafür, daß Progesteron schon vor der Ovulation im sprungreifen Follikel vermehrt gebildet wird.
2. Progesteron wird bis zur Mitte der Corpus luteum-Phase des menstruellen Zyklus vermehrt gebildet. Kommt es nicht zur Implantation eines Eies, so nimmt die Progesteronbildung bis zur Menstruation wieder ab. Findet hingegen die Implantation eines Eies statt, so wird Progesteron bereits kurze Zeit nach der Implantation vermehrt gebildet.
3. Das Corpus luteum graviditatis bildet zumindest in den ersten Schwangerschaftswochen mehr Progesteron als das Corpus luteum menstruationis. Eine genaue Information über den lutealen Anteil der Progesteronproduktion in der Frühschwangerschaft ist jedoch zur Zeit nicht möglich. Es ist unklar, von welchem Zeitpunkt an und in welchem Ausmaß der Trophoblast an der Gesamtsteroidproduktion beteiligt ist.
4. Die Relation zwischen Progesteron und 20α-Dihydroprogesteron verschiebt sich in der Schwangerschaft gegenüber dem Zyklus.
5. Die vorliegenden Befunde deuten hin auf eine vermehrte Bildung von 17α-Hydroxyprogesteron in der Ovulationsphase und in der Lutealphase. Ebenso scheint diese Substanz im Corpus luteum graviditatis vermehrt gebildet zu werden.
6. Östrogene werden in der Ovulationsphase und in der Mitte der Corpus luteum-Phase des menstruellen Zyklus vermehrt gebildet. Eine zuverlässige Differenzierung zwischen dem lutealen und dem plazentaren Anteil der in der Frühschwangerschaft gebildeten Östrogene ist zur Zeit nicht möglich.
7. Über Veränderungen der Androgenproduktion in Corpora lutea des Zyklus und der Schwangerschaft liegen keine ausreichenden Informationen vor.

Über Tagesrhythmen in der Steroidproduktion des Corpus luteum ist bisher wenig bekannt. In eigenen Untersuchungen haben wir in der Gelbkörperphase des
Zyklus und in der Schwangerschaft Plasmaprogesteronbestimmungen in 4-stünd-
lichen Abständen über 24 Stunden vorgenommen. Die Ergebnisse sind in Abb. 10

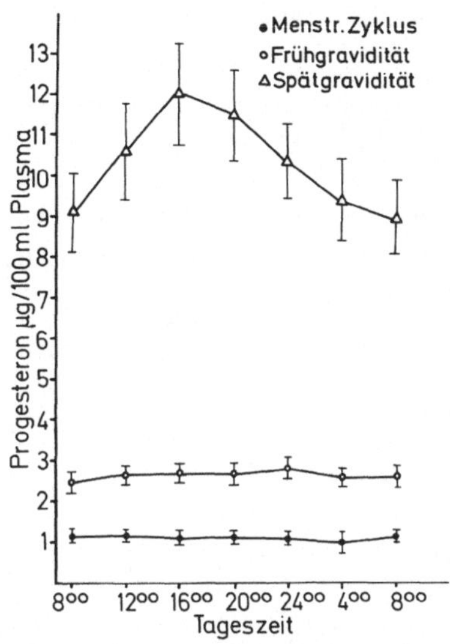


Abb. 10. Durchschnittliches Tagesprofil der Progesteronkonzentrationen im
peripheren Plasma in der Lutealphase des Zyklus (10 Frauen), während der 8.-18. Schwangerschaftswoche (14 Frauen) und während der
20.-41. Schwangerschaftswoche (15 Frauen).

dargestellt. Bei 10 Frauen wurden am 5. und 6. Tag der hyperthermen Phase
des Zyklus keine signifikanten Tagesschwankungen beobachtet. Das gleiche
trifft zu für 14 Frauen in der Frühgravidität (8.-18. Schwangerschaftswoche).
Johansson (1969 a) fand hingegen in der Frühgravidität in den Morgenstunden
geringfügig erhöhte Werte. In der 2. Schwangerschaftshälfte (20.-41. Schwangerschaftswoche) beobachteten wir bei 15 Frauen eine signifikante Tages-
rhythmik der Progesteronkonzentration im Plasma mit maximalen Werten in den
späten Nachmittagsstunden zwischen 16 und 20 Uhr. Wiest (1967) und Craft u.
Mitarb. (1969) konnten jedoch keine signifikanten Tagesschwankungen der
Plasmaprogesteronkonzentration in der Spätschwangerschaft beobachten.
    Über eine Tagesrhythmik der Östrogene in der Corpus luteum-Phase liegen
unseres Wissens bisher keine Mitteilungen vor. In der Spätgravidität sind
nach Selinger u. Levitz (1969) in den Morgenstunden die Östriolwerte um etwa
25% höher als in den Nachmittagsstunden.
    Signifikante Tagesschwankungen für Testosteron im Verlauf des menstruel-
len Zyklus konnten wir in eigenen Untersuchungen ebenso wie andere Autoren
nicht beobachten (Southren u. Mitarb., 1967).


Steroidbiosynthese im Corpus luteum

Unsere derzeitigen Vorstellungen über die Biosynthese der Steroide im Corpus
luteum wurden im wesentlichen aus den Ergebnissen von in vitro-Experimenten
gewonnen.

Auf die viel diskutierte Frage, welche Schlußfolgerungen aus Ergebnissen
von in vitro-Experimenten über die Stoffwechselvorgänge unter physiologischen in vivo-Bedingungen gezogen werden können, soll hier nicht eingegangen
werden. Es werden lediglich die wichtigsten Daten zusammengefaßt.

Die Totalsynthese einer ganzen Reihe von Steroiden unter Einschluß von
Progesteron, Androstendion, Östradiol-17β und Östron aus Azetat wurde unter
in vitro-Bedingungen sowohl im Corpus luteum menstruationis als auch im Corpus luteum graviditatis der Frau erwiesen (Huang u. Pearlman, 1963; Hammerstein u. Mitarb., 1964; Tab. 9). Tab. 10 zeigt, daß die gleichen Steroide

Tab. 9. <u>In vitro-Synthese in menschlichen Corpora lutea</u>
　　　　 <u>Substrat: Na-Acetat-1-$^{14}$ C($-^{3}$H)</u>

| Produkt | Status | Autoren |
|---|---|---|
| Pregnenolon | Gr. | Hammerstein u. Mitarb., 1964; |
| Progesteron | M.Z. und Gr. | Huang u. Pearlman, 1963; |
| | | Hammerstein u. Mitarb., 1964; |
| 20α-OH-Progesteron | M.Z. und Gr. | Hammerstein u. Mitarb., 1964; |
| 17α-OH-Progesteron | M.Z. und Gr. | " |
| Androstendion | M.Z. und Gr. | " |
| Östradiol-17β | M.Z. und Gr. | " |
| Östron | M.Z. und Gr. | " |

Gr.　- Gravidität
M.Z. - Menstrueller Zyklus

Tab. 10. <u>In vitro-Synthese in menschlichen Corpora lutea</u>
　　　　　 <u>Substrat: Pregnenolon -4-$^{14}$C ($-7-^{3}$H)</u>

| Produkt | Status | Autoren |
|---|---|---|
| Progesteron | M.Z. und Gr. | Ryan,1963; Ingiulla u. Mitarb.,1967; |
| 20α-OH-Progesteron | M.Z. | Ingiulla u. Mitarb., 1967; |
| 17α-OH-Progesteron | M.Z. und Gr. | Ryan,1963; Ingiulla u. Mitarb.,1967; |
| 16α-OH-Progesteron | M.Z. | Ingiulla u. Mitarb., 1967; |
| 17,20-OH-Progesteron | M.Z. | " |
| Androstendion | M.Z. | " |
| Testosteron | M.Z. | " |
| Östradiol-17β | M.Z. und Gr. | Ryan,1963; Ingiulla u. Mitarb.,1967; |
| Östron | M.Z. | Inguilla u. Mitarb., 1967; |

M.Z.　- Menstrueller Zyklus
Gr.　　- Gravidität

und darüber hinaus auch Testosteron im Corpus luteum-Gewebe gebildet werden,
wenn Pregnenolon als Substrat eingesetzt wird (Inguilla u. Mitarb., 1967).
Bei Verwendung von Progesteron als Substrat erhöht sich noch die Reihe der
nachgewiesenen Produkte (Tab. 11). Axelrod u. Goldzieher (1970) wiesen u.a.
kürzlich 19-Nor-Testosteron und 19-aldo-Androstendion als Metaboliten des
Progesterons im Corpus luteum-Gewebe des menstruellen Zyklus nach. Von Interesse sind vergleichende Untersuchungen von Hammerstein u. Mitarb. (1964)
über die quantitativen Verhältnisse bei der Steroidbiosynthese aus Azetat in
Corpora lutea des Zyklus und der Gravidität (Tab. 12). Die Autoren fanden in
Corpora lutea der 8.-12. Schwangerschaftswoche eine gegenüber den Corpora
lutea menstruationis deutlich vermehrte Steroidsynthese. Mit Schlußfolgerungen aus solchen Ergebnissen für die physiologischen Verhältnisse wird man
noch vorsichtig sein müssen.

Tab. 11. **In vitro-Synthese in menschlichen Corpora lutea**
         Substrat: Progesteron $-4-^{14}C$ $(-7-^{3}H)$

| Produkt | Status | Autoren |
|---|---|---|
| 20α-OH-Progesteron | M.Z. | Huang u. Pearlman, 1963; Axelrod u. Goldzieher, 1970; |
| 17α-OH-Progesteron | M.Z.und **Gr.** | Huang u. Pearlman, 1963; Ryan, 1963; Inguilla u. Mitarb.,1967; Axelrod u. Goldzieher, 1970; |
| 16α-OH-Progesteron | M.Z. | Huang,1967; Axelrod u. Goldzieher,1970; |
| 17,20-OH-Progesteron | M.Z. | Inguilla u. Mitarb., 1967; |
| 6β,17α-OH-Progesteron | M.Z. | Axelrod u. Goldzieher, 1970; |
| Androstendion | M.Z. | Inguilla u. Mitarb.,1967; Axelrod u. Goldzieher, 1970; |
| 19-aldo-Androstendion | M.Z. | Axelrod u. Goldzieher, 1970; |
| Testosteron | M.Z. | Inguilla u. Mitarb.,1967; Axelrod u. Goldzieher, 1970; |
| 19-Nor-Testosteron | M.Z. | Axelrod u. Goldzieher, 1970; |
| Östradiol-17β | M.Z.und Gr. | Ryan,1963; Inguilla u. Mitarb., 1967; |
| Östron | M.Z. | Inguilla u. Mitarb., 1967; |

M.Z.  - Menstrueller Zyklus
Gr.   - Gravidität


Tab. 12. **In vitro-Steroidsynthese aus Acetat-1-$^{14}$-C in menschlichen**
         **Corpora lutea**[1]

| Produkt | DPM/Corpus luteum | | | |
|---|---|---|---|---|
|  | Zyklus | | Gravidität | |
| Progesteron | 66.750 | (2) | 162.675 | (4) |
| 20α-OH-Progesteron | 1.170 | (1) | 16.420 | (2) |
| 17α-OH-Progesteron | 9.605 | (2) | 83.300 | (3) |
| Androstendion | 902 | (2) | 38.003 | (3) |
| Östradiol-17β | 1.925 | (2) | 21.900 | (3) |
| Östron | 822 | (2) | 7.013 | (3) |

( ) - Anzahl der Corpora lutea

---

[1] nach Hammerstein u. Mitarb., 1964


   Weiter ist unter in vitro-Versuchsbedingungen die Einwirkung von HCG auf
die Steroidbildung im Corpus luteum-Gewebe von Interesse. So fanden Rice u.
Mitarb. (1964, Tab. 13) bei Zusatz von 100 - 175 IE HCG zum Inkubationsge-
misch eine signifikant erhöhte Einbaurate von Azetat in eine Reihe von Steroiden u.a. in Progesteron, Androstendion und Östradiol-17β.
   Le Maire u. Mitarb. (1968) sowie Maeyama u. Mitarb. (1970) bestimmten die
Progesteronkonzentration im Corpus luteum-Gewebe vor und nach der Inkubation
und errechneten daraus die Progesteronproduktion unter in vitro-Bedingungen
(Tab. 14). Nach Zusatz von HCG fanden sie ebenfalls sowohl in Corpora lutea
des Zyklus als auch der Schwangerschaft eine erhöhte Produktion von Progesteron. In Abb. 11 sind derzeitige Vorstellungen über die Steroidbiosynthese

Tab. 13. **Einfluß von Gonadotropinen auf die in vitro-Steroidsynthese aus Acetat-1-[14]-C in menschlichen Corpora lutea[1]**

| Produkt | DPM/g Feuchtgewebe Menstrueller Zyklus | | DPM/g Feuchtgewebe Gravidität | |
|---|---|---|---|---|
| | ohne HCG | mit HCG | ohne HCG | mit HCG |
| Progesteron | 7.400 (1) | 31.700 (1) | 214.550 (4) | 689.000 (4) |
| 20α-OH-Progesteron | Spur (1) | 1.750 (1) | 21.667 (4) | 34.667 (4) |
| 17α-OH-Progesteron | 1.700 (1) | 8.300 (1) | 180.000 (2) | 513.000 (2) |
| Androstendion | 800 (1) | 8.700 (1) | 77.075 (2) | 213.500 (2) |
| Östradiol-17β | 1.200 (1) | .3.300 (1) | 64.250 (2) | 117.950 (2) |

( ) - Anzahl der Corpora lutea

---

[1] nach Rice u. Mitarb., 1964

Tab. 14. **In vitro-Produktion von Progesteron in menschlichen Corpora lutea**

| Status | µg/g Feuchtgewebe | | Autoren |
|---|---|---|---|
| | ohne HCG | mit HCG | |
| C.L.Phase 1.Drittel | 55,3 (9) | 126,4 (9) | Le Maire u. Mitarb., 1968; |
| C.L.Phase | 14,6 (22) | 157,8 (20) | Maeyama u. Mitarb., 1970; |
| Gravidität 1.Drittel | 11,5 (4) | 29,5 (4) | Le Maire u. Mitarb., 1968; |
| " | 20,0 (12) | 218,6 (10) | Maeyama u. Mitarb., 1970; |

C.L. - Corpus luteum
( ) - Anzahl der Corpora lutea

in menschlichen Corpora lutea zusammengefaßt. Der bekannte Syntheseweg über Progesteron, 17α-Hydroxyprogesteron, Androstendion und Östradiol ist durch in vitro-Untersuchungen wiederholt bestätigt worden (Ryan, 1963; Huang u. Pearlman, 1963; Hammerstein u. Mitarb., 1964; Ingiulla u. Mitarb., 1967; Axelrod u. Goldzieher, 1970).

Der vorher beschriebene Nachweis einer vermehrten Konzentration dieser Substanzen im Venenblut von Ovarien mit Corpus luteum gegenüber dem Venenblut von Ovarien ohne Corpus luteum spricht ebenso wie der Nachweis dieser Substanzen im Corpus luteum-Gewebe selbst für die Richtigkeit der aus den in vitro-Untersuchungen abgeleiteten Vorstellungen.

Ergebnisse von in vitro-Untersuchungen geben weiterhin Hinweise für die Annahme, daß auch ein zweiter Syntheseweg unter Umgehung des Progesterons über 17α-Hydroxypregnenolon und Dehydroepiandrosteron möglich ist (Ryan, 1963; Ingiulla u. Mitarb., 1967). Es ist noch ungeklärt, in welchem Ausmaß dieser zweite Syntheseweg unter physiologischen Bedingungen eine Rolle spielt.

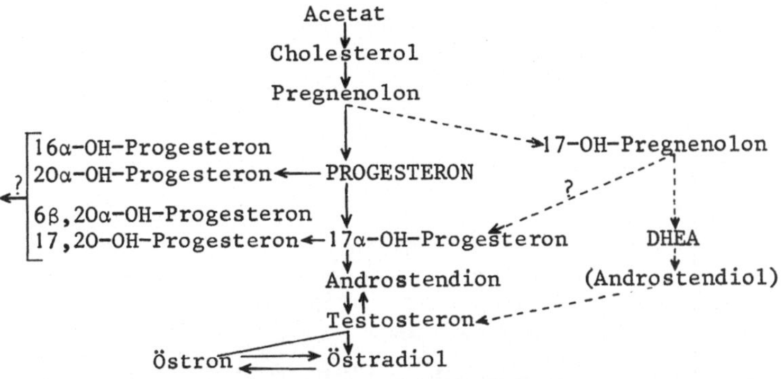


Abb. 11. Biosyntheseweg von Steroiden in menschlichen Corpora lutea nach
in vitro-Untersuchungen von Ryan (1963), Huang u. Pearlman (1963),
Hammerstein u. Mitarb. (1964), Ingiulla u. Mitarb. (1967) und
Axelrod u. Goldzieher (1970). Neben dem bekannten Syntheseweg über
Progesteron, 17α-Hydroxyprogesteron, Androstendion und Östradiol-
17β erscheint ein zweiter Syntheseweg unter Umgehung des Progesterons über 17α-Hydroxypregnenolon und Dehydroepiandrosteron möglich.


## Regulation der Corpus luteum-Funktion

Untersuchungen über die Induktion von Ovulationen mit Gonadotropinen sprechen dafür, daß bei der Frau der Follikelsprung in einem durch FSH stimulierten Follikel sowie die Umwandlung dieses Follikels in einen Gelbkörper
durch HCG bzw. LH hervorgerufen werden kann. Der bekannte LH-Gipfel im Plasma und im Harn während der Ovulationsphase wird dahingehend gedeutet, daß
auch unter physiologischen Bedingungen der Follikelsprung und die Umwandlung
des Graaf'schen Follikels in den Geldkörper durch dieses Gonadotropin stimuliert wird. Bei einer kritischen Untersuchung der bisher vorliegenden Ergebnisse erscheint es jedoch nicht ausreichend gesichert, daß der LH-Gipfel jeweils unmittelbar vor der Ovulation liegt. Auch hier ist wieder auf die
Schwierigkeit hinzuweisen, bei der Korrelation verschiedener Vorgänge das
Ereignis der Ovulation selbst mit ausreichender Sicherheit zu bestimmen.
Die Bedeutung des in der letzten Zeit mit radioimmunologischen Methoden
mehrfach nachgewiesenen FSH-Gipfels in der Ovulationsphase ist noch unklar
(Midgley u. Jaffe, 1968 a; Saxena u. Mitarb., 1968). Untersuchungen von
Geiger in unserem Laboratorium weisen darauf hin, daß diese FSH-Bestimmungen
zum Teil nicht ausreichend spezifisch sind. Schaltet man durch Zugabe von
HCG zum Inkubationsansatz eine Kreuzreaktion mit LH aus, so tritt der FSH-
Gipfel in der Zyklusmitte weniger deutlich in Erscheinung. Dies zeigt
Abb. 12 mit FSH und LH-Bestimmungen in 12-Stunden-Harnportionen während des
menstruellen Zyklus. FSH-Untersuchungen mit biologischen Bestimmungsmetho-
den führten bisher zu recht differenten Ergebnissen.
Es ist zur Zeit noch völlig unklar, welche Mechanismen die sehr konstante Lebensdauer des Corpus luteum von 14 Tagen regulieren. Der verhältnismäßig scharfe Abfall des LH im Plasma und im Harn am Ende der Ovulationsphase
sowie die im allgemeinen recht niedrigen LH-Werte im Verlauf der Lutealphase sind in dieser Richtung nur schwierig zu deuten.
Befunde bei hypophysektomierten Frauen, bei denen eine Ovulation durch
Gonadotropine induziert wurde, sprechen dafür, daß das Corpus luteum auch
ohne weitere LH- bzw. HCG-Stimulation zumindest bis zur Bildung funktionsfä-
higer Trophoblastzellen nach der Implantation existieren kann (Gemzell u.
Kjessler, 1964; Bettendorf u. Mitarb., 1964; Knörr, 1967).

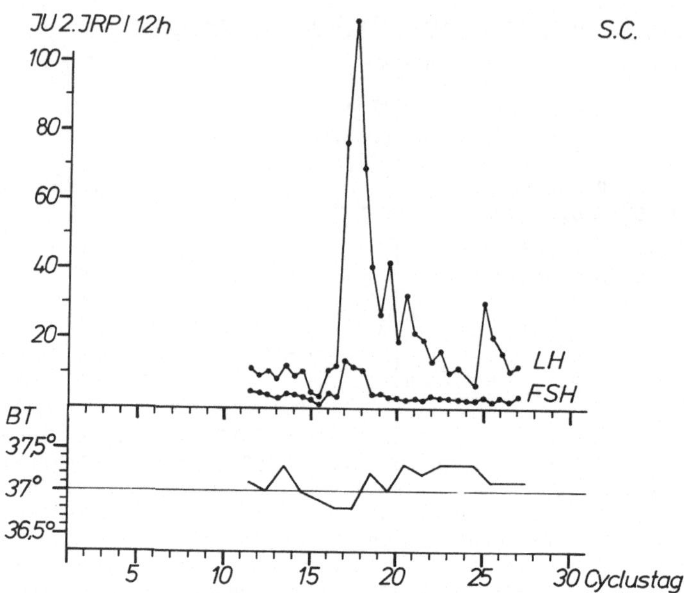


Abb. 12. Radioimmunologische Bestimmungen von FSH und LH in 12-Stunden Harn-
portionen während eines menstruellen Zyklus. Eine Kreuzreaktion mit
LH wurde durch Zugabe von HCG zum Inkubationsansatz ausgeschaltet.
(Nach Geiger, unveröffentlichte Resultate)

Abrams u. Mitarb. (1967) wiesen bei einem 15jährigen hypophysektomierten
Mädchen, bei dem eine Ovulation durch Gonadotropine ausgelöst wurde, eine
Corpus luteum-Funktion von 14 Tagen nach ohne weitere Stimulation mit LH.
In diesem Zusammenhang sind neuere Untersuchungen von Llerena u. Mitarb.
(1969) von Interesse. Sie fanden in der Lutealphase des Zyklus im Venenblut
von Ovarien mit Corpus luteum signifikant niedrigere LH-Konzentrationen als
im peripheren Blut. Unter Ovulationshemmern war dieser Unterschied nicht
erkennbar. Ebenso war für die Konzentration von FSH im Ovarialvenenblut sowie im peripheren Blut kein Unterschied nachweisbar. Die Autoren deuten
diese Befunde hypothetisch dahingehend, daß das Corpus luteum möglicherweise
über spezifische Rezeptoren verfügt, welche LH binden.
Nicht ausgeschlossen ist auf der anderen Seite die Möglichkeit, daß das
Corpus luteum menstruationis für die Erhaltung seiner 14tägigen Lebenszeit
keiner übergeordneten Regulationsmechanismen bedarf. Es ist durchaus denk-
bar, daß diese Lebensdauer einfach durch eine begrenzte Wachstumspotenz der
Corpus luteum-Zellen bedingt ist. Auch für eine Corpus luteum erhaltende
Funktion des bei verschiedenen Tierspezies nachgewiesenen luteotrophen Hormons (LTH) gibt es bei der Frau bisher keine Hinweise. Ebenso konnten die
bei einigen Spezies nachgewiesenen sogenannten luteolytischen Faktoren beim
Menschen bisher nicht demonstriert werden.
Mit dem Eintritt der Schwangerschaft bzw. dem Abschluß der Implantation
des Eis kann mit der Bildung von HCG in den Trophoblastzellen gerechnet
werden. In neueren Untersuchungen wurde HCG im mütterlichen Blut mit radioimmunologischen Methoden etwa 3 Tage nach der angenommenen Implantation
nachgewiesen (Marshall u. Mitarb., 1968; Wide, 1969; Yoshimi u. Mitarb.,
1969). Es ist äußerst wahrscheinlich, daß die weitere Erhaltung des Corpus
luteum und seine Entwicklung zum Corpus luteum graviditatis im wesentlichen
durch das in den Trophoblastzellen gebildete HCG bedingt ist.
Eine Verlängerung der Lebensdauer des Corpus luteum durch HCG wurde
schon in älteren experimentellen Untersuchungen bei der Frau nachgewiesen
(Browne u. Vennig, 1938; Brown u. Bradbury, 1947). Strott u. Mitarb. (1969)

berichteten kürzlich über eine Verlängerung der Corpus luteum-Phase bis zu
21 Tagen bei Verabreichung einer täglichen Dosis von 2000 IE von HCG. Unter
dieser Behandlung stieg gleichzeitig die Progesteron- und die 17α-Hydroxy-
progesteronkonzentration im Blut um das Zweifache an. Kaiser u. Geiger
(1969, 1970) gaben 3 Frauen nach der Ovulation steigende, der Frühgravidität
angepaßte HCG-Dosen von 300 - 80.000 IE täglich bis zum 24. postovulatorischen Tag. Zu diesem Zeitpunkt war die Östrogenausscheidung in 24 Stunden
etwa doppelt so hoch wie in der Lutealphase des Zyklus. Die Pregnandiolausscheidung im Harn stieg unter der Behandlung bis zum 18. postovulatorischen
Tag um etwa 50% an.

In eigenen Untersuchungen haben wir uns kürzlich mit der Einwirkung von
HCG auf die Progesteronkonzentration im peripheren Blut im Verlauf des Zyklus und der Schwangerschaft beschäftigt. Das aus dem Plasma extrahierte Progesteron wurde enzymatisch reduziert und das reduzierte Produkt als Chlor-
monoazetat gaschromatographisch quantitativ bestimmt (Llauró u. Mitarb.,
1968).

In Abb. 13 sind die Ergebnisse bei 12 Zyklen von 10 Frauen am 5. Tag der
hyperthermen Phase dargestellt. 5000 bzw. 20.000 IE HCG wurden 2 Stunden im

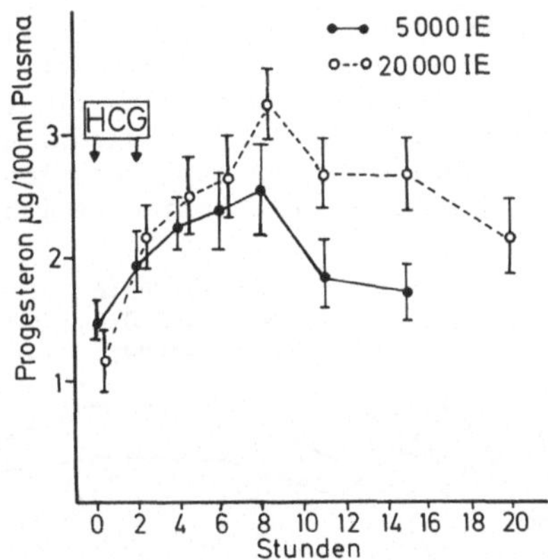


Abb. 13. Durchschnittliche Plasmaprogesteronkonzentrationen mit den entsprechenden Standardabweichungen vor, während und nach der intravenösen
Gabe von 5000 IE und 20.000 IE HCG bei jeweils 6 Frauen.


Dauertropf intravenös infundiert. Es zeigte sich in allen Fällen ein eindeutiger Anstieg der Plasmaprogesteronkonzentration nach der HCG-Infusion. Sowohl bei der niedrigen als auch bei der höheren Dosierung wurden maximale
Progesteronkonzentrationen im peripheren Plasma erst 8 Stunden nach Beginn
der HCG-Infusion beobachtet. Danach kommt es zuerst zu einem schnelleren und
später zu einem langsameren Abfall der Progesteronkonzentrationen. Noch 20
Stunden nach der Infusion von 20.000 IE HCG ist die mittlere Progesteronkonzentration im peripheren Plasma gegenüber dem mittleren Ausgangswert
deutlich erhöht.

Aus diesen Ergebnissen ergibt sich einmal eine deutliche Dosisabhängig-
keit der Plasmaprogesteronkonzentration von HCG. Zum zweiten bleibt die Auswirkung des HCG auf den Plasmaprogesteronspiegel über einen überraschend langen Zeitraum bestehen, wobei das Maximum dieser Auswirkung erst 6 Stunden
nach Beendigung der zweistündigen HCG-Infusion erreicht wird.

Eine Analyse dieser Ergebnisse wird mehrere Faktoren berücksichtigen müssen. Einer der Faktoren dürfte die verhältnismäßig lange Halbwertzeit des
HCG sein (Tab. 15). Von verschiedenen Autoren wurden Halbwertzeiten zwischen
23 und 37 Stunden ermittelt. Demgegenüber sind die Halbwertzeiten für LH um
ein vielfaches geringer.

Tab. 15. <u>Halbwertzeit von menschlichem Plasma-LH und HCG (Werte in Stunden)</u>

| Status | schnelle Komponente | | langsame Komponente | | Autoren |
|---|---|---|---|---|---|
| | LH | HCG | LH | HCG | |
| Eine i.v. Injektion | 1 | 8 | 5 | 24 | Parlow, 1965; |
| Hypophysektomie; Entfernung der Placenta | 1/3 | 11 | 4 | 23 | Yen u. Mitarb., 1968; |
| Post partum; eine i.m. Injektion | – | 8,9 | – | 37,2 32,0 | Midgley u. Jaffe, 1968; |
| Eine i.v. oder i.m. Injektion | – | 5,6 | – | 23,9 | Rizkallah u. Mitarb., 1969; |

In Abb. 14 sind die Ergebnisse von Untersuchungen gleicher Art gegen Ende des ersten Trimesters der Schwangerschaft dargestellt. Lediglich bei 7 von 9 Frauen fand sich ein geringer, jedoch statistisch nicht signifikanter Anstieg der Progesteronkonzentration nach HCG-Infusion. Auch im letzten Trimester der Schwangerschaft war bei 6 Fällen nach Infusion von 20.000 IE HCG kein signifikanter Anstieg der Progesteronkonzentration zu beobachten (Abb. 15).

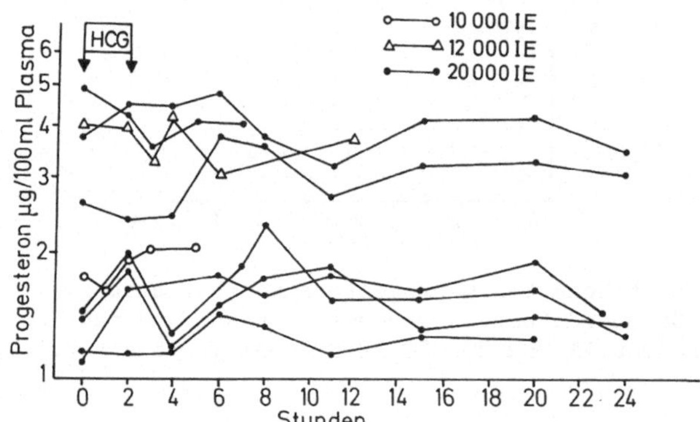


Abb. 14. Plasmaprogesteronkonzentrationen nach intravenöser Gabe von 10.000 IE, 12.000 IE und 20.000 IE HCG bei 9 graviden Frauen in der 6. bis 19. Schwangerschaftswoche.

Zusammengefaßt ergibt sich für die Regulation der Corpus luteum-Funktion unter physiologischen Bedingungen bei kritischer Betrachtung der bisher vorliegenden Daten folgendes Bild:
1. Es ist sehr wahrscheinlich, daß zwischen dem LH-Gipfel in der Ovulationsphase und der Umwandlung des Graaf'schen Follikels in ein Corpus luteum ein ursächlicher Zusammenhang besteht.
2. Es ist unklar, welche Faktoren die Lebensdauer des Corpus luteum menstruationis bestimmen.
3. Die Entwicklung des Corpus luteum graviditatis sowie dessen Erhaltung in der Frühschwangerschaft steht sehr wahrscheinlich in einem ursächlichen Zusammenhang mit dem in den Trophoblastzellen gebildeten HCG.
4. Es ist sehr wahrscheinlich, daß HCG zumindest im frühesten Stadium der Schwangerschaft die Progesteronsynthese im Corpus luteum stimuliert.

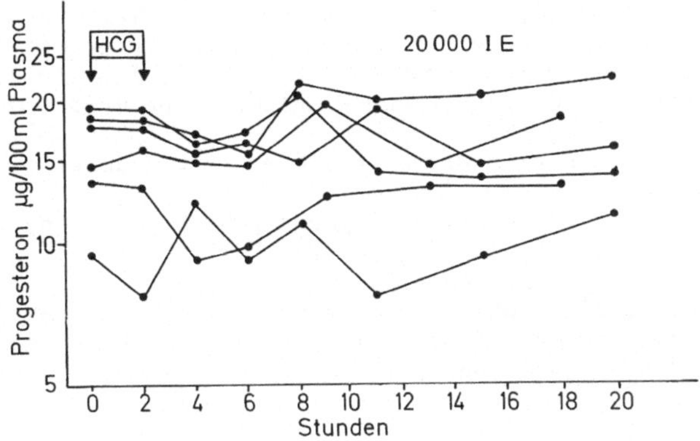


Abb. 15. Plasmaprogesteronkonzentrationen nach intravenöser Gabe von 20.000
IE HCG bei 6 Frauen im letzten Drittel der Schwangerschaft.


## Störungen der Funktion des Corpus luteum

Als letztes möchten wir die Störungen der Corpus luteum-Funktion diskutieren. Insgesamt ist hierüber wenig bekannt.

Zu den häufigsten Anomalien gehört zweifellos die zuerst von Knaus (1929) nachgewiesene verkürzte Lebensdauer des Corpus luteum. Diese kann klinisch für das Problem der Sterilität erhebliche Bedeutung gewinnen.

Kürzlich mitgeteilte Ergebnisse von Strott u. Mitarb. (1970) sind hier von Interesse. Sie bestimmten bei Frauen mit verkürzter Gelbkörperphase täglich die Plasmakonzentration von FSH, LH, Progesteron und 17α-Hydroxyprogesteron. Die durchschnittliche FSH-Konzentration war niedriger als bei Frauen mit normaler Gelbkörperphase. Hingegen lag der präovulatorische LH-Gipfel höher, dagegen die LH-Konzentration im Verlauf der Lutealphase niedriger als bei Frauen mit ungestörter Lutealphase. Diese Befunde sind noch schwierig zu deuten. Sie weisen auf Veränderungen in den regulatorischen Mechanismen bei der Corpus luteum-Insuffizienz hin.

Strott u. Mitarb. (1970) fanden weiterhin bei Frauen mit Corpus luteum-Insuffienz 4-5fach niedrigere Plasmaprogesteronkonzentrationen als bei Frauen mit normal langer Lutealphase.

Ein Wiederanstieg des 17α-Hydroxyprogesterons, welcher gleichzeitig mit dem Progesteronanstieg bei normal langer Corpus luteum-Phase zu beobachten ist, war bei der Corpus luteum-Insuffizienz nicht nachweisbar.

Diese Befunde geben Hinweise für Störungen im Bereich der regulativen Vorgänge sowie im Bereich der Steroidproduktion bei der Corpus luteum-Insuffizienz mit verkürzter Corpus luteum-Phase.

Auffallend sind auch die relativ häufig zu beobachtenden verkürzten Corpus luteum-Phasen nach Auslösung der Ovulation mit Clomiphen. In solchen Fällen wurde mehrfach trotz hyperthermen Verlaufs der Basaltemperaturen eine gänzlich unzureichende Progesteronauswirkung am Endometrium beobachtet.

Es wird heute noch vielfach angenommen, daß die Voraussetzung für die Bildung eines Corpus luteum der Follikelsprung sei. Ausreichende Beweise für diese Annahme können jedoch nicht vorgelegt werden. Es erscheint erlaubt, dahingehend zu spekulieren, daß unter bestimmten Bedingungen eine Lutenisierung des reifen Follikels auch dann stattfindet, wenn keine Ovulation erfolgt ist. Möglicherweise verbergen sich hier auch Ursachen für die Entstehung eines insuffizienten Corpus luteum.

Unklar ist bis heute, ob ohne Graviditätsvorgang auch eine verlängerte Lebensdauer des Corpus luteum menstruationis möglich ist. Man nimmt an, daß gelegentlich zu beobachtende verlängerte Corpus luteum-Phasen in der Regel

durch eine Gravidität bedingt sind, welche vorzeitig zugrundegeht. Für den
Einzelfall können meist entsprechende Beweise nicht vorgelegt werden.

Es war im Rahmen dieser Ausführungen nur möglich, Aspekte der Physiologie
und Pathophysiologie der Corpus luteum-Funktion zu berühren, welche uns besonders wichtig erscheinen.

Es kam uns vor allen Dingen darauf an, Grenzen des derzeitigen Wissens
aufzuzeigen.

Ludwig Fraenkel, dessen Werk wir eingangs gewürdigt haben, sagte bei der
Feier seines 70. Geburtstages in Montevideo: "Meine Kompassion gehört den
Kranken, meine Passion der Biologie."

Dieses Wort durfte ein großer Mann glaubhaft nach einem erfüllten Leben
aussprechen. Man darf jedoch sicher sein, daß sich auch aus der unausgespro-
chenen Intimität dieses Antriebes weitere Erkenntnisse auf den Spuren Ludwig
Fraenkels ergeben.


## Literatur


Aakvaag, A., Fylling, P.: A method for the simultaneous determination of
progesterone, androstenedione, testosterone and dehydroepiandrosterone
sulphate in biological fluids, its application in the analysis of venous
plasma and cyst fluid from human ovaries in situ. Acta Endocr. (Kbh.) 57,
447 (1968).

Abrams, C.A.L., Grumbach, M.M., Dyrenfurth, J., Wiele, R.L. van de: Ovarian
stimulation with human menopausal and chorionic gonadotropins in a pre-
pubertal hypophysectomized female. J. clin. Endocr. 27, 467 (1967).

Axelrod, L.R., Goldzieher, J.W.: Metabolism of (4-$^{14}$C) progesterone by components of normal human ovaries. Acta Endocr. (Kbh.) 65, 607 (1970).

Baird, D.T.: A method for the measurement of estrone and estradiol-17β in
peripheral human blood and other biological fluids using $^{35}$S pipsyl
chloride. J. clin. Endocr. 27, 244 (1968).

- Guevara, A.: Concentration of unconjugated estrone and estradiol in peripheral plasma in nonpregnant women throughout the menstrual cycle,
castrate and postmenopausal women and in man. J. clin. Endocr. 29, 149
(1969).

Bermudez, J.A., Doerr, P., Lipsett, M.B.: Measurement of pregnenolone in
blood. Steroids 16, 505 (1970).

Bettendorf, G., Breckwoldt, M., Knörr, K., Stegner, E.: Gravidität nach
Hypophysektomie und Behandlung mit hypophysärem Human-Gonadotropin.
Dtsch. med. Wschr. 89, 1952 (1964).

Brown, J.B.: The relationship between urinary oestrogens and oestrogens produced in the body. J. Endocr. 16, 202 (1957).

Brown, W.E., Bradbury, J.T.: A study of the physiologic action of human
chorionic hormone. The production of pseudopregnancy in women by chorionic hormone. Amer. J. Obstet. Gynec. 53, 749 (1947).

Browne, J.S.L., Venning, E.H.: The effect of intramuscular injection of
gonadotropic substances on the corpus luteum phase of the human menstrual
cycle. Amer. J. Physiol. 123, 26 (1937).

Craft, J., Wyman, H., Sommerville, J.F.: Serial analysis of plasma progesterone and pregnandiol in human pregnancy. J. Obstet. Gynec. Brit. Cwlth.
76, 1080 (1969).

Eren, S., Reynolds, G.H., Turner, M.E., Schmidt, F.H., Mackay, J.H., Howard,
C.M., Preedy, J.R.K.: Estrogen metabolism in the human: III. A comparison
between females, studied during the first and second halves of the menstrual cycle, and males. J. clin. Endocr. 27, 1451 (1967).

Fotherby, K.: Excretion of pregnanetriol during the normal menstrual cycle.
Brit. med. J. 1960 I, 1545.

Fraenkel, L.: Die Funktion des Corpus luteum. Arch. Gynäk. 68, 438 (1903).
- Zur Histo-Physiologie des Corpus luteum. Arch. Gynäk. 181, 217 (1952).

Geiger, W.: Unveröffentlichte Resultate.

Gemzell, C., Kjessler, B.: Treatment of infertility after partial hypophys-
ectomy with human pituitary gonadotrophins. Lancet 1964 I, 644.
Hammerstein, J., Rice, B.F., Savard, K.: Steroid hormone formation in the
human ovary: Identification of steroids formed in vitro from acetate-1-
$^{14}$C in the corpus luteum. J. clin. Endocr. 24, 597 (1964).
Horton, R.: Estimation of androstenedione in human peripheral blood with
$^{35}$S-thiosemicarbazide. J. clin. Endocr. 25, 1237 (1965).
– Romanoff, E., Walker, J.: Androstenedione and testosterone in ovarian
venous and peripheral plasma during ovariectomy for breast cancer.
J. clin. Endocr. 26, 1267 (1966).
Huang, W.Y.: Conversion of progesterone to 16α-hydroxy-progesterone in nor-
mal human corpus luteum tissue. Steroids 10, 107 (1967).
– Pearlman, W.H.: The corpus luteum and steroid hormone formation. II.
Studies on the human corpus luteum in vitro. J. biol. Chem. 238, 1308
(1963).
Ingiulla, W., Forleo, R., Bruni, V.: Steroid biosynthesis in human ovarian
tissues. Proc. 2$^{nd}$ Int.Congr. on Hormonal Steroids Milan 1966 (eds.
Martini, L., Fraschini, F., Motta, M.) Excerpta Med. (Amst.) Intern.
Congr. Ser. 132, 411 (1967).
Johansson, E.D.B.: Plasma levels of progesterone in pregnancy measured by a
rapid competitive protein binding technique. Acta Endocr. (Kbh.) 61,
607 (1969 a).
– Progesterone levels in peripheral plasma during the luteal phase of the
normal human menstrual cycle measured by a rapid competitive protein bind-
ing technique. Acta Endocr. (Kbh.) 61, 592 (1969 b).
– Wide, L.: Preovulatory levels of plasma progesterone and luteinizing hor-
mone in women. Acta Endocr. (Kbh.) 62, 82 (1969).
Kaiser, R., Geiger, W.: Zur Frage der Oestrogen- und Progesteronproduktion
des Corpus luteum graviditatis in der Frühschwangerschaft. Symp. Dtsch.
Ges. Endokr. 15, 178. Oestrogene – Hypophysentumoren. Berlin-Heidelberg-
New York: Springer 1969.
– – Untersuchungen zur Differenzierung zwischen materner, plazentarer und
embryonaler Hormonbildung. Geburtsh. u. Frauenheilk. 30, 307 (1970).
Kalliala, K., Laatikainen, T., Vihko, R.: Neutral steroid sulfates in human
ovarian vein blood. J. clin. Endocr. 30, 533 (1970).
Knaus, H.: Über den Zeitpunkt der Konzeptionsfähigkeit des Weibes im Inter-
menstruum. Münch.med. Wschr. 76, 1157 (1929).
Knörr, K.: Erfolgreiche Sterilitätsbehandlung mit Humangonadotropinen nach
Hypophysektomie. Med. Welt 18, 3128 (1967).
LeMaire, W.J., Conly, P.W., Moffett, A., Cleveland, W.W.: Plasma progeste-
rone secretion by the corpus luteum of term pregnancy. Amer. J. Obstet.
Gynec. 108, 132 (1970).
– Rice, B.F., Savard, K.: Steroid hormone formation in the human ovary:
V. Synthesis of progesterone in vitro in corpora lutea during the repro-
ductive cycle. J. clin. Endocr. 28, 1249 (1968).
Llauró, J.L., Runnebaum, B., Zander, J.: Progesterone in human peripheral
blood before, during, and after labor. Amer. J. Obstet. Gynec. 101, 867
(1968).
Llerena, L.A., Guevara, A., Lobotsky, J., Lloyd, C.W., Weiss, J., Pupkin,
M., Zanartu, J., Puga, J.: Concentration of luteinizing and follicle-
stimulating hormones in peripheral and ovarian venous plasma. J. clin.
Endocr. 29, 1083 (1969).
Lobotsky, J., Wyss, H.J., Seger, E.J., Lloyd, C.W.: Plasma testosterone in
the normal women. J. clin. Endocr. 24, 1261 (1964).
Maeyama, M., Matuoka, H., Tuchida, Y.: Progesterone biosynthesis by human
corpora lutea in vitro. Steroids 15, 789 (1970).
Marschall, J.R., Hammond, C.B., Ross, G.T., Jacobson, A., Rayford, P., Odell,
W.D.: Plasma and urinary chorionic gonadotropin during early human preg-
nancy. Obstet. and Gynec. 32, 760 (1968).

Midgley, A.R., Jaffe, R.B.: Regulation of human gonadotropins: IV. Correlation of serum concentration of follicle stimulating and luteinizing hormones during the menstrual cycle. J. clin. Endocr. 28, 1699 (1968 a).
- - Regulation of human gonadotropins: II. Disappearance of human chorionic gonadotropin following delivery. J. clin. Endocr. 28, 1712 (1968 b).
Migeon, C.J., Keller, A.R., Lawrence, B., Shepard, T.H.: Dehydroepiandrosterone and androsterone levels in human plasma. Effect of age and sex; day to day and diurnal variations. J. clin. Endocr. 17, 1051 (1957).
Mikhail, G.: Sex steroids in blood. Clin. Obstet. Gynec. 10, 29 (1967).
- Hormone secretion by the human ovaries. Gynec. Invest. 1, 5 (1970).
- Allen, W.M.: Ovarian function in human pregnancy. Amer. J. Obstet. Gynec. 99, 308 (1967).
- Wu, C.H., Ferin, M., Wiele, R.L. van de: Radioimmunoassay of plasma estrone and estradiol. Steroids 15, 333 (1970).
- Zander, J., Allen, W.M.: Steroids in human ovarian vein blood. J. clin. Endocr. 23, 1267 (1963).
Mizuno, M., Lobotsky, J., Lloyd, C.W., Kobayashi, T., Murasawa, Y.: Plasma androstenedione and testosterone during pregnancy and in the newborn. J. clin. Endocr. 28, 1133 (1968).
Neill, J.D., Johansson, E.D.B., Datta, J.K., Knobil, E.: Relationship between the plasma levels of luteinizing hormone and progesterone during the normal menstrual cycle. J. clin. Endocr. 27, 1167 (1967).
Ober, K.G., Klein, J., Weber, M.: Zur Frage einer Progesteronbehandlung. Experimentelle Untersuchungen mit dem Hooker-Forbes-Test und klinische Beobachtungen mit Kristallsuspensionen. Arch. Gynäk. 184, 543 (1954).
Parlow, A.F.: Discussion to Gemzell, C. Induction of ovulation with human gonadotropins. Recent. Pogr. Hormone Res. 21, 198 (1965).
Rice, B.F., Hammerstein, J., Savard, K.: Steroid hormone formation in the human ovary: II. action of gonadotropins in vitro in the corpus luteum. J. clin. Endocr. 24, 606 (1964).
Rivarola, M.A., Saez, J.M., Meyer, W.J., Jenkins, M.E., Migeon, C.J.: Metabolic clearance rate and blood production rate of testosterone and androst-4-ene-3,17-dione under basal conditions, ACTH and HCG stimulation. Comparison with urinary production rate of testosterone. J. clin. Endocr. 26, 1208 (1966).
Rizkallah, T., Gurpide, E., Wiele, R.L. van de: Metabolism of HCG in man. J. clin. Endocr. 29, 92 (1969).
Roy, E.J., Harkness, R.A., Kerr, M.G.: The concentration of oestrogens in the peripheral blood of women during the normal menstrual cycle and in the first trimester of pregnancy. J. Endocr. 31, 177 (1965).
Rubin, B.L., Maralit, M., Kinard, J.H.: A reproducible, reliable method for determination of progesterone (P) and 20α-dihydroprogesterone (20 P) in small volumes of plasma or serum, suitable for a range of concentrations from follicular phase to late pregnancy values. Levels found throughout a normal human pregnancy. J. clin. Endocr. 31, 511 (1970).
Runnebaum, B., Molen, H.J. van der, Zander, J.: Steroids in human peripheral blood of the menstrual cycle. Steroids, Supp. II, 189 (1965).
- Zander, J.: Progesterone in the human peripheral blood in the preovulatory period of the menstrual cycle. Acta Endocr. (Kbh.) 55, 91 (1967).
Ryan, K.J.: The conversion of pregnenolone-7-$^3$H and progesterone-4-$^{14}$C to oestradiol by a corpus luteum of pregnancy. Acta Endocr. (Kbh.) 44, 81 (1963).
Saxena, B.B., Demura, H., Gandy, H.M., Peterson, R.E.: Radioimmunoassay of human follicle stimulating and luteinizing hormone in plasma. J. clin. Endocr. 28, 519 (1968).
Selinger, M., Levitz, M.: Diurnal variation of total plasma estriol levels in late pregnancy. J. clin. Endocr. 29, 995 (1969).
Simmer, H., Voss, H.E.: Androgene im menschlichen Ovarium. Klin. Wschr. 38, 819 (1960).

Southren, A.L., Gordin, G.G., Tochimoto, S., Pinzon, G., Lane, D.R.,
    Stypulkowski, W.: Mean plasma concentration, metabolic clearance and
    basal plasma production rates of testosterone in normal young men and
    women using a constant infusion procedure: Effect of time of day and plas-
    ma concentration of the metabolic clearance rate of testosterone. J. clin.
    Endocr. 27, 686 (1967).
Strott, C.A., Cargille, C.M., Ross, G.T., Lipsett, M.B.: The short luteal
    phase. J. clin. Endocr. 30, 246 (1970).
— Yoshimi, T., Ross, G.T., Lipsett, M.B.: Ovarian physiology: Relationship
    between plasma LH and steroidogenesis by the follicle and corpus luteum;
    Effect of HCG. J. clin. Endocr. 29, 1157 (1969).
Van der Molen, H.J.: Patterns of gonadal steroids in the normal human female.
    Proc. 3rd Int. Congr. Endocr., Mexico, D.F. Excerpta Med. (Amst.) Intern.
    Concr. Ser. No 184, 1968.
— Aakvaag, A.: Progesterone in "Hormones in Blood" (C.H. Gray and A.L.
    Bacharach, eds.) 2nd Ed. p. 221. New York and London: Academic Press,
    1967.
— Groen, D.: Determination of progesterone in human peripheral blood using
    gas liquid chromatography with electron capture detection. J. clin. Endocr.
    25, 1625 (1965).
— — Quantitative determination of submicrogramme amounts of steroids in
    blood using electron capture and flame ionization detection following gas
    liquid chromatography. Memoirs of the Society for Endocr. No 16, p. 155,
    1967.
Wide, L.: Early diagnosis of pregnancy. Lancet 1969 II, 863.
Wiest, W.G.: Estimation of progesterone in biological tissues and fluids
    from pregnant women by double isotope derivative assay. Steroids 10, 279
    (1967).
Woolever, C.A.: Daily plasma progesterone levels during the menstrual cycle.
    Amer. J. Obstet. Gynec. 85, 981 (1963).
Yen, S.S.C., Llerena, O., Little, B., Pearson, O.H.: Disappearance rates of
    endogenous luteinizing hormone and chorionic gonadotropin in man. J. clin.
    Endocr. 28, 1763 (1968).
Yoshimi, T., Lipsett, M.B.: The measurement of plasma progesterone. Steroids
    11, 527 (1968).
— Strott, C.A., Marshall, J.R., Lipsett, M.B.: Corpus luteum function in
    early pregnancy. J. clin. Endocr. 29, 225 (1969).
Yussmann, M.A., Taymor, M.L.: Serum levels of follicle stimulating hormone
    and luteinizing hormone and of plasma progesterone related to ovulation
    by corpus luteum biopsy. J. clin. Endocr. 30, 396 (1970).
Zander, J.: 17α-Oxyprogesteron und Δ-4-Andosten-3,17-dion im menschlichen
    Ovarium. Klin. Wschr. 35, 1101 (1957).
— Steroids in the human ovary. J. biol. Chem. 232, 117 (1958).
— Brendle, E., Münstermann, A.-M. von, Diczfalusy, E., Martinsen, B.,
    Tillinger, K.G.: Identification and estimation of oestradiol-17β and
    estrone in human ovaries. Acta obstet. gynec. scand. 38, 724 (1959).
— Forbes, T.R., Münstermann, A.-M. von, Neher, R.: $\Delta^4$-3-ketopregnene-20β-ol
    and $\Delta^4$-3-ketopregnene-20α-ol, two naturally occurring metabolites of pro-
    gesterone. Isolation, identification, biologic activity and concentration
    in human tissues. J. clin. Endocr. 18, 337 (1958).
— Thijssen, J., Münstermann, A.-M. von: Isolation and identification of
    16α-hydroxyprogesterone from human corpora lutea and placental blood.
    J. clin. Endocr. 22, 861 (1962).

Symp. Dtsch. Ges. Endokrin. _17_, 85-100 (1971)
© by Springer-Verlag

# Stoffwechsel und Wirkung der synthetischen Gestagene

## Metabolism and Effects of Synthetic Gestagens

G. A. OVERBEEK

N.V. Organon, Oss, Holland
Mit 11 Abbildungen[+]

## Summary

It is shown that the effects of various synthetic gestagens on the uterus
and the pituitary gland may differ considerably. Some possible explanations
are:
1. The interference of other activities i.e. oestrogenicity with progestational effects.
2. Physiological differences in various animal species such as the relative
   importance of ovarian and placental function for the maintenance of pregnancy.
3. The presence or absence of central effects.
4. Properties of receptors.
   Metabolic differences between some derivatives of pregnane and retro-
pregnane are discussed.
   The examples mentioned show that in a number of instances molecular biological studies can yield an explanation for the different effects observed
with related substances in the same animal species, and for those of one
substance in various species. More rarely the analyses of metabolites will
serve this purpose. It is hardly to be expected that this kind of study will
allow to predict the presence of desired or undesired effects of new compounds.

Das Thema, welches den Auftrag zu diesem Vortrag bildet, ist sehr breit.
Sogar bei einer Beschränkung auf die in pharmazeutischen Präparaten enthaltenen Gestagene müßten 23 verschiedene Substanzen mit zwei verschiedenen chemischen Grundstrukturen (Pregnane und Oestrane) besprochen werden (Tab. 1).
Der Autor freut sich deshalb, daß wenigstens über den Stoffwechsel der meisten nicht sehr viel bekannt ist. Andererseits haben jedoch die synthetischen Gestagene mehrere Wirkungen, die manchmal noch aufgrund verschiedener
Wirkungsmechanismen zustande kommen. Es ist deshalb klar, daß in der zur Verfügung stehenden Zeit nur einige Beispiele gegeben werden können, die leider
hauptsächlich die Lücken unserer Kenntnisse demonstrieren werden und vielleicht zeigen, in welcher Richtung man die Lösung der Probleme suchen könnte.

Ob eine Aktivität gestagener Art ist und eine Substanz deshalb als ein
Gestagen betrachtet werden sollte, muß immer aufgrund einiger Effekte im
Tierexperiment entschieden werden. Wenn eine Substanz wenigstens einen der
in Tab. 2 erwähnten Effekte auslöst, wird sie als ein "Gestagen"[1] betrachtet.

---

[+] Halbtonbilder s. Anhang S. 191
[1] Ob der Name Gestagen korrekt ist und eine Bezeichnung als Progestativum,
   Progestagen oder sogar eine Untereinteilung in Prägestagen und Gestagen
   zu bevorzugen wäre, wird in diesem Rahmen nicht diskutiert.

Tab. 1. <u>Synthetische Gestagene in Handelspräparaten</u>

<u>Pregnanderivate</u>

| | |
|---|---|
| Chlormadinonazetat | 6-Chloro-17α-azetoxy-pregna-4,6-dien-3,20-dion |
| Dydrogesteron | 9β, 10α-Pregna-4,6-dien-3,20-dion |
| Medrogeston | 6,17α-Dimethyl-pregna-4,6-dien-3,20-dion |
| Medroxyprogesteronazetat | 6α-Methyl-17α-azetoxy-pregn-4-en-3,20-dion |
| Megestrolazetat | 6-Methyl-17α-azetoxy-pregna-4,6-dien,3,20-dion |
| Melengestrolazetat | 6-Methyl-16-methylen-17α-azetoxy-pregna-4,6-dien-3,20-dion |
| Superlutin[a] | 16-Methylen-17α-azetoxy-pregna-4,6-dien-3,20-dion |

<u>Oestranderivate</u>

| | |
|---|---|
| Allylestrenol | 17α-Allyl-estr-4-en-17β-ol |
| Dimethisteron | 6α-Methyl-17α-prop-1-ynyl-17β-hydroxy-androst-4-en-3-on |
| Ethisteron | 17α-Äthynyl-17β-hydroxy-androst-4-en-3-on |
| Ethinodioldiazetat | 3β,17β-Diazetoxy-17α-äthynyl-estr-4-en |
| Lynestrenol | 17α-Äthynyl-estr-4-en-17β-ol |
| Norethisteron | 17α-Äthynyl-17β-hydroxy-estr-4-en-3-on |
| Norethisteronazetat | 17α-Äthynyl-17β-acetoxy-estr-4-en-3-on |
| Norethinodrel | 17α-Äthynyl-17β-hydroxy-estr-5(10)-en-3-on |
| Norgesteron | 17α-Vinyl-17β-hydroxy-estr-5(10)-en-3-on |
| Norgestrel | (dl)-13β-Äthyl-17α-äthynyl-17β-hydroxy-gon-4-en-3-on |
| Norgestrienon | 17α-Äthynyl-17β-hydroxy-estra-4,9,11-trien-3-on |
| Normethandron | 17α-Methyl-17β-hydroxy-estr-4-en-3-on |
| Norvinisteron | 17α-Vinyl-17β-hydroxy-estr-4-en-3-on |
| Quingestanolazetat | 17α-Äthynyl-17β-acetoxy-estr-3,5-dien-3-ol-cyclopentyläther |

[a] Handelsname

   Abhängig von der Tierart und der Substanz kann eine vorherige oder gleichzeitige Behandlung mit einem Oestrogen notwendig sein, um den erwünschten Effekt zu erzielen.
Tab. 2.

GESTAGENE WIRKUNGEN

1 Sekretion Endometrium
2 Deziduombildung
3 Schwangerschaftserhaltung


   Von den vielen in Tab. 1 erwähnten Substanzen sind nur relativ wenige imstande, die Schwangerschaft bei kastrierten Tieren zu erhalten; diese sind in Tab. 3 zusammengefaßt worden. Warum die anderen die Schwangerschaft bei kastrierten Tieren *nicht* erhalten, ist nicht in allen Fällen klar.
   Es erscheint nicht unwahrscheinlich, daß die sonstigen biologischen Aktivitäten dieser Gestagene dafür verantwortlich sein könnten, und schon aus diesem Grunde wurden die wichtigsten nicht gestagenen Wirkungen der Gestagene in Tab. 4 zusammengefaßt.
   Es ist gerade die zuerst erwähnte *oestrogene* Wirkung, die bei den für Oestrogene hochempfindlichen Nagetieren dafür verantwortlich ist, daß manche Gestagene die Schwangerschaft nach vorheriger Kastration dieser Tiere nicht erhalten.
   Auch die Deziduombildung wird von den Oestrogenen oft verhindert. Der Mensch, wie der Hamster, ist viel weniger oestrogenempfindlich als Ratte und Maus. Deshalb ist es wichtig, daß man bei Hamstern auch mit oestrogenen Gestagenen, wie Lynestrenol und Norethinodrel, Deziduome erzeugt hat (Abb. 1).

Tab. 3.

SCHWANGERSCHAFTSERHALTUNG
(KASTRIERTE RATTEN)

Progesteron  (s.c.)

Normethandron  (s.c.)

Norgestrel  (s.c.+or.)

Allylestrenol  (s.c.+or.)

Medroxyprogesteronazetat  (s.c.+or.)

Medrogeston  (s.c.+or.)

Megestrolazetat  (s.c.+or.)

Chlormadinon  (s.c.+or.)


Tab. 4. "Neben"-Wirkungen synthetischer Gestagene

1 Oestrogene und anti-oestrogene Wirkung
2 Androgene und anti-androgene Wirkung
3 Anti-gestagene Wirkung
4 Förderung und Hemmung der gonadotropen Funktion der Hypophyse (Produktion
  oder Abgabe von LH, FSH, LTH)
5 Förderung und Hemmung der endokrinen Funktion des Ovars (Produktion oder
  Abgabe von Oestrogenen und Gestagenen)
6 Hemmung der Befruchtung
7 Hemmung der Nidation
8 Hemmung der Motilität vom Uterus und Eileiter
9 Förderung und Hemmung des Spermatransports

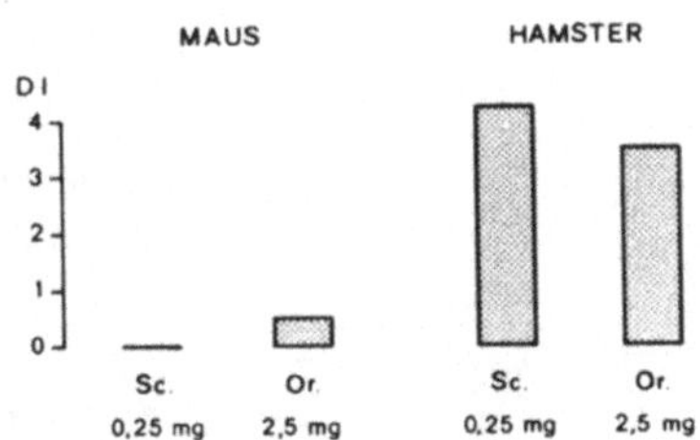


Abb. 1. Deziduombildung durch Norethinodrel


   Wenn bewiesen wurde (Madjerek et al., 1960; Desaulles u. Krähenbühl,
1962; Suchowsky, 1963), daß das Allylestrenol die Schwangerschaft bei kastrierten Ratten erhalten konnte, war dies Anlaß, die Substanz bei drohendem und habituellem Abort zu verwenden. Obwohl es bekanntlich sehr schwierig
oder vielleicht gar fast unmöglich ist, den Erfolg einer derartigen Therapie
zu beweisen, haben wir doch heute, nach einer Erfahrung von etwa 10 Jahren,
die rein klinische Bestätigung, daß das Allylestrenol auch bei der Frau die
bedrohte Schwangerschaft erhält.
   Aus den folgenden Gründen war zu vermuten, daß der Wirkungsmechanismus
bei der Ratte und bei der Frau verschieden sein könnte. Erstens beendet die

Kastration die Schwangerschaft bei der Ratte, nicht jedoch bei der Frau. Bei
der Frau ist die Plazenta das Organ, welches als Gestagenquelle betrachtet
werden soll, während die Rattenplazenta fast gar keine Hormone produziert.
Damit könnte zusammenhängen, daß die therapeutische Dosierung bei der Ratte
relativ viel höher ist (20-40 mg/kg/Tag) als bei der Frau (0.02-0.04 mg/kg/
Tag). Für die Ratte könnte eine solche Menge zur Substitution eines Proge-
steronmangels ausreichen; die wirksame menschliche Dosierung ist für eine
Substitution viel zu niedrig.

Das Problem scheint gelöst zu werden durch die Beobachtungen von Szontagh,
1963, und Toth u. Treit, 1964, daß nach Verabreichung von Allylestrenol die
Ausscheidung von HCG, Oestriol und Pregnandiol im Harn schwangerer Frauen
zunimmt (Abb. 2). Dies kann offenbar als eine Anregung der Hormonproduktion
der Plazenta betrachtet werden.

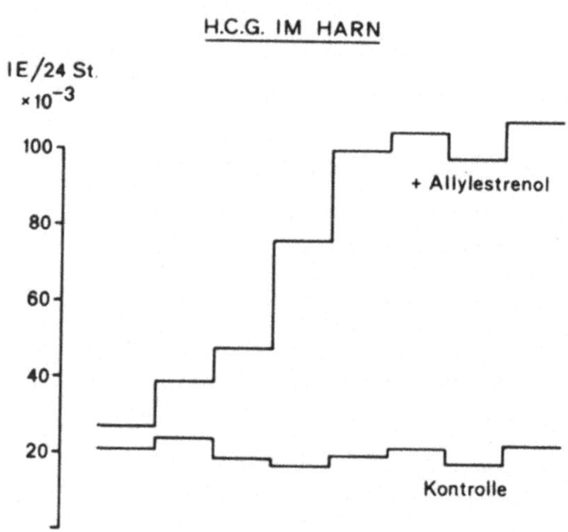


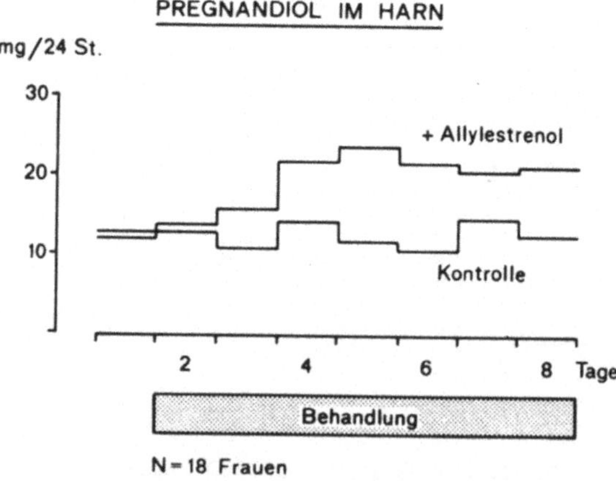


Abb. 2. Nach Toth und Treit, 1964.

Wir haben hier also zwei Beispiele der verschiedenartigen Wirkung der
Gestagene. Bei den Nagetieren ist die quantitativ verschiedene oestrogene
"Neben"aktivität der Gestagene die Ursache verschiedener Effekte auf Deziduombildung. Im zweiten Beispiel wirkt *eine* Substanz (Allylestrenol) bei
zwei Tierarten auf andere Endorgane (resp. Ovar und Plazenta).

Man kann sich kaum vorstellen, wie eine Substanz hier zwar den gleichen
Effekt auslöst, jedoch in ganz verschiedener Weise. Dies ist nun aber gerade

typisch für die Schwangerschaftshormone, wie Progesteron und Prolaktin, die
auf vielerlei Weisen vor, während und nach der Schwangerschaft irgendwie das
neue Leben schützen.

Heutzutage ist es unmöglich, von synthetischen Gestagenen zu sprechen und
dabei die Effekte auf das Zentralnervensystem außer Betracht zu lassen. Deshalb werden die folgenden Effekte erwähnt:
1. Erhöhung der Basaltemperatur.
2. Hemmung der Ovulation.
3. Auslösung der Ovulation.

Wir werden uns auf die drei *Substanzgruppen*, Oestrane, Pregnane und Retro-
pregnane beschränken und nicht die individuellen Substanzen besprechen.

Bekanntlich wird die Erhöhung der Basaltemperatur nicht nur durch Progesteron sondern auch durch die meisten Gestagene ausgelöst. Eine Ausnahme bilden die Retrosteroide wie Retroprogesteron und Dydrogesteron, die nicht
einen solchen Effekt haben. (Das in diese Gruppe gehörende Trengeston
scheint in hohen Dosen doch thermogen zu sein (Onetti, 1968), ist aber höchstens schwach aktiv (Stamm u. Gerhard, 1968).

Nach allgemeiner Ansicht wirkt wenigstens das Progesteron als ein Stimu-
lans des thermoregulatorischen Zentrums im Hypothalamus. Eine Wirkung via
Uterus oder Schilddrüse ist ausgeschlossen, da die Erhöhung der Temperatur
auch beim Fehlen dieser Organe auftritt, z.B. bei normalen Männern und auch
bei weiblichen Kretins (Rothschild u. Rapport, 1952), nicht aber bei Frauen
mit hypothalamischen Defekten (Netter, 1954) oder nach Behandlung mit Pheno-
barbital (Elert, 1951). Wie immer erhebt sich die Frage, ob dies geschieht
durch eine *direkte* Wirkung des Progesterons oder durch einen Metaboliten, da
ja bekanntlich die 5β-Pregnane thermogen sein können, und im Organismus wird
das Progesteron in einen 5β-Pregnan übergeführt. Ehe man dies annehmen könnte, sollte gezeigt werden, daß auch die hormonal inaktiven Abbauprodukte der
Estrene eine solche Wirkung haben (z.B. 17α-Äthinyl-5β-estran-17β-ol oder
17α-Methyl-17β-hydroxy-5β-estran-3-on). Die Metabolittheorie würde eine Erklärung geben für die Wirkung der Retro-steroide.

Bei diesen Substanzen findet nämlich die Reduktion der Doppelbindung (und
der 3-Ketogruppe) im A-Ring nicht statt, weil die diesbezüglichen Enzyme mit
dem Retropregnangerüst, das sterisch von dem der Pregnane so verschieden
ist, sich nicht binden können (Diczfalusy u.a., 1963; Walep u. de Lange,
1966; Breuer u. Knuppen, 1969).

Eine andere Möglichkeit wäre aber, daß die Rezeptoren im Hypothalamus
ähnliche Schwierigkeiten haben wie die reduzierenden Enzyme und sich ebenfalls mit den Retro-steroiden schlecht abfinden. Dafür würde plädieren, daß
auch die zentrale Ovulationshemmung durch Retro-steroide kaum verursacht wird.
Ein ähnlicher Unterschied zwischen Rezeptoraffinität in verschiedenen peripheren Organen für ein normales und ein Retro-steroid findet sich bei Chlormadinon und Retro-Chlormadinon.

Chlormadinon ist im Clauberg-Versuch deutlich aktiver als Retro-Chlormadinon, während im Deziduomtest das Wirkungsverhältnis gerade umgekehrt ist
(Abb. 3). Das wird nach Krause (1970) erklärt durch die Annahme, daß die Rezeptoren im Endometrium das Chlormadinon und die Rezeptoren im Stroma das
Retro-Chlormadinon bevorzugen.

Es erscheint angebracht, jetzt auch einmal eine schon kurz erwähnte andere zentrale Wirkung zu betrachten. Hierzu ist die Wirkung der synthetischen
Gestagene auf den Hypothalamus, die via Hemmung der Gonadotropinabgabe aus
der Hypophyse zum Ausbleiben der Ovulation führt, geeignet. Auch hier nehmen
die Retro-steroide eine Sonderstellung ein, da sie die Ovulation nicht hemmen (Trengeston schwach), wofür die obengenannte Erklärung gelten könnte.

Da es aber seit kurzer Zeit ziemlich genau bekannt scheint, wie die Ovulation zustande kommt und wo die hemmenden Substanzen angreifen, soll dies
kurz besprochen werden.

Nachdem schon die Arbeiten der Sawyer-Gruppe den Hypothalamus, eher als
die Hypophyse selbst, als den Angriffspunkt der Ovulationshemmer angezeigt
hatten, haben Schally et al. (1969) einen sehr wertvollen Beitrag geliefert,
indem sie die Wirkung eines gereinigten Hypothalamusextrakts, der die Abgabe
von luteinisierendem Hormon aus der Hypophyse stimuliert, ausführlich unter-

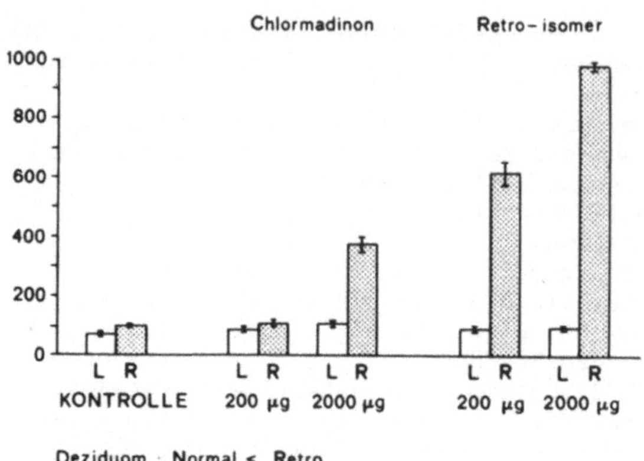



Abb. 3.

suchten. Die aktive Substanz – meistens als LRF, von Schally aber als LH-RH
(= luteinizing hormone releasing hormone) bezeichnet – ist ein Peptid, das im
Hypothalamus gebildet wird. Es wird durch Neurosekret zur Hypophyse geführt,
wo es die Abgabe von LH fördert, wie im Blut kastrierter weiblicher oder
männlicher Ratten nachgewiesen werden konnte.

Schally u. Mitarb. (1969) haben erst gezeigt, daß Progesteron und Oestradiol den Blutspiegel des LH erniedrigen. Wenn aber exogen LH-RH i.v. injiziert wird, steigt der LH-Spiegel zu normalen oder sogar über normale Werte
an. Actinomycin D, das die Bildung von RNA im Hypothalamus hemmt (und dies
auch an anderen Stellen tut) hemmt *nicht* die Wirkung des LH-RH. Jedoch konnte die durch Progesteron-Oestradiol hervorgerufene Ovulations*hemmung* verhindert werden durch Aktinomycin D.

Schally et al. (1969) haben die Hypothese geäußert, daß im Hypothalamus
anwesende Peptidasen das LH-RH abbauen können (Abb. 4). Ohne LH-RH findet

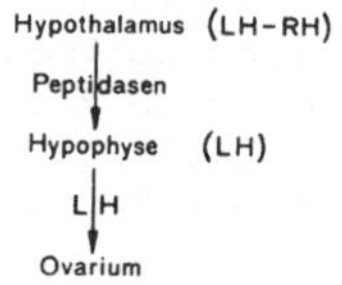

Abb. 4.

keine LH-Abgabe statt, und deshalb bedeutet eine höhere Peptidaseaktivität
auch weniger LH. Es könnte sein, daß die Ovulationshemmer die RNA-Synthese
anregen und daß infolgedessen mehr Peptidasen und weniger LH-RH zur Verfügung
stehen. Hiermit würde übereinstimmen, daß Actinomycin D durch Hemmung der
RNA-Synthese und der Peptidase-Synthese die Wirkung der Ovulationshemmer verhindert. Das würde auch verständlich machen, warum *verabreichtes* LH-RH sowohl bei mit Ovulationshemmern als auch bei mit Actinomycin D behandelten
Tieren gut wirksam ist und auch, daß Actinomycin D die normale Abgabe von LH
nicht hemmt.

Eine weitere rezente Bestätigung der Peptidasetheorie wurde von Frith u.
Hooper (1971) veröffentlich. Sie zeigten, daß verschiedene Ovulationshemmer
(Norethisteron, Chlormadinon, Äthinyloestradiol und Oestron) die Peptidase-
aktivität des Hypothalamus erhöhen (Abb. 5). Eine weitere Arbeit von Schally
et al. (1970) hat bewiesen, daß die LH-Abgabe durch exogen verabreichtes

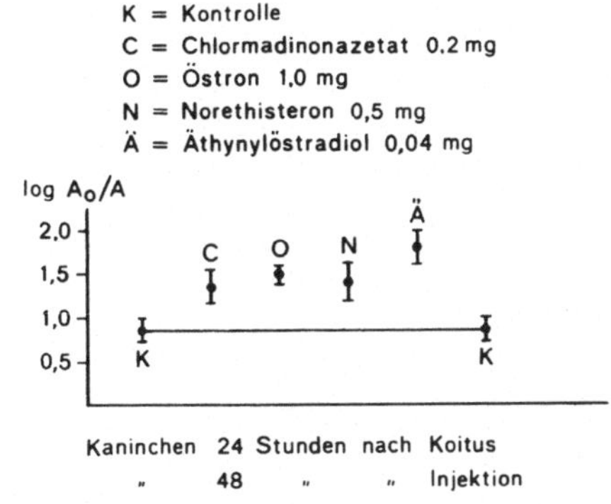



Abb. 5. Peptidase-Aktivität im Hypothalamus (nach Frith u. Hooper, 1971).

LH-RH auftritt bei kastrierten Ratten, die mit hohen Dosierungen verschiedener synthetischer Ovulationshemmer behandelt wurden. Unter den verabreichten Substanzen befanden sich sowohl Oestrene wie Norethisteron, Lynestrenol und Norethinodrel als auch Pregnene wie Megestrolazetat und Ay 11440 (3β, 17α-Diacetoxy-6-chloropregna-4,6-dien-20-on) und die Kombinationen dieser Gestagene mit Mestranol oder Äthinyloestradiol (Tab. 5). Es ist also höchst wahr-

Tab. 5.

Effekt von LH-RH bei kastrierten
mit Ovulationshemmern behandelten Ratten

| | LH | FSH (ng/ml Plasma) |
|---|---|---|
| Kontrollen | 531 | 2800 |
| Megestrolazetat 10 mg + Äthinylöstradiol 0,2 mg | 119 | 1800 |
| Id + LH-RH | 963 | -- |
| Lynestrenol 10 mg + Mestranol 0,3 mg | 119 | 1600 |
| Id + LH-RH | 969 | 1300 |

Behandlung: 5 Tage 10 mg Gestagen/Tag

scheinlich, daß alle synthetischen (und nicht synthetischen) Ovulationshemmer wirken durch Beseitigung des im Hypothalamus gebildeten LH-RH mittels einer Erhöhung der Peptidaseaktivität im Hypothalamus.

Dies bedeutet keineswegs, daß die gleichen Substanzen nicht auch noch periphere Angriffspunkte haben (Ovar, Uterus, Cervix) und daß in dieser Hinsicht nicht große individuelle Unterschiede bestehen!

Nur ganz kurz möchten wir einige Probleme erwähnen, die mit der ovulationsinduzierenden Wirkung der Gestagene zusammenhängen. Für das Progesteron ist insbesondere durch die Untersuchungen von Sawyer u. Kawakami (Effekt von Progesteron auf EEG, "after reaction" und "arousal threshold") ein cerebraler Angriffspunkt bewiesen. Mit Medroxyprogesteronacetat konnten Odell u. Swerdloff (1968) einen LH-Gipfel im Plasma menopausaler Frauen hervorrufen (Abb. 6). Auch Lynestrenol kann die Ovulation fördern oder auslösen (Zeilmaker). In erster Linie ist anzunehmen, daß diese Substanzen ebenso auf den

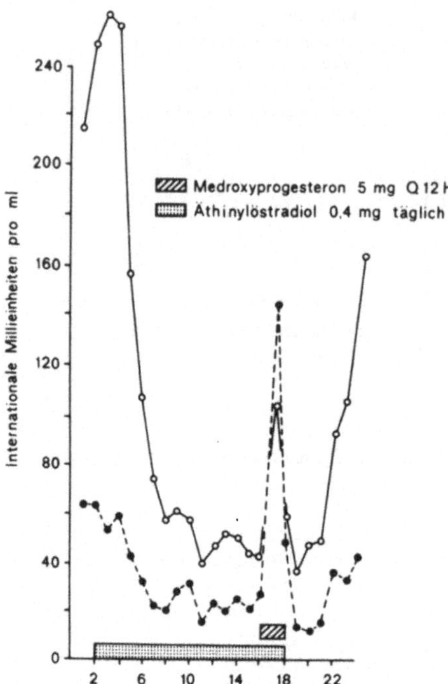



Abb. 6. FSH und LH im Blut einer Kastratin nach oraler Gabe von Äthinylöstradiol und Medroxyprogesteronazetat
(Nach Odell und Swerdloff, Proc. Nat. Acad. Sc. 61 (1968 529)

Hypothalamus wirken, obwohl die publizierten Versuche in vielen Fällen eine
periphere Wirkung durch Sensibilisierung des Ovars keineswegs ausschließen.
Letzteres ist sicher zu berücksichtigen im Falle des Retro-steroids Trengeston, wobei andere zentrale Wirkungen nur schwach sind. Auch Krause (1970)
betrachtet dieses Problem als ungelöst. Ob der Hypothalamus für die Ovula-
tionsinduktion andere Rezeptoren enthält als für die Hemmung, oder aber ob
die Peptidasen nach Trengeston und ovulationsfördernden Dosierungen anderer
Gestagene gehemmt werden (siehe oben unter Ovulationshemmung) ist völlig
unbekannt.

Diesen ersten Teil abschließend sollen noch einige Bemerkungen gemacht
werden über die schon mehrfach hier erwähnte Bindung der Gestagene an Enzyme
und Rezeptoren. So wurde bei der Besprechung der Retro-steroide eine mögliche Verschiedenartigkeit der Bindung bei normalen und Retro-steroiden als
Erklärung für das Fehlen einer zentralen Wirkung bei den letzteren herangezogen. Man bedenke, daß dies zwar eine intelligente und bestechende Theorie
ist, daß aber hierfür vorläufig noch keine experimentellen Beweise vorliegen. Es wurden nämlich viel weniger Arbeiten publiziert über gestagenbinden-
de als über oestrogen- und androgenbindende Enzyme und Rezeptoren, und die
wenigen vorliegenden Arbeiten behandeln meistens Progesteron und nicht die
synthetischen Gestagene. Untersuchungen, wie sie Unjehm (1970 a, b, c) mit
Testosteron ausgeführt hat, wobei die Enzym- und Rezeptorbindungen getrennt
wurden, sollten auch mit Progesteron und den synthetischen Gestagenen wiederholt werden.

So haben O'Malley et al. (1970) die Rezeptoren im Eileiter von Hühnern
untersucht. Sie konnten zeigen, daß es zwei verschiedene Rezeptoren in Cytoplasma und Kern gibt, die für die Bindung von Progesteron verschiedene Temperaturen erfordern.

Eine Vorbehandlung mit Oestrogen erhöht die Menge gebundenen Progesterons. Ähnliche Versuche wurden ausgeführt von Milgrom u. Baulieu (1970),
die den Rattenuterus als Testobjekt wählten. Diese Autoren haben aber auch
einige synthetische Gestagene (Norethisteron, Chlormadinon) geprüft. Das Er-

gebnis, daß Norethisteron eine geringere und Chlormadinon keine Affinität zu
den Rezeptoren hat (siehe Abb. 7, die Verdrängung von $^3$H-Progesteron durch
Progesteron und durch die synthetischen Gestagene) läßt daran zweifeln, ob

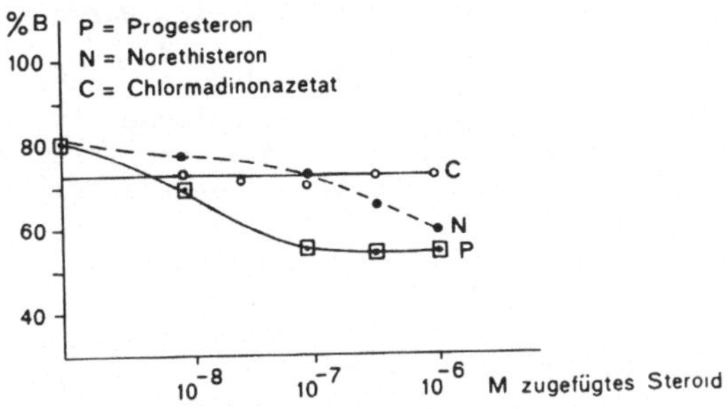


Abb. 7. Verdrängung von $^3$H-Progesteron von Protein in Uterusextrakt. (nach
Milgrom u. Baulieu, 1970)

wirklich der spezifische Gestagenrezeptor in den verwendeten Extrakten vorhanden war. Sonst würde man verschiedene Rezeptoren für die verschiedenen
synthetischen Gestagene annehmen müssen, was unwahrscheinlich ist.

Eine dritte Möglichkeit ist, daß Chlormadinon als solches inaktiv ist und
erst durch ein Enzym, das in den Extrakten möglicherweise fehlt, zu einer
aktiven und bindungsfähigen Substanz metabolisiert wird. Die beschriebenen
Eigenschaften des Chlormadinons geben aber vorläufig keinen Anlaß zu dieser
Annahme.

Diese Überlegungen führen zum zweiten Thema dieses Vortrags, dem Stoffwechsel der synthetischen Gestagene. Hier wollen wir uns wiederum auf einige
Beispiele beschränken.

Da im Vorhergehenden die Retro-steroide schon mehrfach erwähnt wurden,
fangen wir an mit dem weiteren Vergleich der verschiedenen metabolisierenden
Reaktionen. Die Abb. 8 mit den sterischen Strukturen des Progesterons und

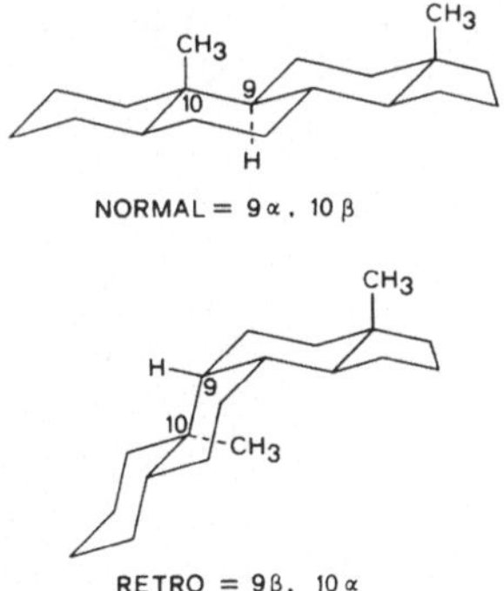


Abb. 8.

des Retroprogesterons zeigt, wo die größten Unterschiede im räumlichen Bau
auftreten. Es ist klar, daß es in den Ringen C und D keine Unterschiede
gibt, und erwartungsgemäß verlaufen die Reaktionen an Kohlenstoffatomen
dieser Ringe gleich. Das gilt für die Reduktion der 20-Ketogruppe und für
die Einführung von Hydroxylgruppen. Dies konnte nicht nur für Reaktionen bei
Mensch und Tier gezeigt werden (Diczfalusy et al., 1963; Krause, 1970), sondern auch für Umsetzungen durch Mikroorganismen, die von de Flines u.
Mitarb. (1966) (Abspaltung der C17-Seitenkette und 11, 15 und 16 Hydroxylie-

rung) beschrieben sind. Interessant ist aber, daß Curvularia lunata, welches
in 9α, 10β-Steroiden in 11β und 11α hydroxyliert, in Retro-steroiden zu 9β-
Hydroxylierung imstande ist. Eine Betrachtung der Formeln in Abb. 8 zeigt,
daß während bei 9α, 10β Steroiden das H an C9 weit von den Wasserstoffatomen
an C11 entfernt ist, bei den Retro-steroiden dieses Wasserstoffatom ganz nahe an die C11-Wasserstoffatome herangerückt ist (van der Syde et al., 1969).

In vitro-Versuche mit menschlichen plazentären Mikrosomen haben gezeigt,
daß eine Aromatisierung des A-Ringes nicht stattfindet in den verschiedenen
10α-Verbindungen; die 9β-Stellung hat hierauf einen geringeren Einfluß
(Walop u. de Lange, 1966) (Tab. 6). Für beide Retro-steroide ist es gelun-

Tab. 6.

AROMATISIERUNG UND KONFIGURATION

I = Androst–4–en–3,17–dion = 100 %

II = Testosteron

| 8 β, 9 α, 10 β. | | 8 β, 9 β, 10 β. | | 8 β, 9 α, 10 α. | | 8 β, 9 β, 10 α. | |
|---|---|---|---|---|---|---|---|
| I | II | I | II | I | II | I | II |
| 100 | 100 | 30 | 30 | <10 | <10 | <10 | <10 |

Mikrosomen menschlicher Plazenta

Nach Walop und de Lange

gen, verständlich zu machen, warum bestimmte Metabolite nicht auftreten. Offenbar fügt sich das Molekül nicht mehr an die Enzyme an, die sich mit dem
A- oder B-Ring befassen.

Wenden wir uns jetzt den für die kontrazeptiven Präparate so wichtigen
Oestrenderivaten zu und vergleichen den Metabolismus der nachfolgenden vier
Äthinylverbindungen beim Menschen (Abb. 9):

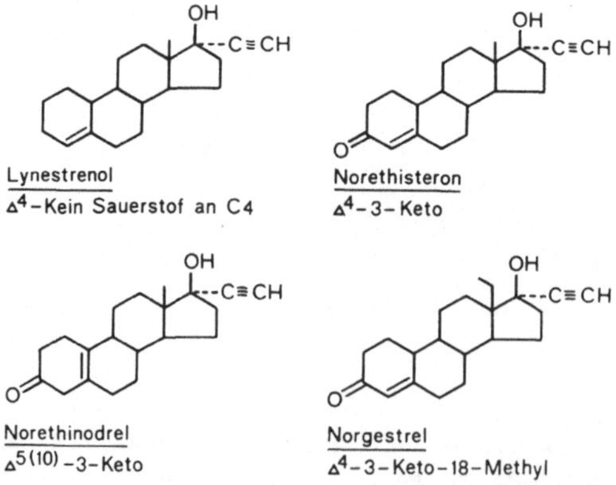

Lynestrenol
Δ⁴–Kein Sauerstof an C4

Norethisteron
Δ⁴–3–Keto

Norethinodrel
Δ⁵(10)–3–Keto

Norgestrel
Δ⁴–3–Keto–18–Methyl

Abb. 9.

I Lynestrenol ($\Delta^4$, kein Sauerstoff an C 3)
II Norethisteron ($\Delta^4$-3-Keto)
III Norethinodrel ($\Delta^{5\,(10)}$-3-Keto)
IV Norgestrel ($\Delta^4$-3-Keto-18-Methyl)

Die auffallendsten Unterschiede der biologischen Aktivitäten in Tierver-
suchen sind:
1. Eine niedrige gestagene Aktivität von III
2. Eine relativ hohe oestrogene Aktivität von III
3. Eine niedrige Antifertilitätswirkung von IV (Ganz anders beim Menschen,
   offenbar wegen eines anderen Angriffspunktes!)

Eine mögliche Erklärung dieser Unterschiede, nicht aufgrund des Stoffwechsels sondern aufgrund der unterschiedlichen Oestrogenizität, wurde schon
im ersten Teil dieses Vortrags behandelt.

Bei der Betrachtung quantitativer Unterschiede zwischen bestimmten Metaboliten und auch der Gesamtradioaktivitäten soll man sich bewußt sein, daß
verschiedene Dosierungen einer Substanz sich unerwartet verschieden verhalten können. So wurde in unserer Forschungsgruppe von Wijmenga u. Wijnand in
einer Zusammenarbeit mit Van der Molen (1969) gefunden, daß bei oraler Verabreichung von 0,5 mg Lynestrenol der Gipfel der Radioaktivität im Plasma
schon nach zwei Stunden vorlag, während nach 5 mg dieser Gipfel nach 4 Stunden auftrat. Auch war der Gipfel nach 0,5 mg relativ höher (relativ = ausge-
drückt als Prozentsatz der verabreichten Dosis). Schließlich waren 3 Tage
nach 0,5 mg noch 1 - 2% und sogar nach 8 Tagen noch immer 0,5 - 0,8% der
verabreichten Dosierung pro Liter Plasma zu finden. Die letzte Menge war
nach Verabreichung der *hohen* Dosierung schon nach 3 Tagen erreicht. Es ist
deutlich, daß diese Tatsachen den Vergleich verschiedener in der Literatur
beschriebenen Resultate sehr erschweren. Deshalb sollen nur grobe Unterschiede als wahrscheinlich betrachtet werden.

Keine der vier Substanzen wird in nennenswerten Mengen unverändert im
Harn ausgeschieden (für Lynestrenol und Norethisteron < 0,1 - 1,0%). Die
Ausscheidung mit dem Stuhl ist schwer zu beurteilen, da einerseits nicht-
resorbiertes Material anwesend sein kann, das andererseits stark durch Darm-
bakterien verändert wird.

Sind hier also keine Unterschiede beweisbar, ist dies bei der Konjugation
anders. Fotherby et al. (1968) haben Lynestrenol, Norethisteron und Norgestrel verglichen und gefunden, daß in allen Fällen < 5% nicht-konjugiert ausgeschieden wird. Während jedoch etwa 45% des Lynestrenols und Norethisterons
mit Glukuronsäure konjugiert waren und nur 10% als Sulfat, wurden von Norgestrel 33% als Glukuronid und 26% als Sulfat ausgeschieden. Auch waren die
Metabolite des Norgestrels in relativ größerer Menge Reduktionsprodukte,
während für die beiden anderen Substanzen die Metabolite einen mehr polaren
Charakter zeigten und das Verhalten bei saurer resp. alkalischer Extraktion
verschieden war (Abb. 10, Fotherby, 1968).

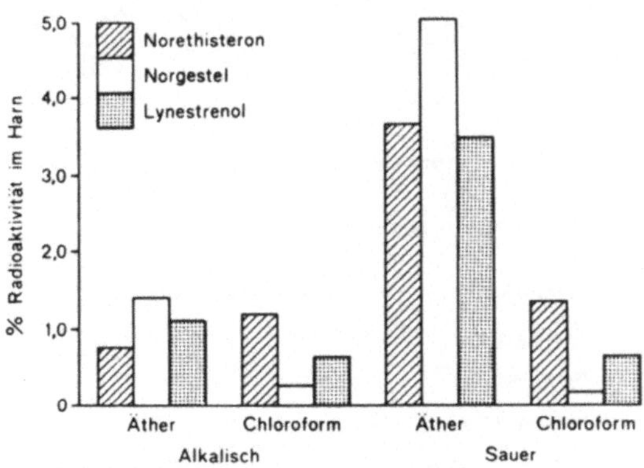



Abb. 10. Extraktion Harn (Mensch) (nach Fotherby, 1968)

Die Harnuntersuchungen ergaben keinen Unterschied im Verhalten der 17α-Äthinylgruppe. In allen Fällen enthielten > 90% der Summe der Metaboliten noch die Äthinylkette (Fotherby et al., 1968; Wiedhaup, unpubliziert).

Wenn aber das Norethinodrel mit einem 10.000 x g Überstand eines Homogenats aus menschlicher Leber inkubiert wurde, so stellte sich heraus, daß die Äthinylgruppe für etwa 40% abgespalten wurde. Kaninchenleber wirkte ebenso, nicht aber Ratten- und Meerschweinchenleber. Die Versuche mit Kaninchenleber zeigten, daß die Abspaltung im Falle von Norethisteron in viel größerem Maße stattfand (Palmer et al., 1969).

Eine weitere Reaktion von einigem Interesse ist die Aromatisierung des A-Ringes, da sie einen Beitrag zur Oestrogenizität und damit zu der gestagenen und hypophysenhemmenden Aktivität liefern könnte. Die beobachtete Aktivität wird aber offenbar einerseits durch das gebildete Oestrogen, andererseits durch die anti-oestrogene Wirkung des ursprünglichen Moleküls oder dessen sonstigen Metaboliten bestimmt.

Es wurde beschrieben, daß bei den genannten vier Steroiden jeweils 2 - 3% aromatisiert werden. Jedoch wird aus Norethisterons das stark oestrogene Äthinyloestradiol gebildet (Brown u. Blair, 1960), während das aus Norgestrel gebildete 18-Äthyl-Homolog viel weniger oestrogen ist (Edgren et al., 1966). Dazu kommt noch, daß das unveränderte Norgestrel stärker anti-oestrogen wirkt als Norethisteron. Dies kann das Fehlen einer oestrogenen Wirkung beim Norgestrel erklären. Die aromatischen Metabolite des Lynestrenols und Norethinodrels wurden noch nicht mit Sicherheit identifiziert, obwohl schon Okado et al. (1964) die mögliche Ausscheidung von Äthinyloestradiol nach Verabreichung dieser beiden Substanzen wahrscheinlich gemacht hat.

Vor kurzem wurde durch Breuer (1970) die Aromatisierung des Norethisterons und damit auch der anderen Substanzen mit guten Gründen in Zweifel gestellt. Erstens bemerkt er, daß die nach Aromatisierung zu erwartenden Oestriolabkömmlinge niemals im Harn gefunden werden (Okado et al., 1964). Zweitens fand er, daß in vitro keine Aromatisierung stattfindet. Er vermutet, daß eine säureempfindliche 1β-Hydroxyverbindung gebildet wird, die durch die Behandlung mit Säuren während der Extraktion in ein phenolisches Steroid umgewandelt wird. Die phenolischen Metabolite wären also reine Artefakte!

Größere Unterschiede zwischen den hier besprochenen Stoffen findet man bei der Reduktion der Doppelbindung und der 3-Ketogruppe. Beim Lynestrenol, bei welchem die letztere Gruppe fehlt, wird diese eingeführt und dadurch Norethisteron gebildet (Murata, 1968; Mahazari et al., 1970). Hauptsächlich finden die nachfolgenden Reaktionen statt. Aus Lynestrenol (Wiedhaup) und Norethisteron (Murata, 1968) werden meistens 5αH-3β-Hydroxy- und Ketoverbindungen gebildet, in Norethinodrel wird die $\Delta^5$ $(^{10})$-Doppelbindung nur teilweise und die 3-Ketogruppe hauptsächlich zu 3α-Hydroxy reduziert (Palmer et al., 1969). Letzten Endes wird Norgestrel zum größten Teil zu 5β-H und 3α-Hydroxy metabolisiert (Tab. 7) (de Jongh et al., 1968).

Tab. 7.

REDUKTIONSPROZESSE

Wichtigste Ausscheidungsprodukte beim Menschen

| | |
|---|---|
| Lynestrenol | 3 = O, 5α-H; 3β-OH, 5α-H |
| Norethisteron | 3 = O, 5α-H; 3β-OH, 5α-H |
| Norethinodrel | 3α-OH, $\Delta^{5\,(10)}$ |
| Norgestrel | 3α-OH, 5β-H |

Ein weiterer Unterschied zwischen Norethinodrel und Norgestrel wurde gefunden in der Hydroxylierung am Kohlenstoffatom 10. Aus Norethinodrel wird eine 10-Hydroxyverbindung gebildet, wobei die Doppelbindung nach $\Delta^4$ verschiebt. Beim Norgestrel wird eine 10-Hydroxylierung nicht gefunden.

Zusammen führen die obengenannten Reaktionen zu einem Ausscheidungspatron,
wobei Lynestrenol und Northisteron einander am meisten ähnlich sind, und wo-
von die beiden anderen Verbindungen stärker abweichen, als aufgrund der geringen Wirkungsunterschiede erwartet werden konnte.

Es könnte sinnvoll sein, hier einige Beispiele zu erwähnen, wobei der
Metabolismus bei Mensch und Laboratoriumstieren deutlich verschieden ist.
Erstens die Verteilung der Metabolite des Lynestrenols und des Norgestrels
in Harn und Stuhl.
Beim Menschen: 40 – 60% im Harn und 20 – 30% in Faeces.
Bei der Ratte: 20% im Harn und 70% in Faeces.

Auch die Konjugierung ist verschieden. Die folgenden Werte wurden für
Lynestrenol gefunden:
Beim Menschen: etwa 4% nicht konjugiert.
Bei der Ratte: etwa 30% nicht konjugiert.

Die nachfolgenden Werte aus einer schon erwähnten Arbeit von Palmer
et al. (1969) zeigen die großen Unterschiede in der Bildung von 3α und 3β-
Hydroxyverbindungen des Norethinodrels bei vier Arten (Tab. 8).

Tab. 8.

α -UND β-OH METABOLITE DES NORETHINODRELS
NACH INKUBATION MIT $10^4$ x G ÜBERSTAND VON LEBER

| | | % 3-OH-Norethinodrel | |
| | | 3α | 3β |
|---|---|---|---|
| Ratte | ♂ | 96 | 4 |
| " | ♀ | 90 | 10 |
| Meerschweinchen | ♀ | 9 | 91 |
| Kaninchen | ♀ | 36 | 64 |
| Mensch | ♀ | 70 | 30 |

Nach Palmer et al

Als letztes unserer Beispiele sei der Einfluß der Veresterung gezeigt.
Schon Kimbel et al. (1960) fanden, daß die Kapronsäure-Ester des 17α-Hydroxy-
progesterons und des 17α-Hydroxy-19-nor-progesterons im Tierkörper nicht
hydrolysiert werden, und daß ihr Abbau anders verläuft als bei den freien
Steroiden. Wie aus chemischen Gründen zu erwarten war, sind die Ester dieser
tertiären Alkohole viel stabiler als die der sekundären Alkohole, wie z.B.
das Oestradiol und das Testosteron.

Coert fand beim Lynestrenolacetat eine Hydroxylierung im D-Ring, vermut-
lich an C-15 oder C-16, eine Reaktion, die bei unverestertem Lynestrenol
nicht beobachtet wurde (Abb. 11). Nach dieser Hydroxylierung folgt die Abspaltung der Acetatgruppe.

Es ist wahrscheinlich, daß Ähnliches nach oraler Verabreichung stattfindet, da Lynestrenolacetat von Magen- und Darmsaft nicht hydrolysiert und
auch als solches vom Darm resorbiert wird. Zwar wird in der Leber Essigsäu-
re abgespalten, jedoch offenbar langsam genug, um die Bildung der oben erwähnten Metabolite zu ermöglichen. Auch im peripheren Blut wurde noch Lynestrenolacetat angetroffen (Tab. 9).

Die beschriebenen Beispiele sollen zeigen, wie wenig wir wissen und was
wir gerne wissen möchten. Man soll sich aber auch fragen, *weshalb* wir die
Auskunft haben möchten und wozu wir sie brauchen. Hierzu ergibt sich sofort
*eine* Antwort. Aus rein *wissenschaftlichen* Gründen möchte und muß man alles
wissen.

Im Falle einer wirksamen Substanz bedeutet dies die Kenntnis ihrer Wirkungen, des Wirkungsmechanismus im ganzen Organismus und auf molekularem
Niveau und die Kenntnis dessen, was im Körper mit der Substanz geschieht.

Tab. 9.

LYNESTRENOLAZETAT NACH ORALER GABE VON
LYNESTRENOLAZETAT-1'-$^{14}$C AN RATTEN

Portalvenenblut   : + + +
Intestinale Lymphe : +
Peripheres Blut   . +

HYDROLYSE-GESCHWINDIGKEIT

|  |  | % | Stunden |
|---|---|---|---|
| Leberhomogenat | ( Ratte ) | 50 | $^3/_4$ |
| Plasma | ( ·· ) | 50 | 24 |
| Darmsaft | ( ,, ) | 10 | 24 |
| Magensaft | ( ,, ) | 0 | 24 |
| Plasma | ( Mensch ) | 15 | 24 |
| Darmsaft | ( ,, ) | 0 | 24 |

Viel schwieriger ist die Frage der *praktischen* Bedeutung zu beantworten.
Praktisch wünscht man eine Antwort zu den Fragen, warum eine Substanz nur
auf bestimmte Organe eine Wirkung ausübt oder warum sie bei einer Tierart so
und bei einer anderen anders wirkt. Man möchte die Eigenschaften schätzen
aufgrund einfacher in vitro-Versuche über Bindung an gereinigte Proteine.
Man möchte anhand pharmakokinetischer und metaboler Experimente Versuchstie-
re wählen, die dem Menschen am meisten ähneln. Man möchte am Schreibtisch
chemische Strukturen entwerfen, die ganz bestimmte Wirkungen haben oder
nicht haben, wozu höchstens physikalische Messungen helfen sollten ("ratio-
nal design of drugs").
Kurz gesagt: man möchte zuverlässig prophezeien mit einem minimalen Ar-
beitsaufwand, insbesondere unter Vermeidung der langwierigen und kostspieli-
gen Tierversuche, die von manchen vielleicht sogar als unethisch angesehen
werden. Was ist möglich?
Offenbar liegen die Aussichten besser bei den retrospektiven als bei den
prospektiven Studien. Wenn man eine Erklärung für ein unerwartetes Ergebnis
sucht, weiß man wenigstens, wo man suchen soll, und eine intelligente Analyse auf molekularer Ebene oder die des Stoffwechsels *kann* den erwünschten Erfolg haben. Dazu soll bemerkt werden, daß man heutzutage oft in erster Linie
meint, daß jedes unterschiedliche Verhalten einer Substanz durch einen unterschiedlichen Stoffwechsel verursacht werde, was aber nur ziemlich selten der
Fall ist.
Die An- oder Abwesenheit von Rezeptoren in bestimmten Organen, meistens
festgestellt anhand der Anhäufung markierter Substanz, *kann* erklären, warum
das Organ reagiert oder nicht. Dies trifft aber keineswegs immer zu, da man
schon längst bevor von Rezeptoren die Rede war, wußte, daß in den reagieren-
den Organen oft ziemlich wenig der verabreichten Substanz gefunden wird oder
daß diese sogar vollständig fehlen kann, z. B. infolge eines raschen Abbaus.
Nur in sehr seltenen Fällen wurde festgestellt, daß die Wirkung nicht durch
die verabreichte Substanz selbst, sondern durch einen ihrer Metabolite ausgelöst wurde und hat dieser Befund zu der weiteren Verwendung dieses aktiven
Metabolits in einem Präparat geführt. Besser liegen die Möglichkeiten für
die Pharmakokinetik. Die Chancen sind sicher gegeben, daß man aufgrund theo-
retischer oder praktischer Kenntnis eine Substanz derart modifiziert, daß
sie langsamer resorbiert oder ausgeschieden wird, höhere Blutspiegel erreicht oder sogar ihren Abbau verändert.
Aber damit sind dann wenigstens vorläufig die praktischen prospektiven
Möglichkeiten erschöpft. Denn wer wird die unabsehbare Reihe Rezeptorpro-
teine beschaffen, die eine zur vollständigen Charakterisierung einer völlig

unbekannten Substanz erforderliche in vitro-Untersuchung ermöglichen würde?

Was soll man machen, wenn sich herausstellt, daß nach Verabreichung einer Substanz bei den üblichen Laboratoriumtieren der Stoffwechsel sich von dem beim Menschen gefundenen unterscheidet, und welche Unterschiede sind praktisch wichtig?

Die Absicht dieser Vorlesung sollte keineswegs sein, die neuen Entwicklungen und Auffassungen als wertlos zu verwerfen, sondern nur zu warnen vor zu hohen Erwartungen. Die alte medizinisch-biologische Analyse des Wirkungsmechanismus soll mit den bisher zu sehr vernachlässigten Stoffwechseluntersuchungen ergänzt werden. Aber die beiden im Anfang erwähnten Beispiele, die keineswegs Ausnahmen bilden, zeigen, daß man die Forschung bei Tier und Menschen noch immer nicht entbehren kann, wenn man wissen will, ob und wie eine Substanz eine biologische Wirkungs ausübt.

Der Autor dankt den Herren Prof. Dr. M. Tausk, Dr. J.H.H. Thijssen, Dr. K. Wiedhaup, Drs. A. Coert, J. de Visser und J. van der Vies für ihre freundliche Hilfe bei der Vorbereitung dieses Manuskripts.


Zusammenfassung

Anhand einiger Wirkungen auf den Uterus und die Hypophyse, welchen zwei der praktisch wichtigsten Anwendungen (Schwangerschaftserhaltung und Schwangerschaftsverhütung) zugrunde liegen, wird gezeigt, wie unterschiedlich die Wirkung verschiedener Gestagene erklärt werden könnte. Die folgenden Möglichkeiten werden erwähnt:
1. Das Interferieren anderer Aktivitäten, wie die Oestrogenizität, mit gestagenen Wirkungen.
2. Unterschiede in der Physiologie bei verschiedenen Tierarten (Bedeutung der Ovar- resp. Plazentafunktion für die Erhaltung der Schwangerschaft).
3. Die An- oder Abwesenheit einer zentralen Wirkung.
4. Eigenschaften der Rezeptoren.

Unterschiede im Metabolismus zwischen einigen Oestranderivaten und zwischen einigen Pregnanen und Retro-Pregnanen werden besprochen.

Die gegebenen Beispiele zeigen, daß die Molekularbiologie manchmal Unterschiede der Wirkung verschiedener Substanzen bei einer Tierart, oder einer Substanz bei verschiedenen Tierarten, erklären kann. Viel seltener wird die Analyse der Metaboliten in dieser Hinsicht von Nutzen sein. Es ist kaum zu erwarten, daß man aus Untersuchungen dieser Art erwünschte oder unerwünschte Wirkungen prophezeien könnte.


Literatur

Breuer, H.: Lancet 1970 II, 615.
- Knuppen, R.: Z. physiol. Chem. 350, 581 (1969).
Brown, J.B., Blair, H.A.F.: Proc. roy. Soc. Med. 53, 433 (1960).
Coert, J.: Unveröffentlicht.
Desaulles, P.A., Krähenbühl, C.: Acta endocr. (Kbh.) 40, 217 (1962)
Diczfalusy, E., Tillinger, K.G., Esser, R.J.E.: Nature (Lond.) 200, 79 (1963).
Edgren, R.A., Peterson, D.L., Jones, R.C., Nagra, C.L., Smith, H., Hughes, G.A.: Recent Progr. Hormone Res. 22, 305 (1966).
Elert, R.: Geburtsh. u. Frauenheilk. 11, 325 (1951).
Flines, J. de, Syde, D. van der, Waard, W.F. van der: Rec. Trav. chim. (Pays-Bas) 85, 701, 712, 721 (1966).
Fotherby, K., Kamyab, S., Littleton, P., Klopper, A.I.: J. Reprod. Fertil. Suppl. 5, 51 (1968).
Frith, D.A., Hooper, K.C.: Acta endocr. (Kbh.) 66, 221 (1971).
Jongh, D.C. de, Hribar, J.D., Littleton, P., Fotherby, K., Rees, R.W.A., Schrader, S., Foell, T.J., Smith, H.: Steroids 11, 649 (1968).

Kimbel, K.H., Willenbrink, J., Schulze, P.E.: Acta endocr. (Kbh.) Suppl. 51, 737 (1960).
Krause, R.: Vortrag III World Congress Horm. Steroids (1970).
Madjerek, Z., Visser, J. de, Vies, J. van der, Overbeek, G.A.: Acta endocr. (Kbh.) 35, 8 (1960).
Mahzeri, A., Fotherby, K., Chapman, J.R.: J. Endocr. 47, 251 (1970).
Milgrom, E., Baulieu, E.E.: Endocrinology 87, 276 (1970).
Molen, H.J. van der, Hart, P.G., Wijmenga, H.G.: Acta endocr. (Kbh.) 61, 255 (1969).
Murata, S.: Nippon Naibunpi Gakkai Zasshi 43, 1083 (1968).
Netter, A., Lumbroso, P., Salomon, Y.: La fonction luteale. Paris: Masson et Cie., 1954, p. 315.
Odell, W.D., Swerdloff, R.S.: Proc. nat. Acad. Sci. (Wash.) 61, 529 (1968).
Okado, H., Amutsa, M., Ishihara, S., Gen-Ichi, T.: Acta endocr. (Kbh.) 46, 31 (1964).
O'Malley, B.W., Sherman, M.R., Toft, D.O.: Proc. nat. Acad. Sci. (Wash.) 67, 501 (1970).
Onetti, E.B., zitiert von Dapunt, O., Windbichler, H.: Wien. klin. Wschr. 81, 785 (1969).
Palmer, K.H., Ross, F.T., Rhodes, L.S., Baggett, B., Wall, M.E.: J. Pharmacol. exp. Ther. 167, 207, 217 (1969).
Rothschild, J., Rapport, R.L.: Endocrinology 50, 580 (1952).
Sawyer, C.H., Kawakami, M.: Endocrinology 65, 622, 652 (1959).
Schally, A.V., Bowers, C.Y., Carter, W.H., Arimura, A., Redding, T.W., Saito, M.: Endocrinology 85, 290 (1969).
- Parlow, A.F., Carter, W.H., Saito, M., Bowers, C.Y., Arimura, A.: Endocrinology 86, 530 (1970).
Stamm, O., Gerhard, J.: Geburtsh. u. Frauenheilk. 28, 483 (1968).
Suchowsky, G.K.: Acta endocr. (Kbh.) 42, 533 (1963).
Szontagh, F.E., Sas, M.: Gynaecologia (Basel) 154, 81 (1962).
Syde, D. van der, Flines, J. de, Waard, W.F. van der, Smit, A.: Rec. Trav. chim. (Pays-Bas) 88, 1437 (1969).
Toth, F., Treit, S.: Geburtsh. Gynäk. 162, 140 (1964).
Unjehm, O.: Acta endocr. (Kbh.) 65, 517, 525, 533 (1970).
Walop, J.N., Lange, N. de: Biochim. biophys. Acta (Amst.) 130, 249 (1966).
Wiedhaup, K.: Unveröffentlicht.
Zeilmaker, G.H.: Unveröffentlicht.

Symp. Dtsch. Ges. Endokrin. 17, 101-129 (1971)
© by Springer-Verlag

# Neue Entwicklungen in der Gestagentherapie

## Recent Developments in the Therapeutical Application of Gestagens

J. HAMMERSTEIN

Abteilung für Gynäkologische Endokrinologie der Frauenklinik im Klinikum
Steglitz der Freien Universität Berlin

Mit 15 Abbildungen

Summary

A survey is given of recent developments in the field of the therapeutical
application of progestagens. Introductory remarks deal with the chemical
structure, the profile of actions and the non-endocrine side effects of progestational agents. It is emphasized that the systemic actions of progestagens on skin, connective tissue, skeleton, blood clotting, vascular system,
psyche etc. are poorly understood at present and need further elucidation.

With regard to the clinical application of progestagens, some new indica-
tions outside the gynecological and obstetrical speciality, i.e. hormone-
dependent tumours, pubertas praecox, migraine and dermatosclerosis, are
reported.

Recent progress in hormonal contraception with progestagens alone is the
main topic of the review. For introductory information, some basic aspects
concerning the influence of progestational agents on the endocrinology of
the menstrual cycle in dependence on timing and dosage of the application
are discussed. Special attention is paid to the contraceptive action of con-
tinuous low doses of progestagens both from the theoretical and clinical
point of view. Some recent advances in the field of contraception such as
long-acting injections of depot-preparations, subdermal implants of proges-
tagen-containing silastic capsules, intrauterine, intracervical and intra-
vaginal devices with progestagen fillings and the postcoital intake of progestational agents in low doses are also mentioned.

Bei den ersten und bisher einzigen Verhandlungen unserer Gesellschaft
über das Hauptthema Gestagene vor 12 Jahren in Kiel konzentrierte sich die
Thematik auf pharmakologische und experimentelle Fragen. Die klinischen
Aspekte traten demgegenüber in den Hintergrund. Das mag daran gelegen haben,
daß der Indikationskatalog für die Gestagenanwendung damals wohl weitgehend
als abgeschlossen galt. Nicht einmal die hormonale Kontrazeption fand auf
diesem Symposion besondere Beachtung, obwohl der erste Bericht zu diesem
Thema von Rock, Pincus u. Garcia (1956) schon 3 Jahre zurücklag.

War die Anwendung der Gestagene Ende der fünfziger Jahre noch streng auf
gynäkologische und geburtshilfliche Indikationen abgestellt und in ihrem Umfang insgesamt ziemlich begrenzt, so sind die Gestagene heute dank der wei-
ten Verbreitung der modernen Methoden der hormonalen Empfängnisverhütung zu
den am meisten verwendeten Arzneimitteln überhaupt geworden. Hinzu kommt,
daß im letzten Jahrzehnt mit der Prostatahypertrophie, der Pubertas praecox
und der Sklerodermie neue Indikationsgebiete für die Gestagenbehandlung er-
schlossen wurden.

In Anbetracht der Fülle neuartiger wissenschaftlicher Informationen aus
jüngster Zeit ist eine Begrenzung der Thematik auf die Entwicklungstendenzen

der letzten Jahre unter besonderer Berücksichtigung der in stürmischer Entwicklung begriffenen Kontrazeption auf reiner Gestagenbasis unumgänglich. Altbewährte Behandlungsformen können daher ebensowenig Berücksichtigung finden wie die kombinierte Verabfolgung von Gestagenen mit anderen Hormonen.

## Chemie und Wirkungen der therapeutisch wichtigen Gestagene

### A. Chemische Struktur

Von den annähernd 1000 Verbindungen mit gestagenen Haupt- bzw. Partialwirkungen (Neumann, 1968) haben nur knapp 30 therapeutische Bedeutung erlangt. Die Strukturformeln der wichtigsten von ihnen sind in den Tabellen 1 - 3 zusammengestellt. Vom chemischen Aufbau her haben sich zwei für die hormonale Kontrazeption besonders bedeutsame Hauptgruppen herauskristallisiert: Die 17α-Alkyl-19nor-androstan-Derivate (Tab. 1) und die 17α-Acetoxyprogesteronabkömmlinge (Tab. 2). Innerhalb dieser Gruppen sind die Strukturunterschiede oft außerordentlich gering; trotzdem können die Wirkungsspektren erheblich voneinander abweichen (s. u.). Andererseits sind metabolische Umwandlungen ineinander in vivo möglich, wie das von Mazaheri u. Mitarb. (1970) für die Transformation von Lynestrenol (VI) in Norethisteron (I) bei Inkubationsexperimenten mit Leberhomogenaten nachgewiesen worden ist. Durch Veresterung des 17-Hydroxyprogesterons an $C_{17}$ mit mittellangen Fettsäuren (XVIII-XX) bzw. Acetophenidbildung des 16α,17α-Dihydroxyprogesterons (XVIII) entstehen intramuskulär applizierbare Gestagene mit ausgesprochener Depotwirkung. Das gleiche gilt auch für Medroxyprogesteronacetat (XIII). Cyproteronacetat (XV) aus derselben Gruppe übt sogar bei oraler Applikation protrahierte Wirkungen aus (s. u.).

Die Retroprogesterone Dydrogestron (XXIV) und Trengeston (XXV) sind dadurch charakterisiert, daß ihre Ringe B und C nicht in cis- sondern in trans-Stellung zueinander angeordnet sind. Diese andersgeartete räumliche Struktur verleiht den Retroprogesteronen ein besonders interessantes Wirkungsprofil.

### B. Wirkungsstärke

Die Wirkungsintensität der einzelnen synthetischen Gestagene ist sehr unterschiedlich. Zu ihrer Ermittlung werden bei der Frau die Transformationsdosis und die Menstruationsverschiebungsdosis als approximative Kriterien herangezogen. Diese beiden Größen stimmen allerdings in ihren quantitativen Relationen zueinander nur sehr bedingt überein (Tab. 4). Bei den 19nor-Androstanen pflegt der Unterschied zwischen den beiden Meßwerten größer zu sein als bei den 17α-Acetoxyprogesteronen; besonders gering ist er beim Medroxyprogesteronacetat (XIII). Ähnliche Verhältnisse liegen auch bei dem in Tab. 4 nicht mit aufgeführten Allylöstrenol (VII) und Dydrogestron (XXIV) vor. Ob es auf mehr als nur auf Zufall beruht, daß gerade die drei letztgenannten Gestagene bevorzugt zur Schwangerschaftserhaltung bei drohendem und habituellem Abort angewandt werden, muß dahingestellt bleiben.

Auffallend gering sind die quantitativen Korrelationen zwischen den vorgenannten beiden Kriterien und den in der hormonalen oralen Kontrazeption angewandten Gestagendosen, eine Beobachtung, auf die hier nicht näher eingegangen werden kann (Hammerstein, 1970).

### C. Depotwirkung

Daß Progesteron im Fettgewebe gespeichert wird, haben Kaufmann u. Zander schon vor 15 Jahren zeigen können. Am Ende der Gravidität liegt dieses Gestagen im Fettgewebe in ähnlichen Konzentrationen vor wie in der Placenta. In Übereinstimmung dazu hat Plotz (1961) nach Injektion von markiertem Progesteron 1/3 der Radioaktivität im Fettgewebe angetroffen.

Tab. 1. **Zusammenstellung der wichtigsten Gestagene. I.17α–Alkyl–19nor–androstan–Derivate.**

Östrogenkurzbezeichnungen:

| | |
|---|---|
| EE | = Äthinylöstradiol |
| EEME | = Mestranol |
| QE | = Quinestrol |
| EOE | = Östradiol–17β–Oenanthat |

I. 17α – ALKYL – 19 NOR – ANDROSTANDERIVATE

| Trivialname | Formel | Präparat* | Gesta-gen mg | Östrogen µg | Östrogen Kurzbez. | Anmerkungen |
|---|---|---|---|---|---|---|
| I Norethin-dron (Norethi-steron) | OH ···C≡CH | ORTHO-NOVUM 1/50 | 1 | 50 | EEME | In Deutsch-land übliche Gestagen-komponenten der konven-tionellen hormonalen Kontrazeptiva |
| | | ORTHO-NOVUM 2 mg | 2 | 100 | EEME | |
| II Norethindron-acetat (Norethisteron-acetat) | OAc ···C≡CH | ANOVLAR | 4 | 50 | EE | |
| | | GYNOVLAR | 3 | 50 | EE | |
| | | ETALONTIN 21 | 2,5 | 50 | EE | |
| | | ORLEST | 1,0 | 50 | EE | |
| III Norgestrel | CH₃ CH₂ OH ···C≡CH | EUGYNON 21 | 0,5 | 50 | EE | |
| | | NEOGYNON | 0,25 | 50 | EE | |
| | | STEDIRIL | 0,5 | 50 | EE | |
| | | STEDIRIL-D | 0,25 | 50 | EE | |
| IV Ethyno-dioldi-acetat | 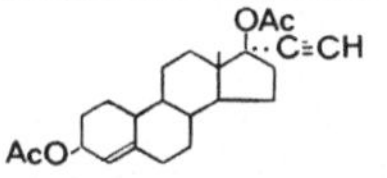 OAc ···C≡CH AcO | OVULEN | 1 | 100 | EEME | |
| V Lynestrenol | 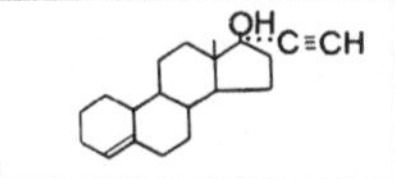 OH ···C≡CH | LYNDIOL 2,5 | 2,5 | 75 | EEME | |
| | | NORACYCLIN 22 | 2,5 | 75 | EEME | |
| | | NORACYCLIN | 5 | 150 | EEME | |
| | | ANACYCLIN | 1 | 100 | EEME | |
| VI Allyloestre-nol | 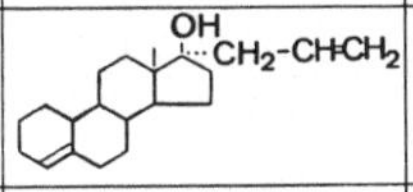 OH ···CH₂-CH=CH₂ | GESTANON* | 5 | – | – | Indikation: Schwanger-schafts-erhaltung |
| VII Norethyno-drel | OH ···C≡CH | ENOVID | 10 | 150 | EEME | ursprüngl. "PINCUS-pille" |
| | | GYNORM | 2,5 | 100 | EEME | |
| VIII Norvino-drel | 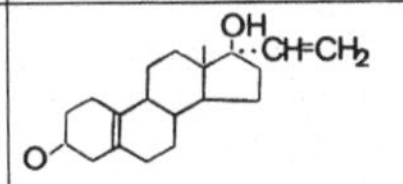 OH ···CH=CH₂ | VESTALIN | 5,0 | 75 | EEME | In Deutsch-land nicht erhältlich |
| IX Quingesta-nol-acetat | OAc ···C≡CH | REGLOVIS | 0,5 | 50 | EE | In Deutsch-land nicht erhältlich Ausländisches Versuchs-präparat |
| | | Einmonatspille | 2,5 | 2000 | QE | |
| X Ethyneron | 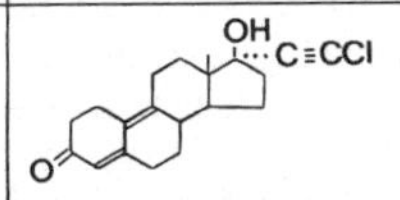 OH ···C≡CCl | Ausländisches Versuchspräparat | | | | Wegen meta-plastischer Mam-maverände-rungen beim Affen 1967 zu-rückgezogen |
| XI Norgest-rienon | 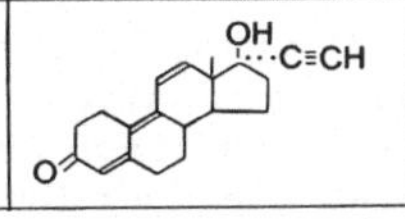 OH ···C≡CH | PLANOR | 2,0 | 50 | EE | In Deutsch-land nicht erhältlich |

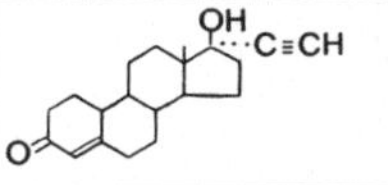

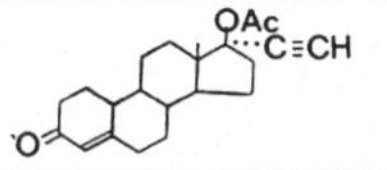

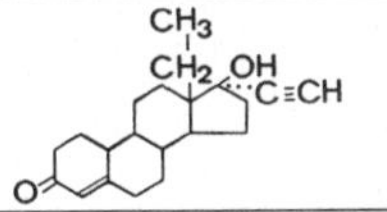

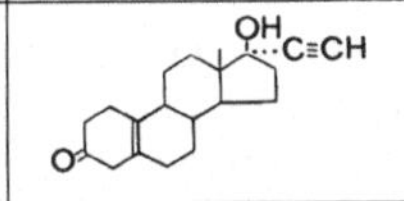

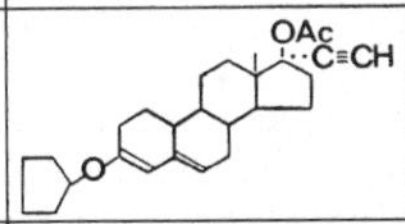

+ Bei den Präparatangaben konnten im allgemeinen nur Einphasenkontrazeptiva berücksichtigt werden. Gestagene, die nicht in der Kontrazeption Verwendung finden, sind durch einen Stern markiert.

Tab. 2. Zusammenstellung der wichtigsten Gestagene. II. 17α-Acetoxy-progesteron-Derivate. Erklärungen s. Tab. 1

| Trivialname | Formel | Präparat* | Zusammensetzung | | | Anmerkung |
|---|---|---|---|---|---|---|
| | | | Gestagen mg | Östrogen µg | Östrogen Kurzbez. | |
| XII Megestrolacetat | 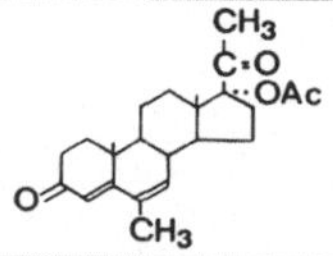 | AGENORAL<br>PLANOVIN<br>WERADYS<br>CO-ERVONUM | 4<br><br>2 | 50<br><br>100 | EE<br><br>EE | Als einziges Gestagen dieser Gruppe noch als Kombinationspille i. Handel |
| XIII Medroxyprogestronacetat | 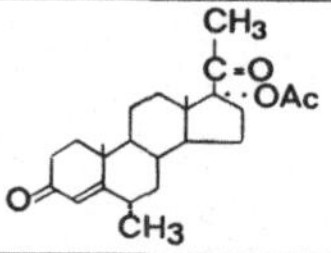 | ZYKLO-FARLUTAL<br>DEPOT-CLINOVIR | 5<br>150 | 75<br>--- | EE | Kürzlich aus dem Verkehr gezogen<br>"3-Monatsspritze" |
| XIV Chlormadinonacetat | 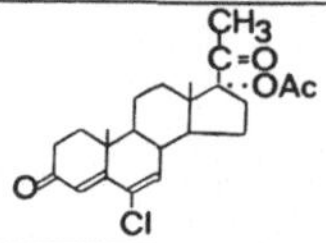 | ACONCEN<br>DISUT 21 | 3<br>2 | 0,1<br>0,08 | EEME<br>EEME | Wegen Auftreten von Mammatumoren beim Beagle kürzlich aus dem Verkehr gezogen |
| XV Cyproteronacetat | 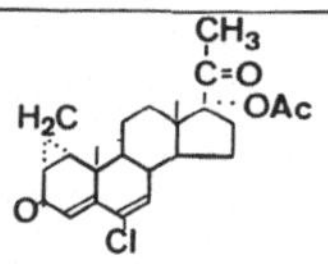 | SH 80714* (Versuchspräparat) | 50<br>100 | – | – | Gleichzeitig stärkstes aller bisher bekannten Antiandrogene |
| XVI Flumedroxonacetat | 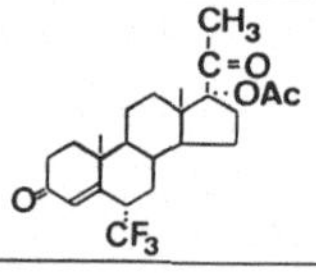 | DEMIGRAN* | 10 | – | – | Einzige Indikation: Migräne-Prophylaxe |

* s. Tab. 1

Bei der oralen Anwendung von Gestagenpräparaten spielen derartige Speichervorgänge keine Rolle, auch wenn die Pharmakokinetik der verschiedenen Stoffe beträchtlich voneinander abweicht. Eine bemerkenswerte Ausnahme von dieser Regel macht Cyproteronacetat (XV), das zusammen mit Östrogenen nur in der ersten Hälfte des Intermenstruum nach Art einer umgekehrten Zweiphasentherapie gegeben werden darf, wenn ein regelmäßiger Blutungszyklus aufrechterhalten bleiben soll (Hammerstein u. Cupceancu, 1969).

Zu ausgesprochenen Fernwirkungen, die auf einer derartigen Fettspeicherung beruhen dürften, kann es bei Depotpräparaten (s. o.) kommen. In diesem Zusammenhang ist besonders das Medroxyprogesteronacetat (XIII) zu nennen. Nach seiner intramuskulären Applikation während der Schwangerschaft kommt es postpartal bei zwei Drittel aller Frauen zu Blutungsstörungen der verschiedensten Art, die bis zur Dauer eines Jahres anhalten können (Turner, 1964; Luby, 1965). Orale Verabfolgung desselben Steroids läßt dergleichen Wirkungen vermissen. Ebenso ist die von Piver u. Mitarb. (1967) angegebene Quote von 38% "missed abortion" nach parenteraler Applikation von Medroxyprogesteronacetat bei Patientinnen mit drohendem Abort ungewöhnlich hoch. Schließlich verdienen auch die sich oft über viele Monate hinziehenden Zyklusstörungen in Verbindung mit Anovulation und Infertilität nach Absetzen der "Dreimonatsspritze" von Medroxyprogesteronacetat hier der Erwähnung (Abb. 1).

Auch andere Depotpräparate scheinen nicht ganz frei von solchen Fernwirkungen zu sein. Nach Verabfolgung von 17-Hydroxyprogesteronkapronat (XVIII) in der Schwangerschaft kommt es immerhin auch bei 16,1% der Fälle im Spätwochenbett zu dysfunktionellen Blutungsstörungen (Turner, 1964). Anderer-

Tab. 3. <u>Zusammenstellung der wichtigsten Gestagene. III. Sonstige synthetische Gestagene. Keine dieser Substanzen ist Bestandteil der in Deutschland erhältlichen Einphasen-Kontrazeptiva. Östrogen-Kurzbezeichnungen s. Tab. 1</u>

III. SONSTIGE SYNTHETISCHE GESTAGENE

| Trivialname | Formel | Präparat | Zusammensetzung | | | Indikation Anmerkungen |
|---|---|---|---|---|---|---|
| | | | Gestagen mg | Östrogen µg | Kurzbez. | |
| XVII Droxone | 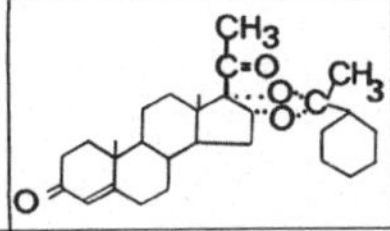 | DELADROXAT | 150 | 10.000 | EOE | "Einmonatsspritze" In Deutschl. nicht erhältl. |
| XVIII Hydroxyprogesteronkapronat | 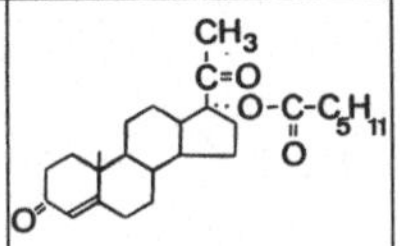 | PROLUTON-DEPOT. | 250–500 | | | Gynäkolog. Indikationen |
| XIX Norhydroxyprogesteronkapronat | 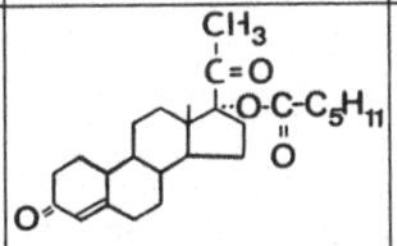 | DEPOSTAT | 200 | | | Prostatahypertrophie — Endometr. Karzinom |
| XX Norethisteronönanthat | 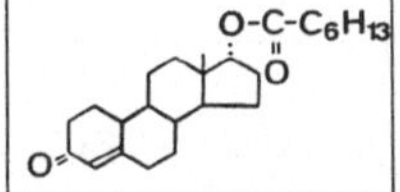 | SH 393 (Versuchspräparat) | 200–300 | | | "3-Monatsspritze" |
| XXI Dimethisteron | 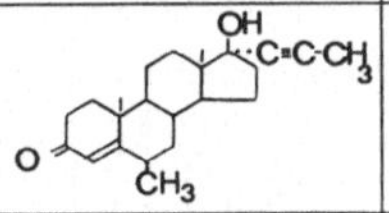 | SECRODYL | 10 | 50 | EE | Einphasenpräparat In Deutschl. nicht erhältl. |
| XXII Anagestonacetat | 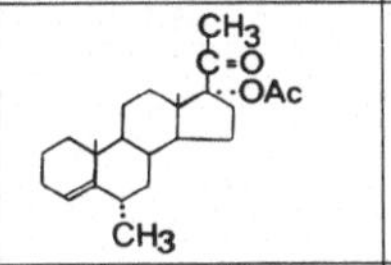 | NEONOVUM | 2 | 0,1 | EEME | Zweiphasenpräparat. Wegen Auftreten von Mammatumoren beim Beagle 1969 aus d. Verkehr gezogen |
| XXIII Quingestron | 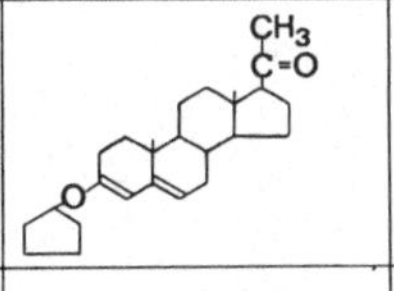 | VERSUCHS-PRÄPARAT | | | | In Deutschland nicht erhältlich |
| XXIV Dydrogestron | 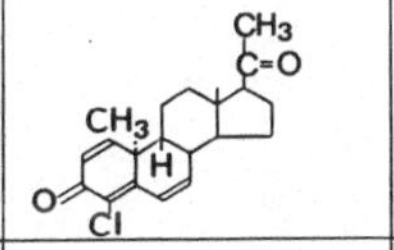 | DUPHASTON | 5 | | | Gynäkolog. Indikationen Schwangerschaftserhaltung |
| XXV Trengeston | 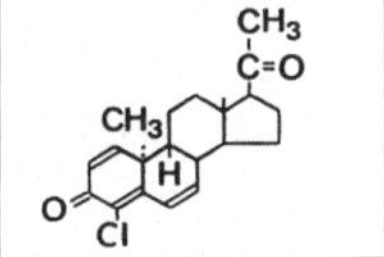 | RETROID Ro4 - 8347 (Versuchspräparat | 4 - 8 | | | Ovulationsauslösung |

seits ist das ebenfalls als "Dreimonatsspritze" an einer großen Patientinnenzahl erprobte Norethisteronänanthat (XX) frei von Fernwirkungen auf Zyklus und Fertilität (Larrañaga u. Kesserü, 1967).

Tab. 4. <u>Wirkungsstärke und Wirkungsspektrum der 9 in der hormonalen Kontrazeption wichtigsten Gestagene (Hammerstein, 1970).</u>

| | Befunde beim Menschen | | | Partialwirkungen im Tierexperiment | | | | | | |
|---|---|---|---|---|---|---|---|---|---|---|
| | Transformations-dosis mg/Zyklus | Menstruations-verschiebungs-test n. Green-blatt mg/Tag | angew.Dosis b. d. hormonalen Kontrazeption mg/Tag | Schwangerschafts-erhaltung | Östrogene (Vaginal smear) | Androgene | Anti-Androgene | Anti-Östrogene | Hemmung d. Go-nadotropinaus-schüttung | Beeinflussung d. ACTH-Ausschüttg. |
| Progesteron | 200 i.m. | 1000 | Ø | + | Ø | Ø | Ø | + | + | ↑ |
| VII. Norethynodrel | 150-200 | 14 | 2,5-10 | Ø | ++ | Ø | Ø | Ø | ++ | ↑ |
| I. Norethindron | 100-150 | 15 | 1-10 | Ø | + | + | Ø | ++ | ++ | ↑ |
| II. Norethindronacetat | 50-60 | 7,5 | 1-4 | (+) | + | + | Ø | ++ | ++ | ↑ |
| XIII. Medroxyprogsteronacetat | 40-70 | 20-30 | 5-10 | ++ | Ø | (+) | (+) | + | ++ | (↓) |
| V. Lynestrenol | 35-70 | 10 | 1-5 | Ø | (+) | + | Ø | ++ | + | ↑ |
| XII. Megestrolacetat | 35-70 | 5-10 | 2-4 | + | Ø | Ø | (+) | (+) | + | ↓ |
| XIV. Chlormadinonacetat | 20-30 | 4 | 2-3 | + | Ø | Ø | + | (+) | + | ↓ |
| IV. Ethynodioldiacetat | 10-15 | 1 | 1 | Ø | + (+) | + | Ø | + | ++ | ↑ |
| III. d-Norgestrel | 12 | 0,5 | 0,25 | + | Ø | ++ | Ø | + | ++ | Ø |

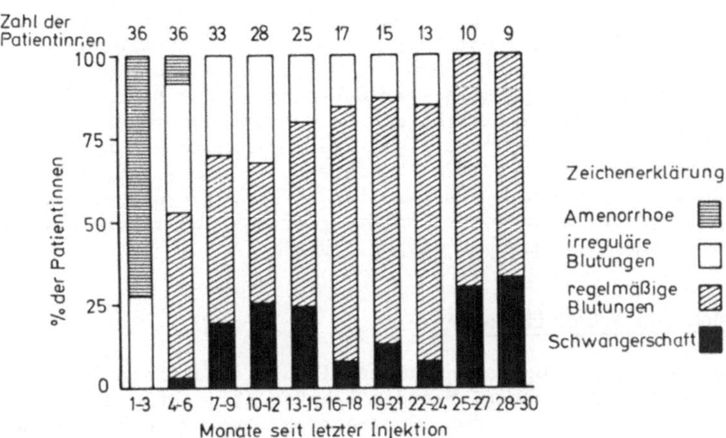


Abb. 1. Zyklusverhältnisse und Auftreten von Schwangerschaften nach Beendigung der Kontrazeption mit der "Dreimonatsspritze" (Medroxyprogesteronacetat XIII) nach Gardner u. Mishell (1970.

## D. Endokrines Wirkungsspektrum

Nur mit größter Zurückhaltung wird man davon ausgehen dürfen, daß die im Tierexperiment ermittelten endokrinen Wirkungsspektren der synthetischen Gestagene (Tab. 4) auf den Menschen übertragbar sind; bestehen doch in dieser Hinsicht schon zwischen verschiedenen Tierspezies erhebliche Unterschiede. Kompliziert wird die Situation noch durch den Umstand, daß viele der endokrinen Partialwirkungen beim Menschen entweder gar nicht oder nur sehr unvollkommen getestet werden können.

Sieht man von diesen grundsätzlichen Bedenken ab, dann ist darüber hinaus auch die häufig geübte Pauschalisierung der Wirkungsspektren der beiden Gestagen-Hauptgruppen nur mit erheblichen Vorbehalten statthaft (Tab. 4). Um nur zwei Beispiele herauszugreifen, so besitzt d-Norgestrel (III) im Gegensatz zur Mehrzahl der 19nor-Androstane keine östrogene, dafür aber ausgeprägte schwangerschaftserhaltende Eigenschaften; es verhält sich in dieser Hinsicht wie die meisten 17-Acetoxyprogesterone. Andererseits ist es im Tierexperiment ebenso wie die meisten anderen 19nor-Androstane ein relativ starkes Androgen, ohne daß sich dies klinisch unangenehm bemerkbar machen würde. Diese Eigenschaft fehlt nun wiederum dem Allylestrenol (VII) und dem Norethynodrel (V). Letzteres ist dafür ein ausgesprochen starkes Östrogen. Diese Wirkung ist dem Norethynodrelmolekül selbst zuzuschreiben, nachdem H. Breuer (1970) kürzlich nachgewiesen hat, daß keine Aromatisierung der 19nor-Androstanderivate in vivo in nennenswerten Mengen erfolgt.

Von klinischer Relevanz ist die schwache antiandrogene Partialkomponente des Chlormadinonacetat (XIV); infolge dieser Eigenschaft war seine Anwendung bei mehr oder weniger stark virilisierten Frauen bis zu seiner Rücknahme vom Markt besonders vorteilhaft. Durch Einführung einer Methylenbrücke zwischen $C_1$ und $C_2$ in das Chlormadinonacetat-Molekül wird der antiandrogene Effekt bei unveränderter gestagener Aktivität bemerkenswert gesteigert. Dieses Cyproteronacetat (XV) ist das stärkste klinisch anwendbare Antiandrogen, das wir z. Zt. kennen (Hammerstein u. Cupceancu, 1969).

Medroxyprogesteronacetat (XVIII) besitzt eine so ausgesprochene Glucocorticoid-Partialwirkung, daß es Cortisol bei der Substitution adrenalektomierter Patienten ersetzen kann (Camanni u. Mitarb., 1963). Gleichzeitig wird

die zentrale ACTH-Ausschüttung im Zuge des negativen Rückkoppelungsmechanis-
mus wie durch Cortisol gebremst. Die von Mathews u. Mitarb. (1970) nachgewiesene verminderte NNR-Reaktion auf ACTH spricht darüber hinaus auch für
einen peripheren Angriffspunkt dieser Substanz an der NNR. Aufgrund der
starken glukogenetischen Aktivität des Medroxyprogesteronacetats ist Zurückhaltung bezüglich seiner Anwendung bei Prädiabetikern und Diabetikern am
Platze (Spellacy u. Mitarb., 1970). Andere Gestagene scheinen in den Kohle-
hydrathaushalt nicht negativ einzugreifen; 17α-Hydroxyprogesteronkapronat
(XVIII) hat sogar bei sehr hoher Dosierung eine Verbesserung der Kohlehydrat-
toleranz zur Folge (Benjamin u. Casper, 1966).


## E. Nichtendokrine Nebenwirkungen

Über Nebenwirkungen der Gestagene außerhalb des endokrinen Systems sind wir
nur unzureichend unterrichtet. Was die Leberfunktion anlangt, so scheint
höchstens einigen Vertretern der 19nor-Androstanreihe gelegentlich eine ne-
gative Wirkung anzuhaften (Eisalo u. Mitarb., 1968). Ganz ausnahmsweise wird
das Auftreten eines Ikterus beobachtet (Somayaji u. Mitarb., 1968). Megestrolacetat (XII) beeinflußt dagegen die Leberfunktion nicht (Eisalo u.
Mitarb., 1968; Clinch u. Tindall, 1969); ähnliches scheint auch für die anderen 17-Acetoxyprogesteron-Derivate zuzutreffen.

Als noch keineswegs geklärt muß der Einfluß der Gestagene auf Gerinnungssystem und Venenapparat angesehen werden. Nach Ludwig (1970) kann es bei der
Behandlung mit Gestagenen zu einer "plasmatischen Tendenz zur Übergerinnbar-
keit und zu einer Erschlaffung des Venentonus" kommen. Zur Vermeidung von
thromboembolischen Konflikten seien daher hormonale Kontrazeptiva mit niedriger Gestagenkomponente zu bevorzugen. Im übrigen haben Poller u. Mitarb.
(1969) gezeigt, daß Chlormadinonacetat (XIV) im Gegensatz zu Norethisteron
und seinem Acetat (I u. II) keine Veränderungen der Gerinnungsfaktoren VII
und X hervorruft. Die Blutlipide werden durch 17-Hydroxyprogesteronkapronat
(XVIII) nicht beeinflußt (Svanborg u. Virkot, 1966).

Verschiedentlich ist behauptet worden, der Gestagenanteil in den hormona-
len Kontrazeptiva sei für das gelegentliche Auftreten von Depressionen und
Libidoverlust unter dieser Therapie verantwortlich. Die Östrogene sollen dem
bis zu einem gewissen Grade entgegenwirken können. Nach der Hypothese von
Grant u. Pryse-Davies (1968) stehen diese psychischen Veränderungen in engem
Zusammenhang mit der Gestagenabhängigkeit der Monoaminoxydase-Aktivität in
Blut und Geweben. Daraus erkläre sich auch die befriedigende klinische Wir-
kung von Monoaminoxydase-Hemmern bei der Behandlung derartiger psychischer
Alterationen.

Noch völlig offen ist die Frage, ob es tatsächlich eine Progesteron-
Allergie gibt, wie verschiedentlich behauptet worden ist. Prämenstruelle
Exacerbationen eines Erythema exsudativum multiforme bzw. eines Pruritus
simplex sind von Shelley u. Mitarb. (1964) und von Hoischen u. Steigleder
(1966) aufgrund verschiedener immunologischer Teste mit einer Autoimmunre-
aktion gegen körpereigenes Progesteron zu erklären versucht worden. Verabfolgung von Ovulationshemmern führte zur Besserung der Beschwerden; eine
Kreuzimmunität mit synthetischen Gestagenen bestand demnach nicht. Umgekehrt
ist kürzlich das Auftreten einer Schwangerschaft bei einer 27jährigen Frau
unter hormonaler Kontrazeption auf die Existenz von Antikörpern gegen Norethisteron (I) zurückgeführt worden (Magré u. Mitarb., 1969).

Klinische Hinweise für eine Kanzerogenität der Gestagene beim Menschen
gibt es nicht. Die diesbezüglichen Hundeexperimente, die zur voreiligen
Rücknahme von Ethyneron (X), Anagestonacetat (XXII), Chlormadinonacetat
(XIV) und partiell auch von Medroxyprogesteronacetat (XIII) geführt haben,
besitzen ganz offensichtlich keine Relevanz für den Menschen. Dies kam auf
zwei kürzlich abgehaltenen Spezialtagungen, deren Verhandlungsberichte dem-
nächst erscheinen werden, deutlich zum Ausdruck. Auf hiermit zusammenhängen-
de Fragen wird auch im anschließenden Rundtischgespräch eingegangen werden.

## Gestagentherapie in der Gynäkologie und Geburtshilfe

In der Frauenheilkunde im engeren Sinne, also dem ursprünglichen Gebiet der
Gestagenanwendung, haben sich in den letzten Jahren kaum neue Aspekte abge-
zeichnet. Zykluseinregulierung speziell bei anovulatorischen Tempoanomalien,
Corpus luteum-Insuffizienz in Verbindung mit Sterilität, Dysmenorrhoe, Endo-
metriose, Uterus myomatosus, Mastopathie, drohender und habitueller Abort
sowie vorzeitige Wehentätigkeit sind die häufigsten Anlässe zur alleinigen
Gestagenapplikation in unserem Fachgebiet.
    Durch die Entwicklung des Trengeston (XXV) sind die auf Rust (1956) sowie
Napp u. Rothe (1958) zurückgehenden Verfahren zur Ovulationsauslösung mit
kleinen Gestagenmengen wieder aufgelebt. Erste Erfahrungsberichte von Stamm
u. Mitarb. (1970), Dapunt (1970) u. a. klingen verheißungsvoll. Die eigenen
Ergebnisse mit diesem Retroprogesteron-Derivat waren indessen enttäuschend.


## Gestagenbehandlung hormonabhängiger Tumoren und Hypertrophien

Seit über 20 Jahren sind günstige Palliativeffekte einer hochdosierten, nach
g/Woche zählenden Gestagenmedikation in etwa 30-40% der Fälle mit fortge-
schrittenem Endometriumkarzinom bekannt. Injektionspräparate wie Medroxyprogesteronacetat (XIII), 17-Hydroxyprogesteronkapronat (XVIII) bzw. sein 19nor-
Analogon (XIX) sind die bevorzugten Steroidpräparate auf diesem Indikationsgebiet. Aber auch mit der oralen Verabfolgung von Gestagenen lassen sich vergleichbare Ergebnisse erzielen, wie unsere Erfahrungen mit Chlormadinonacetat zeigen (Kreuzer u. Bouquoi, 1970). Ob es die antiproliferative Eigenschaft oder ein direkter cytostatischer Effekt der Gestagene ist, der diesen
Behandlungserfolg bedingt, ist noch nicht eindeutig entschieden. Immerhin hat
R. Kaiser schon vor 12 Jahren gezeigt, daß die Mitoserate von Endometrium-
karzinomen nach oraler Verabfolgung von Norethisteronacetat (II) auf fast
die Hälfte abnimmt.
    Die ersten Ansätze zu einer Gestagenbehandlung des fortgeschrittenen
Mammakarzinoms reichen bis in die vierziger Jahre zurück. Trotzdem hat sich
diese Therapieform bis auf den heutigen Tag nicht durchsetzen können. Manche
Gestagene wie z. B. das 17-Hydroxyprogesteronkapronat (XVIII) haben sich als
unwirksam herausgestellt. Beim Medroxyprogesteronacetat (XIII) wird der oralen Applikationsform der Vorzug gegeben. Bewährt hat sich ferner das oral in
einer Höhe von 30-60 mg täglich zu verabfolgende Norethisteronacetat (II).
Weichteil- und Hautmetastasen reagieren am besten auf diese Therapie (Stoll,
1968). Gestagene sind im übrigen auch dann noch indiziert, wenn die Patienten auf andere Arten der Hormontherapie resistent geworden sind. Der Voll-
ständigkeit halber sei noch erwähnt, daß seit einigen Jahren auch eine kombinierte Östrogen-Gestagen-Applikation auf diesem Indikationsgebiet empfohlen wird (Berndt u. Stender, 1970).
    Zum Stellenwert der Behandlung des Prostata-Adenoms mit Gestagenen wird
auf den Beitrag von Klosterhalfen in diesem Band verwiesen. Hier sei nur erwähnt, daß vorwiegend 17-Hydroxyprogesteronderivate zur Anwendung gelangen,
unter ihnen bevorzugt Hydroxyprogesteronkapronat (XVIII), sein 19nor-Analo-
gon (XIX) und Cyproteronacetat (XV). Wie bei dieser Erkrankung erwünscht,
sinkt der Plasma-Testosteronspiegel unter einer solchen hochdosierten
Gestagentherapie beträchtlich ab (Geller u. Mitarb., 1967).


## Pubertas praecox

Bis vor kurzem hat es eine wirksame Therapie zur Abbremsung der Akzelera-
tion von Wachstum und somatischer Reifung bei der idiopathischen Pubertas
praecox nicht gegeben. Von einem wirksamen Mittel für diese Indikation sollte man erwarten, daß es die Gonadotropinausschüttung sicher unterdrückt,
ohne selbst östrogene oder androgene Wirkungen zu entfalten. Die 19nor-

Androstanderivate scheiden unter diesen Umständen von vornherein für diese
Zwecke aus (vgl. Tab. 4).

Dagegen hat sich Medroxyprogesteronacetat (XIII) bisher durchaus bewährt
(Kupperman u. Epstein, 1962). Wenn hoch genug dosiert wird, d. h. 100–200 mg
wöchentlich i.m., dann sistieren bei Mädchen die Menses und die Brüste bilden sich mehr oder weniger vollständig zurück. Bei Jungen kommt es zur Verkleinerung der äußeren Genitalien und zum Verschwinden nächtlicher Pollutio-
nen (Collipp u. Mitarb., 1964; Mathews u. Mitarb., 1970).

Umstritten ist dagegen, ob durch diese Medikation auch das Knochenwachs-
tum nachhaltig beeinflußt und der aus dem vorzeitigen Epiphysenfugenschluß
resultierende Kleinwuchs verhindert werden kann. In dieser Hinsicht gibt es
zweifelsohne resistente Fälle (Teller u. Mitarb., 1969), die keineswegs immer einer zu niedrigen Dosierung angelastet werden können. Ebenso bleibt
auch der angestrebte Abfall der FSH- und LH-Werte in Harn und Blut unter der
Medroxyprogesteronacetat-Medikation nicht selten aus (Rifkind u. Mitarb.,
1969; Kenny u. Mitarb., 1969). Desgleichen weisen Plasma-Testosteronspiegel
und Ausscheidung von $C_{19}$- und $C_{21}$-Steroiden verschiedentlich keine Beeinflussung auf (Rivarola u. Mitarb., 1968; Teller u. Mitarb., 1969).

Nur wenige andere Gestagene sind auf diesem Indikationsgebiet erprobt worden. Während Chlormadinonacetat (XIV) insgesamt enttäuscht hat (Teller u.
Mitarb., 1969; Helge u. Mitarb., 1969 und 1971), geben unsere bisherigen Erfahrungen mit Cyproteronacetat zu einigen Hoffnungen Anlaß. So führte die
tägliche Einnahme von 40 mg dieses Steroids bei 2 Mädchen mit idiopathischer
Pubertas praecox zur Normalisierung der Wachstumsrate (Abb. 2). Bemerkens-
werterweise wurde dabei die Skelettreifung noch stärker gebremst als das

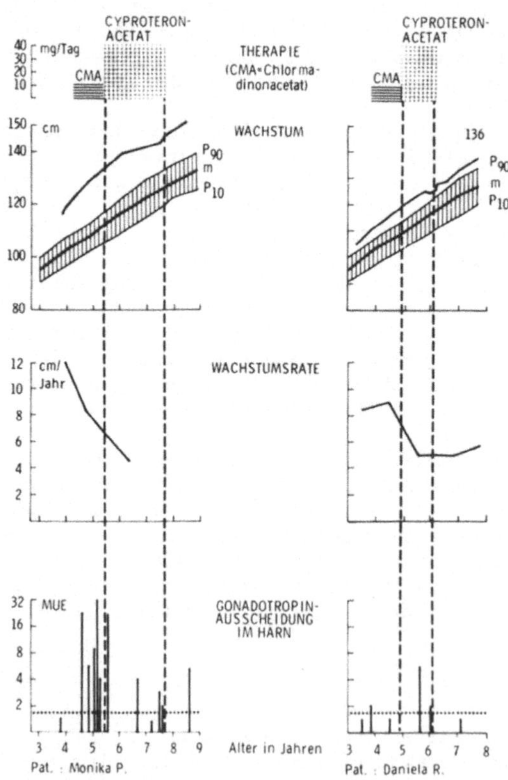


Abb. 2. Einfluß von Cyproteronacetat (XV) auf Wachstum und Gonadotropinausscheidung bei zwei Mädchen mit Pubertas praecox (Helge u. Mitarb.,
1971).

Längenwachstum. Wenn diese Behandlungsergebnisse erzielt wurden, ohne daß
gleichzeitig die Gonadotropinausscheidung auf präpubertale, also nicht mehr
meßbare Werte absank, so könnte das ein Hinweis darauf sein, daß nicht nur
die antigonadotrope, sondern auch die antiandrogene Komponente des Cyprote-
ronacetats für die klinischen Wirkungen Bedeutung besitzt.


## Migräne

Daß eine Migräne während der Behandlung mit Sexualhormonen auftreten oder
intensiviert werden kann, ist sehr viel bekannter als der umgekehrte Fall
einer Besserung der Migräne im Verlauf einer Gestagenmedikation. In dieser
Hinsicht pflegen aber nur solche Patientinnen anzusprechen, deren Beschwerden
zyklusabhängig sind und vor bzw. während den Menses exacerbieren. Charakte-
ristischerweise sind sie während der Gravidität frei von Anfällen. Unter der
üblichen hormonalen Kontrazeption kommt es bei ihnen häufig zu einer unge-
wohnten Verschlimmerung des Leidens um den Menstruationstermin herum, und
zwar speziell bei Verwendung von Präparaten mit hohem Östrogen- und niedri-
gem Progesteronanteil (Whitty u. Mitarb., 1966). Die Überbrückung des pil-
lenfreien Intervalls mit kleinsten Gestagendosen, z. B. 0,5 mg Chlormadinonacetat (XIV) täglich bringt häufig Linderung.

Bei hartnäckiger zyklusabhängiger Migräne führt die reine Gestagentherapie zu den besten Ergebnissen. Wir verwendeten bisher 4–6 mg Chlormadinonacetat (XIV) täglich während des ganzen Intermenstruum und sind jetzt auf
10 mg Megestrolacetat (XII) täglich übergegangen. Auch hierbei erweist sich
eine Überbrückung der Menstruationsphase mit kleinsten Gestagenmengen oft
als nützlich. Einziger Nachteil dieser Therapie ist, daß die Blutungen häufig unregelmäßig werden und nicht mehr richtig gesteuert werden können. Die
Patientinnen nehmen das im allgemeinen gern in Kauf, wenn sie dafür von ihren Migräneanfällen dauerhaft befreit werden.

Nicht eindeutig geklärt ist die klinische Wirksamkeit des Flumedroxon-
acetats (XVI), das ausschließlich zum Zwecke der Migräne-Behandlung in den
Handel gebracht worden ist. Der anfängliche Optimismus ist einer gewissen
Skepsis gewichen, nachdem Bradley u. Mitarb. (1968) im doppelten Blindver-
such keine eindeutige Überlegenheit dieses Präparats gegenüber Placebos
feststellen konnten. Auch 19nor-Androstanderivate scheinen sich auf diesem
Indikationsgebiet nicht zu bewähren.


## Sklerodermie

Auf die Gestagenbehandlung der Sklerodermie braucht hier nicht näher eingegangen zu werden, da dieses Thema von K. Winkler bereits eingehend abgehandelt worden ist. Nur soviel sei erwähnt, daß Holzmann, Kortin u. Morscher
(1965) im Tierexperiment zeigen konnten, daß die Gestagene einen fördernden
Einfluß auf den Abbau der neutralsalzlöslichen Kollagene der Haut, die bei
der Sklerodermie stark vermehrt sind, besitzen. Bei Patienten mit Sklerodermie kommt es dementsprechend unter Gestagentherapie zu einer Zunahme der
Ausscheidung von Hydroxyprolin und Mucopolysacchariden. Diese relativ neuen
Erkenntnisse mögen als Beispiel dafür dienen, wie wenig wir im Grunde genommen noch über die Allgemeinwirkung der Gestagene auf den menschlichen Organismus wissen.


## Hormonale Kontrazeption mit Gestagenen

## A. Methodenübersicht

In Tab. 5 sind alle hormonalen Kontrazeptionsverfahren, bei denen Gestagene
Verwendung finden, zusammengestellt. Die ersten 4 Methoden werden uns im

Tab. 5. Prinzipien der hormonalen Kontrazeption mit Gestagenen

| Kontrazeptionsverfahren | Kurzbezeichnung | bereits eingeführt | Behandlungs – Schemata |
|---|---|---|---|
| Orale kombinierte Östrogen/Gestagen-Medikation | "Pincus-Pille" Klassische oder Kombinations-Pille | Anovlar etc. | menses 5 10 15 20 25 — 0,25 – 5 mg Gestagene täglich; 50 – 150 Östrogene täglich |
| Intramuskuläre Injektion von Östrogen-Gestagen-Depot-Präparaten | "Einmonatsspritze" | Deladroxate (U.S.A.) | 10 mg Östrogenester; 150 mg Gestagenester — 1 x monatlich |
| Orale Depot-Östrogen/Gestagen-Medikation | "Einmonats-pille" | im Versuchs-stadium | langwirkender Östrogenester und kurzwirkendes Gestagenderivat |
| Orale Zweiphasenmethode: 1.Östrogene allein 2. Östrogene + Gestagene | Sequential-Therapie | Ovanon | 2,5 mg Gestagene; 50 – 150 Östrogene täglich |
| Intramuskuläre Gestagen-Depot-Injektion | "Dreimonats-spritze" | Depot-Clinovir | 150 mg Gestagene i.m.; 1 Inj. f. 3 Monate |
| Niedrig dosierte orale Gestagen-Dauer-Medikation | "Minipille" | | 0,03–0,5 mg Gestagene täglich |
| Subkutane Implantation von Gestagenen in Silastik-Kapseln | | im Versuchs-stadium | 4–6 Kapseln für Monate bis Jahre |
| Intrauterine bzw. intravag. Pessare mit Gestagenzusätzen | | im Versuchs-stadium | 1 Pessar für Monate bis Jahre |
| Postcoitale Gestagen-Medikation | "Pille danach" | im Versuchs-stadium | 0,8 mg Gestagen: ● ● ● ● ● ● Verkehr: V V V V V V; menses 5 10 15 20 25 Cyclustage |

Note: The center column spanning "Ovulationshemmer" applies to the rows from "Orale Depot-Östrogen/Gestagen-Medikation" through "Intramuskuläre Gestagen-Depot-Injektion".

Folgenden nicht weiter beschäftigen, da die Gestagene bei ihnen kombiniert
mit Östrogenen verabfolgt werden. Die verbleibenden 5 Verfahren basieren auf
der alleinigen Applikation von Gestagenen; sie sind alle neueren oder neuesten Datums. Während bei der Kombinationstherapie die Ovulation mit größter
Zuverlässigkeit unterdrückt wird, besitzt dieser Effekt bei alleiniger
Gestagenmedikation nur im Falle der "Dreimonatsspritze" Bedeutung. Bei den
übrigen Formen kommt es zwar auch in einem gewissen Prozentsatz zur Ovulationshemmung; diese stellt aber nicht das entscheidende kontrazeptive Prinzip dar.
Nur die "Dreimonatsspritze" (Depot-Clinovir) ist trotz Zurückziehung aller
medroxyprogesteronacetathaltiger Kombinationspräparate bisher noch im Handel.
Dagegen mußte die in einigen Ländern bereits eingeführte chlormadinonacetat-
haltige Minipille unter dem Druck der amerikanischen FDA kürzlich aus dem
Verkehr gezogen werden (s. o.). Die letzten drei Verfahren befinden sich noch
im Versuchsstadium.


## B. Beeinflussung des ovariellen Zyklus durch Gestagene in Abhängigkeit von Dosierung und Zeit

Ebenso wie die kombinierte Verabfolgung von Östrogenen und Gestagenen in
Form der "Pincus-Pille", so führt auch die alleinige intermenstruelle Applikation von Gestagenen in entsprechender Dosierung zur Unterdrückung von Ovulation und Gelbkörperbildung. Apostolakis u. Mitarb. (1966) haben beispielsweise gezeigt, daß der wellenförmige Zyklusverlauf der Östrogen-, Pregnan-
diol- und Testosteron-Ausscheidung durch die tägliche Einnahme von 5 mg
Lynestrenol (V) weitgehend nivelliert wird. Diese Beobachtung steht in eng-
stem Kausalzusammenhang mit der Unterdrückung des LH-Ovulationsgipfels, wie
das von Vorys u. Mitarb. (1965) bei einer Patientin unter 2 mg Ethynodiol-
acetat anhand von Harnanalysen beobachtet worden ist (Abb. 3). Interessanter-

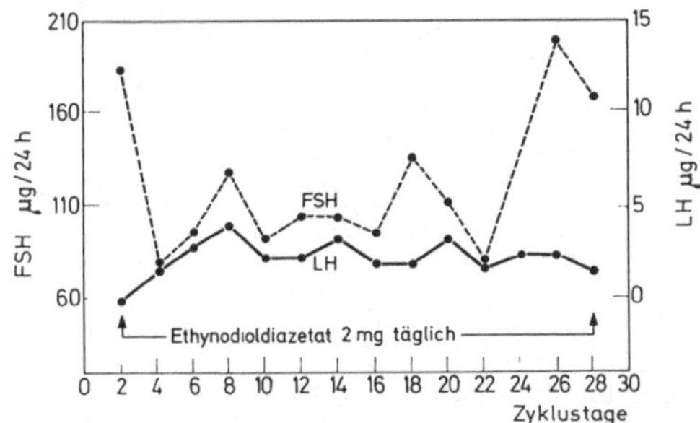


Abb. 3. Einfluß von Ethynodioldiacetat (IV) auf die FSH- und LH-Ausscheidung
im Harn (Vorys u. Mitarb., 1965).


weise wurde die FSH-Ausscheidung bei diesem Therapieregime nicht beeinträch-
tigt. LH- und FSH-Bestimmungen im Blut von Patientinnen unter der Sequential-
Therapie (s. Tab. 5) haben darüber hinaus ergeben, daß die LH-Werte während
der Östrogeneinnahmephase nicht nur nicht absinken, sondern verschiedentlich
sogar kurzfristig ovulationsartig ansteigen; zu einem Abfall auf niedrige
Blutspiegel kommt es erst in der zweiten Phase, also bei zusätzlicher
Gestagenverabfolgung (Swerdloff u. Odell, 1969; Abb. 4). An der spezifischen
Bremswirkung der Gestagene auf die hypophysäre LH-Ausschüttung kann nach
diesen und ähnlichen Untersuchungen nicht mehr gezweifelt werden.
Ein interessanter Beitrag zur Frage, ob die Zyklusregulation auch dann in
Mitleidenschaft gezogen wird, wenn mit der Gestagenmedikation nicht direkt

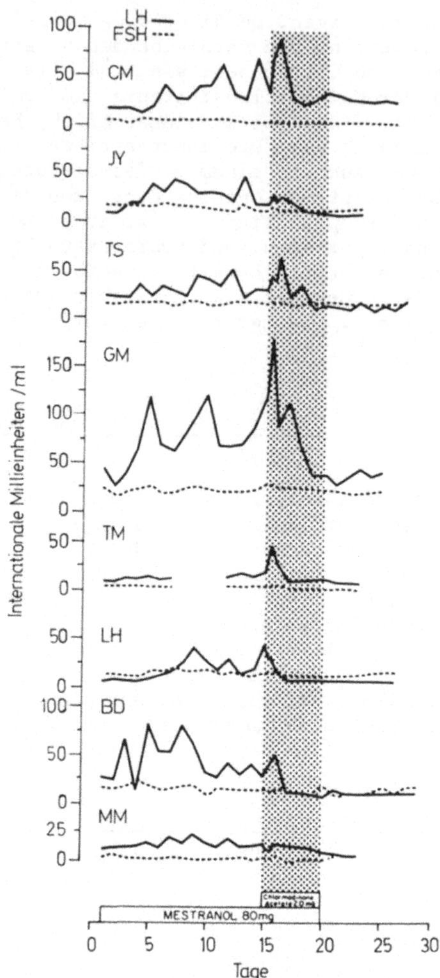



Abb. 4. Tägliche FSH- und LH-Spiegel im Serum von 8 Frauen unter der Zweiphasentherapie mit Mestranol und Chlormadinonacetat (XIV). Das angewandte Präparat C-Quens entspricht dem kürzlich in Deutschland aus
dem Handel gezogenen Estirona. Pat. BD, LH und TM hatten vorher unter Einphasen-Kontrazeption gestanden (Swerdloff u. Odell, 1969).

postmenstruell sondern erst später, z. B. in der präovulatorischen Phase begonnen wird, stammt von Østergaard u. Starup (1968). Diese Autoren konnten
nachweisen, daß die Kombinationspille (Megestrolacetat (XII) und Mestranol)
um so unzuverlässiger ovulationshemmend wirkt, je später die Medikation nach
dem 7. Zyklustag einsetzt. Bei Therapiebeginn nach dem 9. Zyklustag ließ
sich der Follikelsprung sogar in keinem einzigen Fall verhindern, wie der
Nachweis frischer Gelbkörper anläßlich von Laparotomien im selben Zyklus
ergab.

Nach den niedrigen Pregnandiolwerten im Harn zu urteilen, sind diese Corpora lutea jedoch in ihrer Steroidproduktion entscheidend gestört. Dementsprechend ließ sich bei gemeinsam mit Østergaard u. Arends durchgeführten

Inkubationsversuchen solcher Gelbkörper keine Inkorporation von markiertem
Acetat in Progesteron in nennenswerten Mengen, wie das sonst üblich ist
(Hammerstein u. Mitarb., 1964), nachweisen. Dieser suppressive Effekt auf
die steroidbildende Potenz solcher Gelbkörper geht, wie Abb. 5 erkennen läßt,

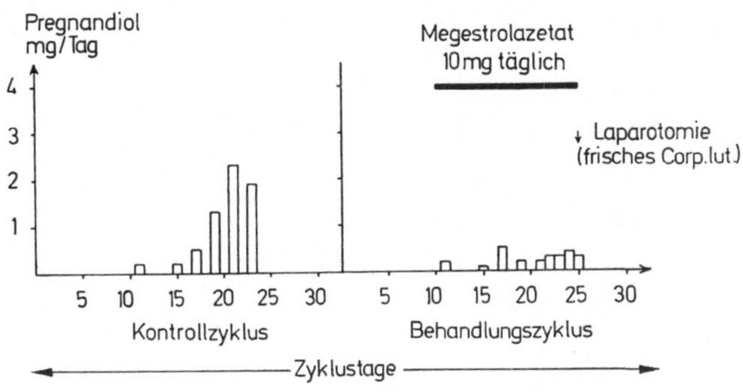


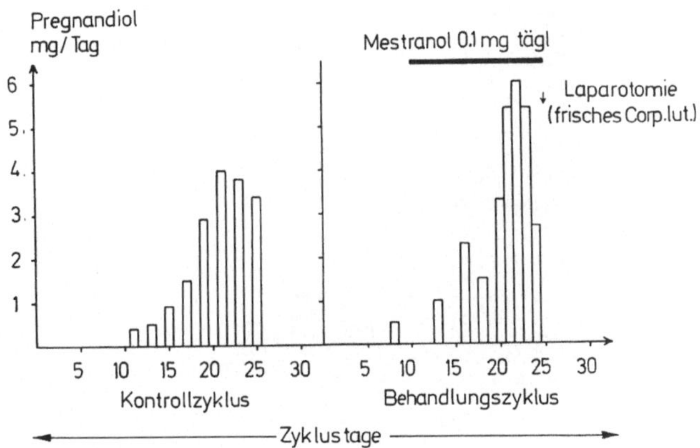


Abb. 5. Beeinflussung der Gelbkörperfunktion durch Verabfolgung von Gestagenen
oder Östrogenen ab 10. Zyklustag. Oben: Pregnandiolausscheidung
einer 34jährigen Frau vor und während einer Behandlung mit 10 mg
Megestrolacetat (XII) täglich. Unten: Pregnandiolausscheidung einer
35jährigen Frau vor und während der täglichen Einnahme von 0,1 mg
Mestranol (Østergaard u. Starup, 1968).

vom Megestrolacetat (XII) aus, nicht dagegen von der Östrogenkomponente der
Pille. Wurde mit der Einnahme der beiden Hormone erst nach der Ovulation begonnen, dann kam es trotz einer Tendenz zur verkürzten Gelbkörperphase zu
keiner Beeinträchtigung der Pregnandiolausscheidung.
In jüngster Zeit ist Johansson (1970) der Frage nachgegangen, ob sich die
Gelbkörperfunktion durch Gestagene in sehr hoher Dosierung beeinflussen läßt.
Die mit Norethyndron (I), Chlormadinonacetat (XIV) und d-Norgestrel (III)
durchgeführten Experimente haben übereinstimmend ergeben, daß der Progeste-
ron-Spiegel im Blut unter dieser Therapie prompt auf sehr niedrige Werte ab-
sinkt. Die Unterdrückung der lutealen Steroidproduktion ist jedoch nicht
irreversibel; sie kann mit Hilfe von HCG-Injektionen durchbrochen werden.
Insofern ist die von Johansson gebrauchte Bezeichnung eines luteolytischen
Effektes der hochdosierten Gestagenapplikation nicht ganz korrekt. Ob diesem
Phänomen eine Bedeutung für die Empfängnisverhütung zukommt, bleibt abzuwarten.

Bei kontinuierlicher Verabfolgung kleinster Gestagenmengen ("Minipille")
wird der Zyklus nach außen hin im allgemeinen nur unwesentlich beeinflußt.
Er pflegt mindestens in 50% der Fälle ovulatorisch zu bleiben; in dieser
Hinsicht gibt es auch keine Abhängigkeit von der Dauer der Behandlung
(Abb. 6). Unter der Minipille entstandene frische Corpora lutea sind mit
Hilfe der Culdospie bzw. anläßlich von Laparotomien verschiedentlich nachge-
wiesen worden (Gutièrrez-Najar u. Mitarb., 1968).

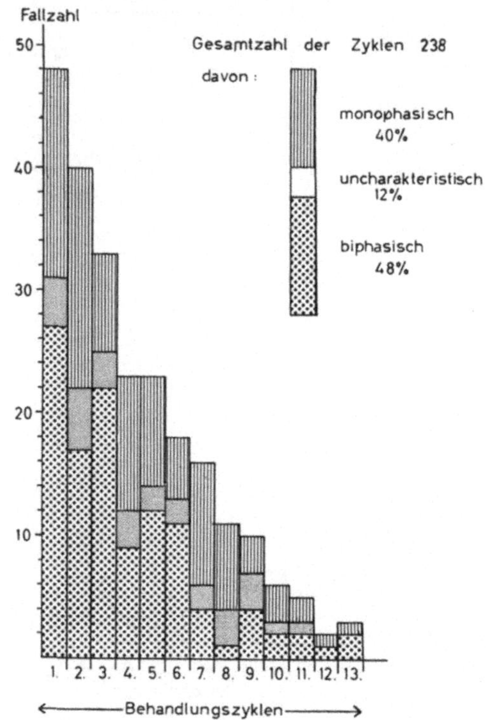


Abb. 6. Aufschlüsselung der Basaltemperaturverläufe von 238 Zyklen unter der
"Minipille" Chlormadinonacetat (XIV) in Abhängigkeit von der Einnahme-
dauer (Hammerstein, 1969 a).


Hormonale Langzeitstudien verschiedener Arbeitskreise haben prinzipiell
zu sehr ähnlichen Resultaten geführt. Bei dem in Abb. 7 gezeigten Fall einer
22jährigen Probandin blieb die Östrogenausscheidung vor, während und nach
der Verabfolgung von 0,3 bzw. 0,5 mg Chlormadinonacetat (XIV) praktisch unverändert; ebensowenig wurde die Ausscheidung der Gesamtgonadotropine im
Harn durch die Gestagenmedikation beeinflußt. Lediglich beim Pregnandiol
fällt eine Abnahme während der beiden Behandlungszyklen auf. Auf mehr oder
weniger ausgeprägte Zeichen einer Corpus luteum-Insuffizienz bei den meisten
Patientinnen unter der "Minipille" haben besonders Fotherby u. Mitarb. (1968)
und Geller u. Scholler (1971) hingewiesen. Dabei gibt es bei gleichzeitiger
beträchtlicher Zunahme der Östrogenausscheidung fließende Übergänge zum
anovulatorischen Zyklus, wie Larsson-Cohn u. Mitarb. (1970) eindrucksvoll
belegen konnten (Abb. 8). Letzteres haben wir besonders ausgeprägt bei
einer 22jährigen Patientin im 6. Behandlungszyklus beobachten können
(Abb. 9). Der Östrogengehalt im 24-Stunden-Harn stieg dabei prämenstruell
auf 180 μg an; das ist der höchste Wert, den wir bisher außerhalb der Gravidität gefunden haben!

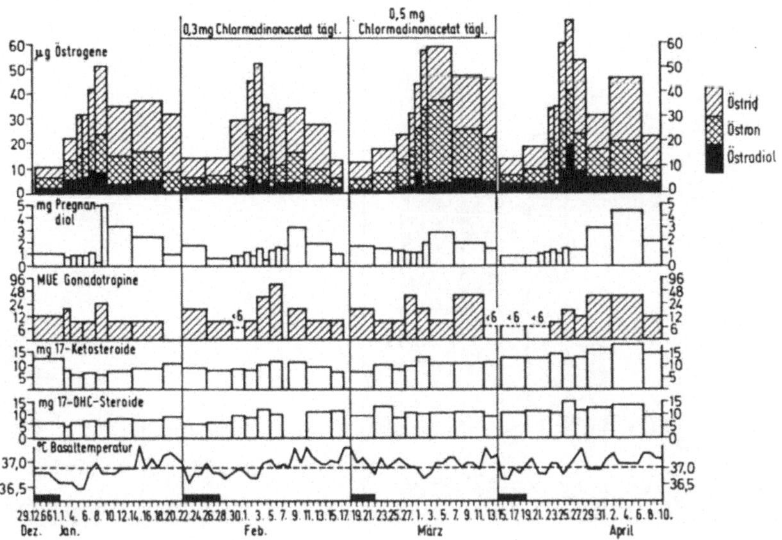

Abb. 7. Ausscheidung von Steroidhormonen und Gonadotropinen vor, während
und nach 0,3 bzw. 0,5 mg Chlormadinonacetat (XIV) täglich bei 22jäh-
riger Probandin (Hammerstein, 1969 a).

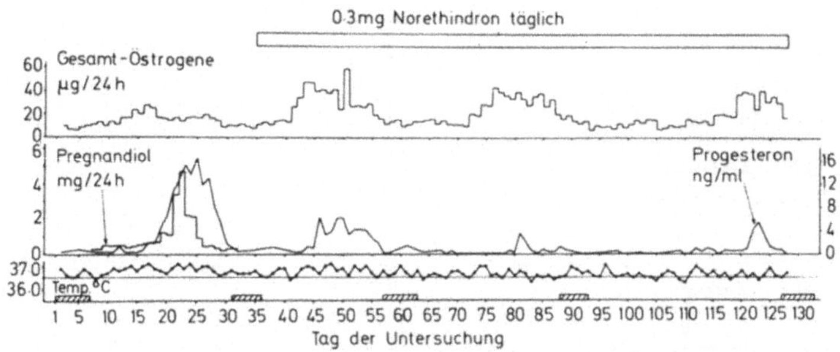

Abb. 8. Östrogenausscheidung und Progesteronspiegel im Plasma vor und während kontinuierlicher Einnahme von 0,3 mg Norethindron (I) täglich
Übergang vom biphasischen zum monophasischen Zyklus bei gleichzeitiger Zunahme der Östrogenwerte im Harn. Die Blutungen sind durch
schraffierte Rechtecke markiert (Larsson-Cohn u. Mitarb., 1970).


Auf eine interessante Beobachtung von Diczfalusy u. Mitarb. (1969) sei abschließend noch hingewiesen: Bei 5 von 6 Frauen verschwand der LH-Ovulations-
anstieg im Harn bei täglicher Einnahme von 0,5 mg Chlormadinonacetat, ohne
daß der für die Gelbkörperphase typische Verlauf der Östrogen- und Pregnandiolausscheidung entscheidend in Mitleidenschaft gezogen worden wäre
(Abb. 10). Der Einfluß von 0,1 mg Norethisteron (I) pro Tag auf die LH-Ausscheidung war weniger ausgeprägt.


C. Kontrazeption mit kleinsten Gestagenmengen ("Minipille")

Nicht nur Gonadotropininkretion und Gelbkörperfunktion, auch Tubenmotilität
und Endometrium haben immer wieder als primärer Angriffspunkt der "Mini-

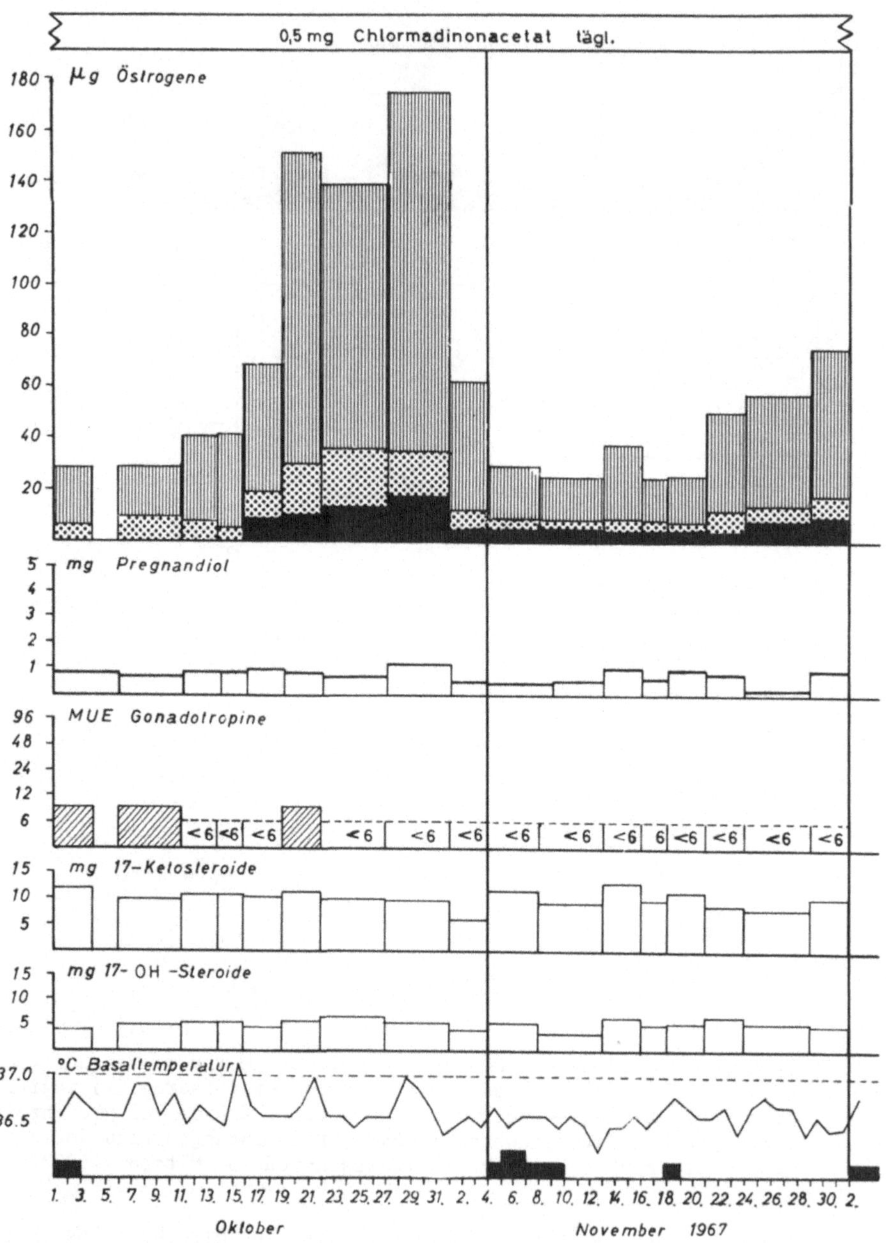



Abb. 9. Harnausscheidung von Steroidhormonen und Gonadotropinen bei 22jähri-
ger Patientin im 6. und 7. Zyklus unter 0,5 mg Chlormadinonacetat
(XIV)-Dauermedikation ("Minipille"): Anovulatorischer Zyklus mit
stark vermehrter prämenstrueller Östrogenausscheidung (Hammerstein,
1969 a).

118

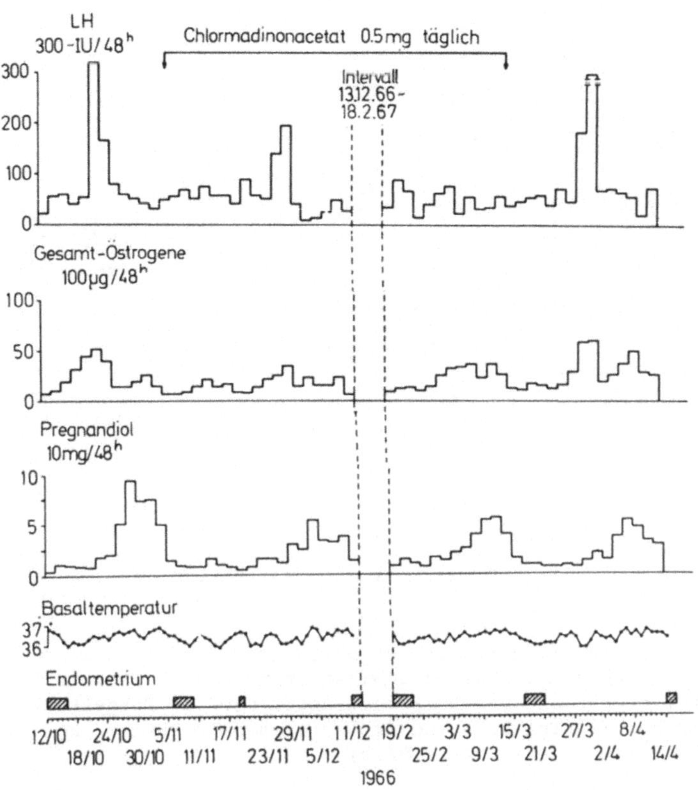


Abb. 10. Normale Ausscheidungsverhältnisse für die Gelbkörperphase bezüglich Östrogenen und Pregnandiol trotz fehlendem Ovulationsanstieg von LH im Harn bei Pat. unter 0,5 mg Chlormadinonacetat (XIV) Dauermedikation ("Minipille") nach Diczfalusy u. Mitarb. (1969).

pille" zur Diskussion gestanden. Vor allem ist aber der Penetrierbarkeit des Cervixschleims für Spermien in diesem Zusammenhang besondere Bedeutung beigemessen worden. Allerdings hat die ursprüngliche Behauptung verschiedener Autoren, daß der Cervixfaktor unter Chlormadinonacetat (XIV) so verändert würde, daß beim Postkoitaltest keine Spermien mehr im Cervixsekret anzutreffen wären, Nachprüfungen nicht standgehalten (Lübke, 1969). Dafür haben aber Kesserü (1969/70) und Roland (1969) unabhängig voneinander gefunden, daß die Sperma-Aszension trotz eines meist intakten Sims-Huhner-Tests durch Norgestrel (III) entscheidend beeinträchtigt wird. Die diesbezüglichen Untersuchungsergebnisse von Kesserü (1969/70) gehen aus Abb. 11 hervor.

Was die kontrazeptive Zuverlässigkeit der "Minipille" anlangt, so haben Martinez-Manautou u. Mitarb. (1967) überzeugend demonstrieren können, daß beim Chlormadinonacetat (XIV) eine geradlinige umgekehrte Korrelation zwischen der Schwangerschaftsrate und der Dosierung im Bereich zwischen 0,1 und 0,25 mg täglich besteht. Trotzdem variieren die Angaben des Pearl-Index bei der üblichen täglichen Dosierung von 0,5 mg Chlormadinonacetat (XIV) in der Weltliteratur zwischen 0,8 und 15%. Unsere eigenen Erfahrungen liegen mit 8% deutlich in der schlechteren Hälfte. Interessant sind in diesem Zusammenhang auch die vergleichenden Untersuchungen über Norethisteron (I), Norgestrel (III), Chlormadinonacetat (XIV) und Megestrolacetat (XII) von Mears u.

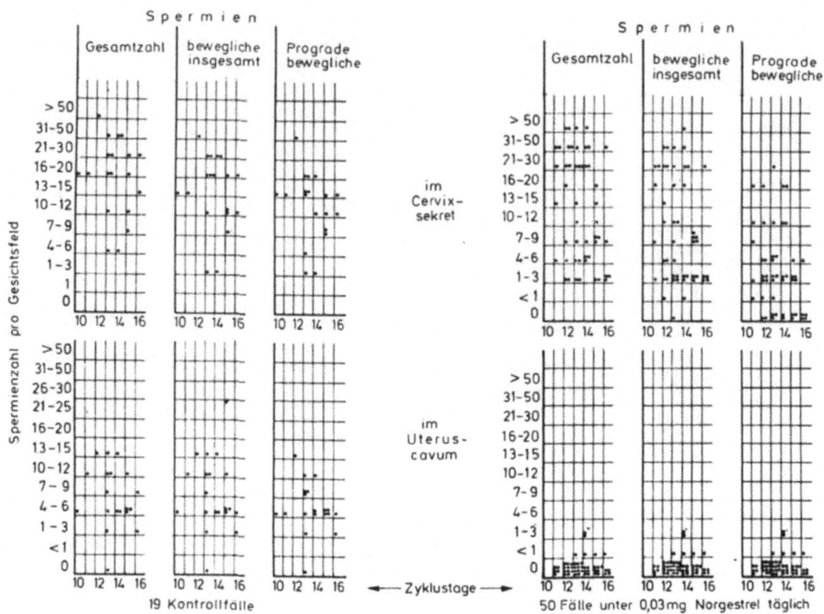

Abb. 11. Nachweis und Differenzierung von Spermien im Cervixsekret und im
Uteruscavumaspirat post cohabitationem. Links 19 Kontrollfälle;
rechts Ergebnisse von 50 Frauen unter 0,03 mg Norgestrel (III).
Unter der "Minipille" sind im Uteruscavum praktisch keine Spermien
nachzuweisen (Kesserü, 1969/70).

Mitarb. (1969). Die in Tab. 6 zusammengestellten Ergebnisse lassen erkennen,
daß Megestrolacetat (XII) zumindest in der angewandten Dosierung mit einer
Schwangerschaftsquote von 50% für die Kontrazeption mit der Minipille undis-
kutabel ist, während Norethisteronacetat (II) am besten abschneidet. Die
Fehlermöglichkeit der kleinen Zahl mahnt jedoch zur Zurückhaltung vor zu
weitgehenden Schlüssen aus diesen vorläufigen Ergebnissen.

Vom klinischen Standpunkt aus ist die Minipille durch das Auftreten grö-
ßerer Zyklustemposchwankungen und häufiger Zwischenblutungen belastet.
Ersteres soll aus Abb. 12 deutlich werden, in der die Zykluslängen von 20
Frauen mit zuvor normal langem und stabilen Zyklus zusammengestellt sind.
Bei Frauen ohne Zwischenblutungen unter der Minipille waren die Abweichungen
von der Norm relativ gering. Dagegen fanden sich erhebliche Zyklustempo-
anomalien bei Patientinnen mit einer oder mehreren Zwischenblutungen unter der
Medikation. Grundsätzlich der gleiche Trend fand sich auch bei Frauen mit
von vornherein unregelmäßigem Zyklus. Was die Zwischenblutungen anlangt, so
soll aus Abb. 13 deutlich werden, daß Metrorrhagien auch noch nach 15 bzw.
16 Behandlungszyklen erstmalig auftreten können; sie sind der häufigste Anlaß zum Abbruch dieser Form der Kontrazeption.

Im übrigen ist die Minipille aber von fast keinen allgemeinen Nebenwirkungen begleitet, wenn man von dem gelegentlichen Auftreten einer Mastodynie ab-
sieht. Darin liegt der hauptsächliche Vorteil dieses kontrazeptiven Verfah-
rens.

## D. Subdermale Implantate von Gestagen-Silastikkapseln

Aufbauend auf der Erkenntnis, daß Steroidhormone durch Silicongummi nach bestimmten Gesetzmäßigkeiten diffundieren können (Dziuk u. Cook, 1966), haben
Segal u. Croxatto 1967 mit Versuchen begonnen, die Versorgung des weiblichen

Tab. 6. Kontrazeptive Wirksamkeit von 4 verschiedenen "Minipillen" (Mears u. Mitarb., 1969).

| | Behandlugsgruppen | | | |
| --- | --- | --- | --- | --- |
| | Norethiste-ronacetat | Norgestrel | Chlormadinon-acetat | Megestrol-acetat |
| Dosierung    mg/pro Tag | 0,3 | 0,05 | 0,5 | 0,25 |
| Zahl der Frauen, die die Behandlung begannen ........ | 41 | 45 | 46 | 43 |
| Zahl der Frauen, die die Behandlung unterbrachen wegen: | | | | |
|   Schwangerschaften ........ | 1 | 3 | 4 | 21 |
|   Zyklusstörungen .......... | 4 | 9 | 13 | 0 |
|   Anderer Nebenwirkungen ... | 5 | 5 | 4 | 1 |
|   Gründe, die nichts mit der Behandlung zu tun haben .. | 12 | 5 | 7 | 7 |
|   Unterbrechung auf ärzt-lichen Rat .............. | 0 | 0 | 0 | 14[+] |
| Zahl der Frauen, die 1 Jahr lang die Behandlung durch-führten ................. | 19 | 23 | 18 | 0 |

[+] Alle nach wenigstens 6 Behandlungsmonaten

Organismus mit kleinsten Gestagenmengen nicht auf oralem Wege sondern durch subkutane Implantate megestrolacetat (XII)-haltiger Silastikkapseln sicher-zustellen.

Die ersten klinischen Berichte über dieses neue kontrazeptive Verfahren klingen durchaus ermutigend (Croxatto u. Mitarb., 1969; Tatum u. Mitarb., 1969; Tejuja, 1970). Für eine zufriedenstellende kontrazeptive Wirkung müs-sen allerdings mehr als nur eine Kapsel implantiert werden; die Empfehlungen liegen zwischen 4–6 Kapseln. Die klinischen Begleiterscheinungen entsprechen erwartungsgemäß denjenigen der "Minipille". Bei 5 implantierten Kapseln ist der Amenorrhoeanteil mit 7% allerdings unerwartet hoch. Wenn sich in der kürzlich bekanntgewordenen Publikation von Coutinho u. Mitarb. (1970) der Hinweis findet, daß die Schwangerschaftsrate ab 9. Monat nach der Implanta-tion steil zunimmt, so paßt das gut zu den Resorptionsstudien am Tier von Benagiano u. Mitarb. (1970), die klar erkennen lassen, daß die Abgabe von Megestrolacetat (XII) aus diesen Kapseln mit der Zeit stetig abnimmt.


## E. Intrauterine und intravaginale Pessare mit Gestagenzusätzen

Angesichts der zunehmenden Gewißheit, daß die Minipille ihre kontrazeptiven Wirkungen über eine Beeinflussung der Spermaaszension ausübt (s. o.), war der Gedanke naheliegend, Gestagene in Silastikkapseln in Verbindung mit einem Intrauterinpessar direkt an ihrem Wirkungsort zu deponieren. Daran knüpften sich drei Erwartungen:
1. Eine Änderung der Endometriumbiologie in kontrazeptivem Sinne ohne syste-mische Nebenwirkungen,
2. eine Senkung der Uteruskontraktibilität und damit eine Verringerung der Ausstoßungsrate der Intrauterinpessare, und
3. eine Verringerung der mit den Intrauterinpessaren verbundenen azyklischen Blutungsepisoden speziell während der ersten Monate.

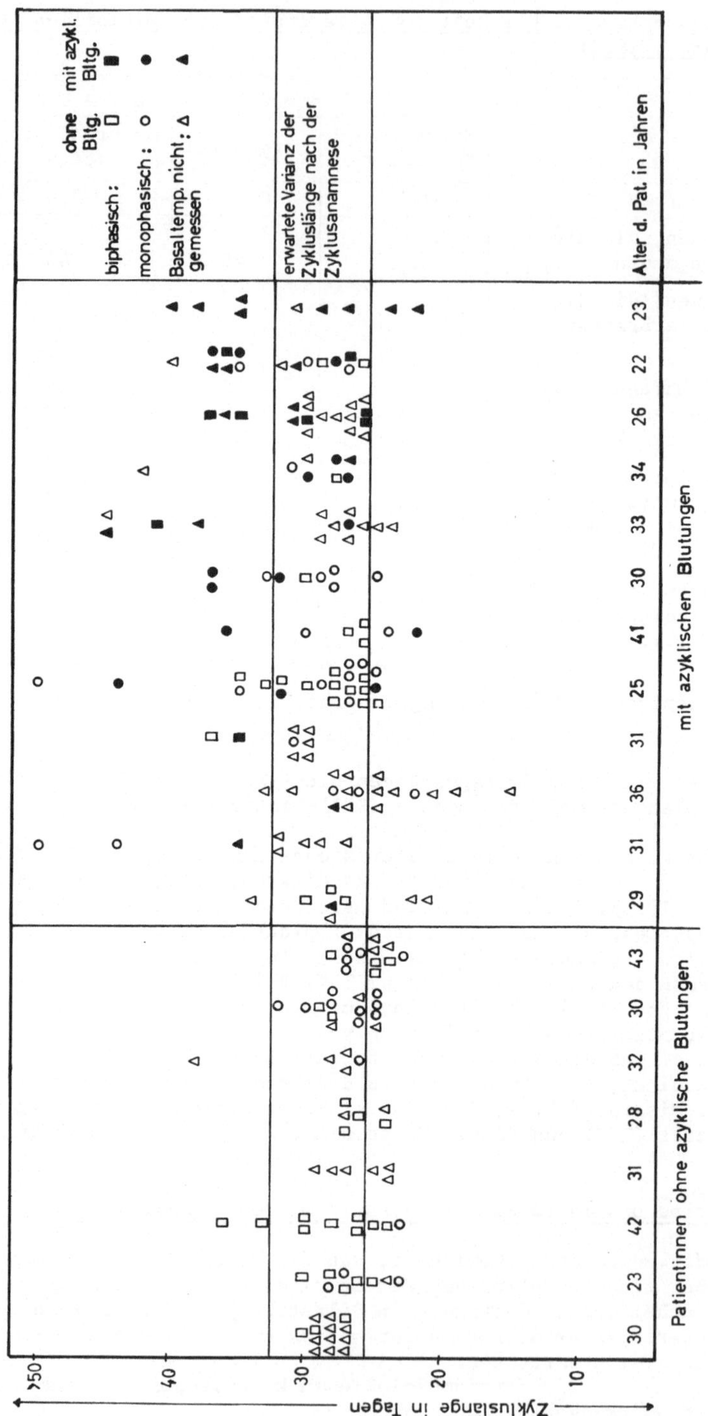


Abb. 12. Varianz der Zykluslänge bei 20 Frauen mit prätherapeutisch stabilem
und normal langem Zyklus unter 0,5 mg Chlormadinonacetat (XIV). Im
linken Teil der Darstellung Pat. ohne Zwischenblutungen, im rechten
solche mit Zwischenblutungen unter der "Minipille" (Hammerstein,
1969 a).

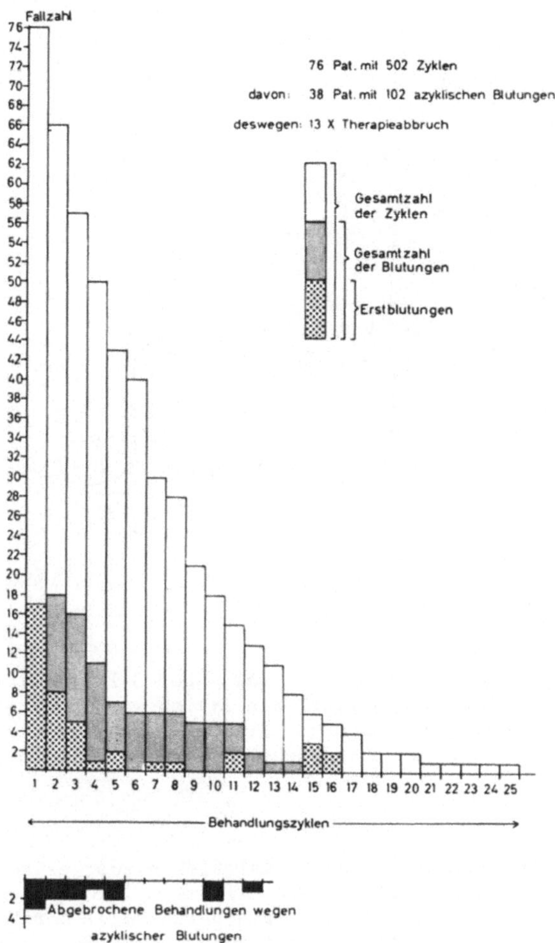


Abb. 13. Häufigkeit von Zwischenblutungen unter 0,5 mg Chlormadinonacetat (XIV)-Dauermedikation ("Minipille") in Abhängigkeit von der Behandlungslänge (Hammerstein, 1969 a).

Ob diese von Scommegna u. Mitarb. (1970) gehegten Hoffnungen in Erfüllung gehen werden, bleibt abzuwarten. Hinsichtlich der Ausstoßungsrate haben die bisherigen Beobachtungen enttäuscht; es konnten keine Unterschiede zwischen Pessarträgerinnen mit und ohne progesteronhaltigen Kapseln festgestellt werden.

Ebenfalls noch im Entwicklungsstadium befinden sich gestagenhaltige Silastikkapseln, die durch eine besondere Vorrichtung in der Cervix uteri gehalten werden (Cohen u. Mitarb., 1970).

Schließlich sind in jüngster Zeit Silastikringe mit Medroxyprogesteronacetat (XIII)-Füllung entwickelt worden, die in die Vagina wie übliche Ringpessare eingeführt werden und dort laufend kleine Progesteronmengen freigeben (Mishell u. Mitarb., 1970); die Scheidenschleimhaut eignet sich bekanntlich besonders gut als Resorptionsorgan. Wie Abb. 14 zeigt, üben die resor-

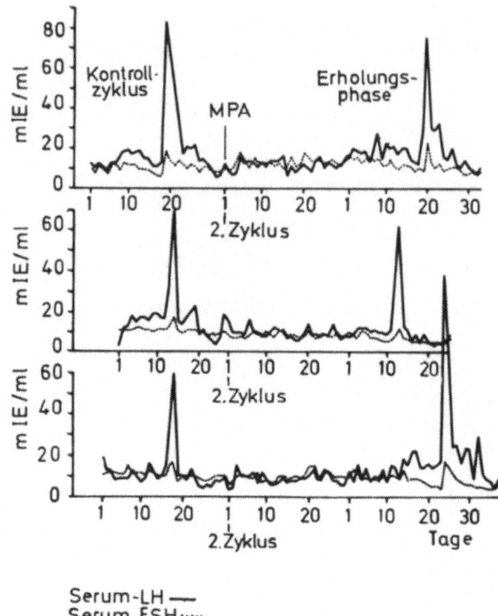


Abb. 14. Tägliche LH- und FSH-Spiegel im Serum bei 3 Probandinnen jeweils
während 3 Zyklen. In den mittleren Zyklen war ein vaginaler Silas-
tikring mit Medroxyprogesteron (XIII)-Füllung eingelegt; die Ovulationen wurden dadurch unterdrückt (Mishell u. Mitarb., 1970).


bierten Gestagene dabei keineswegs nur Lokaleffekte aus; es kommt bei der
untersuchten Ringart auch zu systemischen Allgemeinwirkungen und hier speziell zur Unterdrückung des Zyklus und zur Hyperthermie. Der Vorteil dieses
Modells besteht in seiner einfachen Einbringung und seiner leichten Entfernung durch die Patientin selbst zu jedem beliebigen Zeitpunkt.


## F. Gestagen-Depotinjektionen ("Dreimonatsspritze")

Im Gegensatz zur "Minipille" werden Ovulationen nach Injektion von 150 mg
Medroxyprogesteronacetat (XIII) mindestens für die Dauer von 3 Monaten zuverlässig unterdrückt. Frische Gelbkörper konnten bei keiner von 27 Frauen,
die sich in diesem Zeitraum einer Laparotomie unterziehen mußten, nachgewiesen werden (Zanartu u. Mitarb., 1970). Immerhin verläuft die Follikelreifung
in solchen Ovarien bis zum Graaf'schen Tertiärstadium ungestört. Auch Theka-
luteinisierungen können angetroffen werden. Der aus diesen histologischen
Befunden gezogene Schluß, daß die FSH- und LH-Ausschüttung durch die "Dreimonatsspritze" nicht entscheidend gehemmt würde, fand seine volle Bestätigung
durch Stichprobenbestimmungen des FSH- und LH-Spiegels im Plasma auf radio-
immunologischem Wege durch Goldzieher u. Mitarb. (1970) (Abb. 15 a und 15 b).
Es hat danach den Anschein, als ob die Basalsekretion der beiden Gonadotropine durch das Depotgestagen kaum modifiziert wird, und daß als einzige Abwei-
chung der Ovulationsanstieg der beiden Hormone ausbleibt. Hierauf hatte
Mishell schon 1967 aufgrund immunologischer LH-Bestimmungen im Harn aufmerksam gemacht.

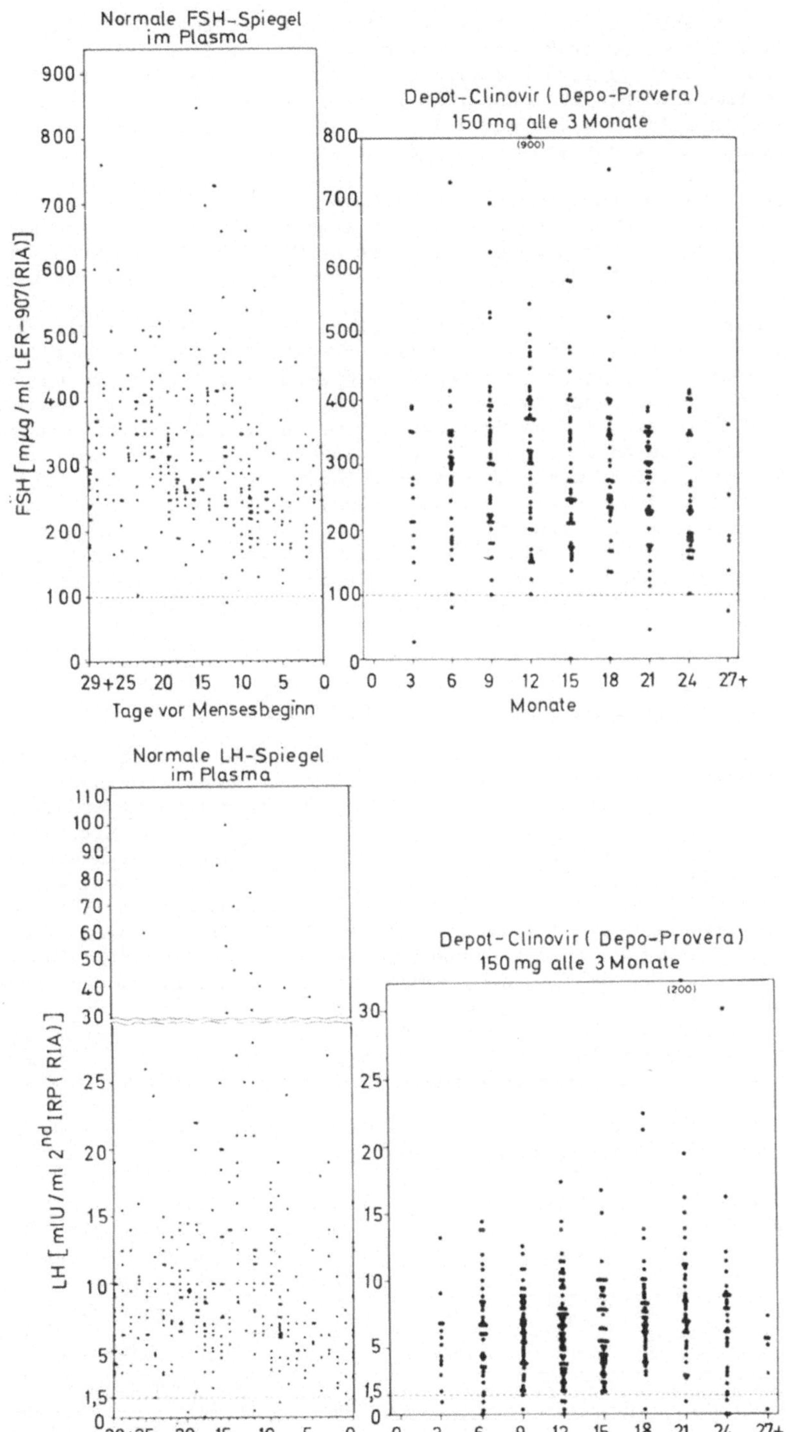



Abb. 15. Stichprobenbestimmungen von FSH (Abb. 15 a) und LH (Abb. 15 b) im Plasma unter der "Dreimonatsspritze" mit Medroxyprogesteronacetat (XIII): Keine wesentlichen Unterschiede zu den Befunden bei unbehandelten Patientinnen; es fehlen lediglich LH-Werte wie zur Zeit der Ovulation (Goldzieher u. Mitarb., 1970).

Klinisch bieten die Injektionen von Medroxyprogesteronacetat in 3monati-
gen Intervallen einen praktisch 100%igen Schutz vor Schwangerschaft. Die
Frauen müssen dafür allerdings eine völlige Regellosigkeit ihrer Blutungen
in Kauf nehmen. 10 hintereinander folgende Blutungen können in einem Zeitraum zwischen 5 1/2 Monaten und 1 1/2 Jahren erfolgen, wie Taylor u. Mitarb.
(1970) kürzlich berichteten. Es ist nicht anzunehmen, daß sich unter diesen
Umständen viele Frauen für die "Dreimonatsspritze" entscheiden werden, auch
wenn Nebenwirkungen selten sind. Zu berücksichtigen ist schließlich, daß bis
zur Wiederherstellung von ovulatorischen Zyklusverhältnissen und Fertilität
viele Monate verstreichen können (Gardner u. Mishell, 1970), worauf bei der
Besprechung der Depotwirkung von Medroxyprogesteronacetat bereits hingewiesen wurde (Abb. 1).

Ein anderes, noch in Erprobung befindliches Depot-Präparat mit Dreimonats-
wirkung, das Norethisteron-Önanthat (XX) sei abschließend erwähnt, weil
es sich in mancher Hinsicht vom Medroxyprogesteronacetat (XIII) unterscheidet. So ist die kontrazeptive Sicherheit mit einem Pearl-Index von 1,58%
nicht so günstig wie die des Medroxyprogesteronacetats. Dafür kommt es nicht
in gleichem Ausmaß zu Blutungsstörungen. Auch ist die Fertilität nach Ablauf
der drei Monate sofort wiederhergestellt (Larrañaga u. Kesserü, 1967). Interessanterweise fanden Achari u. Prasad (1969) anläßlich von Laparotomien, die
2-3 Monate nach der letzten Injektion bei 40 Frauen durchgeführt wurden,
34mal frische Gelbkörper, die sich allerdings durch ihre geringe Größe deutlich von normalen Corpora lutea unterschieden. Norethisteronönanthat hat im
übrigen Wirkungen auf die Plasmaproteine und das Gerinnungssystem (Gilfrich
u. a., 1970).


## G. Postcoitale Gestagenmedikation ("Pille danach")

Offenbar ist ein kontinuierlicher Gestagenstrom, wie er durch die "Minipille",
durch subkutane gestagengefüllte Silastikkapseln, Gestagenpessare bzw. Depot-
gestagene aufrechterhalten wird, zur Kontrazeption gar nicht erforderlich.
Für einen sicheren Empfängnisschutz genügt es nach Rubio u. Mitarb. (1970)
schon, wenn die Frau innerhalb der ersten 24 postcoitalen Stunden ein kurz-
wirkendes, niedrig dosiertes Gestagen einnimmt. Von 200 Frauen, die sich im
Verlauf von 1004 Zyklen regelmäßig 0,8 mg Quingestanolacetat (IX) - das ist
etwa die dreifache Dosis der entsprechenden "Minipille" - postcoital zuführ-
ten, wurde nicht eine einzige schwanger. Bei Reduktion der Einzeldosis auf
0,5 mg stieg der Pearl-Index aber schon auf 6,5% an. Einzeldosen unter 0,4 mg
übten praktisch keinen Empfängnisschutz mehr aus (Tab. 7).
Die Verträglichkeit dieses neuesten aller kontrazeptiven Verfahren war im
Hinblick auf Zwischenblutungen und Hypermenorrhoen verglichen mit der "Minipille" erstaunlich gut. Mit zunehmender Dauer der Einnahme von 0,8 mg Quin-

Tab. 7. Kontrazeptive Wirksamkeit von Quingestanolacetat (IX) als "Pille
danach" in Abhängigkeit von der Dosierung (Rubio u. Mitarb., 1970).

| Dosierung | Zahl der Schwan-gerschaften | Zahl der Zyklen | Schwangerschaftsquote pro 100 Frauen-jahre (Pearl-Index) mit Angabe der 95%igen Vertrauensgrenze |
|---|---|---|---|
| 200 | 7 | 50 | 168.0 (69.8-320.9) |
| 300 | 3 | 100 | 36.0 (7.5-102.2) |
| 400 | 1 | 72 | 16.6 (0.4-90.0) |
| 500 | 5 | 927 | 6.5 (2.1-15.0) |
| 750 | 0 | 28 | 0 (0-121.8) |
| 800 | 0 | 1004 | 0 (0-3.6) |

gestanolacetat sank die Zwischenblutungsrate von anfangs 20% auf 10,6% im
dritten Einnahmezyklus ab. Gleichzeitig nahm die Kohabitationsfrequenz und
damit die Zahl der Einzeldosen pro Zyklus deutlich zu. Man wird weitere Berichte abwarten müssen, ehe man sich über dieses neueste Prinzip der hormonalen Kontrazeption mit Gestagenen ein zuverlässiges Bild machen kann. Von
Interesse ist insbesondere, ob sich auch andere Gestagene für diesen Zweck
eignen.


## Schlußbemerkungen

Diese Übersicht über die jüngsten Entwicklungen auf dem Gebiet der Gestagentherapie dürfte hinreichend deutlich gemacht haben, daß in den letzten Jahren vieles - speziell im Bereich der hormonalen Kontrazeption - wieder in
Bewegung geraten ist. Während vor 12 Jahren auf dem Symposion in Kiel noch
der Eindruck entstehen konnte, der Indikationskatalog für Gestagene sei
praktisch abgeschlossen und neue Impulse könnten bestenfalls von der Entwicklung besser verträglicher und noch wirksamerer Präparate ausgehen, darf
man heute sicher sein, daß die Zukunft auf diesem Gebiet noch manche Überra-
schung bereit hält.

Es ist zu hoffen, daß die den Hundeexperimenten voreilig zum Opfer gefal-
lenen Gestagene möglichst bald wieder rehabilitiert werden, und daß wir -
besser als bisher - lernen, die Gestagene differenziert einzusetzen. Auch
bleibt noch vieles zu tun, bis wir über die Nebenwirkungen dieser Hormone,
über ihre Einflüsse auf Haut, Bindegewebe, Skelet, Gerinnungssystem, Gefäß-
apparat und Psyche sowie über ihre immunologischen Implikationen hinreichend informiert sind. Es besteht also aller Grund, daß den Gestagenen das
wissenschaftliche Interesse erhalten bleibt.


## Literatur


Achari, K., Prasad, V.: Bihar Obstet. u. Gynaec. Soc., 2nd Ann. Conf. Patna
    Oct. 1969.
Apostolakis, M., Becker, H., Voigt, K.D.: Steroids $\underline{7}$, 146 (1966).
Benagniano, G., Ermini, M., Chang, C.C., Sundaram, $\overline{K}$., Kincl, F.A.: Acta
    endocr. (Kbh.) $\underline{63}$, 29 (1970).
Benjamin, F., Casper, D.J.: Amer. J. Obstet. Gynec. $\underline{94}$, 991 (1966).
Berndt, G., Stender, H. St.: Dtsch. Med. Wschr. $\underline{95}$, $\overline{2399}$ (1970).
Bradley, W.G., Hudgson, P., Foster, J.B., Newell, D.J.: Brit. Med. J.$\underline{1968\ III}$,
    531.
Breuer, H.: Lancet $\underline{1970\ II}$, 615.
Camanni, F., Massara, F., Molinatti, G.M.: Acta endocr. (Kbh.) $\underline{43}$, 477
    (1963).
Clinch, J., Tindall, V.R.: Brit. Med. J. $\underline{1969\ I}$, 602.
Cohen, M.R., Pandya, G.N., Scommegna, A.: Fertil. and Steril. $\underline{21}$, 715 (1970).
Collip, P.J., Kaplan, S.A., Boyle, D.C., Plachte, F., Kogut, M.D.: Amer. J.
    Dis. Child. $\underline{108}$, 399 (1964).
Coutinho, E.M., Mattos, C.E.T., Sant'Anna, A.R., Filho, I.A., Silva, M.C.,
    Tatum, H.J.: Contracept. $\underline{2}$, 313 (1970).
Croxatto, H., Diaz, S., Vera, R., Etchart, M., Atria, P.: Amer. J. Obstet.
    Gynec. $\underline{105}$, 1135 (1969).
Dapunt, O.: Bull. Schweiz. Akad. med. Wiss. $\underline{25}$, 481 (1970).
Diczfalusy, E., Goebelsmann, U., Johannisson, E., Tillinger, K.G., Wide, L.:
    Acta endocr. (Kbh.) $\underline{62}$, 679 (1969).
Dziuk, P.J., Cook, B.: Endocrinology $\underline{78}$, 208 (1966).
Eisalo, A., Heino, A., Räsänen, V.: Acta obstet. gynec. scand. $\underline{47}$, 58 (1968).
Fotherby, K., Svendsen, K.E., Foss, G.L.: J. Reprod. Fertil. Suppl. $\underline{5}$, 155
    (1968).

Gardner, J.M., Mishell, D.R.: Fertil. and Steril. 21, 286 (1970).
Geller, J., Fruchtman, B., Meyer, C., Newman, H.: J. clin. Endocr. 27, 556 (1967).
Geller, S., Scholler, R.: Path. et Biol. 19, 157 (1971).
Gilfrich, H.J., Nieschlag, E., Duceck, J., Overzier, C.: Dtsch. Med. Wschr. 94, 2473 (1969).
Goldzieher, J.W., Kleber, J.W., Moses, L.E., Rathmacher, R.P.: Contracept. 2, 225 (1970).
Grant, E., Pryse-Davies, J.: Brit. Med. J. 1968 III, 777.
Greenblatt, R.B.: Med. Sci. 37 (1967).
Gutièrrez-Najar, A., Márquez-Monter, H., Cortéz-Gallegos, V., Giner-Veláz-quez, J., Martinez-Manautou, J.: Amer. J. Obstet. Gynec. 102, 1018 (1968).
Hammerstein, J.: Arch. Gynäk. 207, 48 (1969 a).
- 15. Symp. Dtsch. Ges. Endokrin. 15, 57, 1969 b.
- Symp. der "Beratergruppe für Arzneimittelsicherheit beim Bundesminister für Jugend, Familie und Gesundheit", Berlin, Dez. 1970 (im Druck).
- Cupceancu, B.: Dtsch. Med. Wschr. 94, 829 (1969).
- Rice, B.F., Savard, K.: J. clin. Endocr. 24, 597 (1964).
Helge, H., Weber, B., Hammerstein, J., Neumann, F.: Acta paediat. scand. 58, 672 (1969) und (unveröffentlicht) 1971.
Hoischen, W., Steigleder, G.K.: Dtsch. med. Wschr. 91, 398 (1966).
Holzmann, H., Korting, G.W., Morsches, B.: Hautarzt 16, 456 (1965).
Johansson, E.D.B.: Excerpta med. (Amst.) Intern. Congr. Ser. 210, 215 (1970).
Kaiser, R.: Symp. Dtsch. Ges. Endokrin. 6, 64 (1959).
Kaufmann, C., Zander, J.: Klin. Wschr. 34, 7 (1956).
Kenny, F.M., Midgley, A.R. jr., Jaffe, R.B., Garces, L.Y., Vasquez, A., Taylor, F.H.: J. clin. Endocr. 29, 1272 (1969).
Kesserü, E.: II. Congreso Bolivariano de Endocrinologia 1969. Symposia Lito-grafica del Peru S.A. Lima S. 120, 1970.
Kreuzer, G., Boquoi, E.: Dtsch. med. Wschr. 95, 1863 (1970).
Kupperman, H.S., Epstein, J.A.: J. clin. Endocr. 22, 456 (1962).
Larrañaga, A., Kesserü, E.: Ginec. y Obstet. 8, 49 (1967).
Larsson-Cohn, U., Johansson, E.D.B., Gemzell, C.: Acta endocr. (Kbh.) 64, 38 (1970).
Luby, R.J.: Obstet. and Gynec. 26, 467 (1965).
Ludwig, H.: Ergebn. Angiologie 4, 81 (1970).
Lübke, F.: Arch. Gynäk. 207, 49 (1969).
Magré, J., Leroux, P., Guimbretiere, J., Cadudal: Bull. Féd. Soc. Gynéc. Obstét. franç. 21, 60 (1969).
Martinez-Manautou, J., Giner-Velasquez, J., Rudel, H.: Fertil. and Steril. 18, 57 (1967).
Mathews, J.H., Abrams, C.A.Ö., Morishima, A.: J. clin Endocr. 30, 653 (1970).
Mazaheri, A., Fotherby, K., Chapman, J.R.: J. Endocr. 47, 251 (1970).
Mears, E., Vessey, M.P., Andolšek, L., Oven, A.: Brit. Med. J. 1969 II, 730.
Mishell, D.R.: Amer. J. Obstet. Gynec. 99, 86 (1967).
- Talas, M., Parlow, A.F., Moyer, D.L.: Amer. J. Obstet. Gynec. 107, 100 (1970).
Napp, J.H., Rothe, A.: Dtsch. med. Wschr. 83, 325 (1958).
Neumann, F.: "Handbuch d. experimentellen Pharmakologie" Bd. XXII "Die Gesta-gene" Teil 1, 680. Berlin-Heidelberg-New York: Springer, 1968.
Østergaard, E., Starup, J.: Acta endocr. (Kbh.) 57, 386 (1968).
Piver, M.S., Bolognese, R.J., Feldman, J.D.: Amer. J. Obstet. Gynec. 97, 579 (1967).
Plotz, J.: In: Brook Lodge Symp. Progesterone. Hrsg. A.C. Barnes, Augusta/Mich.: Brook Lodge Press, 1961.
Poller, L., Thomson, J.M., Tabiowo, A., Priest, C.M.: Brit. Med. J. 1969 I, 554.

Rifkind, A.B., Kulin, H.E., Cargille, C.M., Rayford, P.C., Ross, G.T.: J.
    clin. Endocr. 29, 506 (1969).
Rivarola, M.A., Camacho, A.M., Migeon, C.J.: J. clin. Endocr. 28, 679 (1968).
Rock, J., Pincus, G., Garcia, C.R.: Science 124, 891 (1956).
Roland, J.: Bull. N.Y. Acad. Med. 45, 681 (1969).
Rubio, B., Berman, E., Larrañaga, A., Guiloff, E., Aguirre, J.J.: Contra-
    cept. 1, 303 (1970).
Rust, W.: Zbl. Gynäk. 78, 1363 (1956).
Scommegna, A., Pandya, G.N., Christ, M., Lee, A.W., Cohen, M.R.: Fertil. and
    Steril. 21, 201 (1970).
Segal, S.J., Croxatto, H.: Vortrag, 27. Meeting of the Amer. Fertil. Soc.,
    Washington D.C., April 1967.
Shelley, W.B., Preucel, R.W., Spoont, S.S.: J. Amer. med. Ass. 190, 35 (1964).
Somayaji, B.N., Paton, A., Price, J.H., Harris, A.B., Flewett, T.H.: Brit.
    med. J. 1968 II, 281.
Spellacy, W.N., McLeod, A.G.W., Buhi, W.C., Birk, S.A., McCreary, S.A.:
    Fertil. and Steril. 21, 457 (1970).
Stamm, O., Gerhard, I., Zarro, D.: Bull. schweiz. Akad. med. Wiss. 25, 472
    (1970).
Stoll, B.A.: Brit. med. J. 1969 II, 293.
Svanborg, A., Vikrot, O.: Acta med. scand. 179, 615 (1966).
Swerdloff, R.S., Odell, W.D.: J. clin. Endocr. 29, 157 (1969).
Tatum, H.J., Coutinho, E.M., Filho, J.A., Sant'Anna, A.R.: Amer. J. Obstet.
    Gynec. 105, 1139 (1969).
Tejuja, S.: Amer. J. Obstet. Gynec. 107, 954 (1970).
Teller, W.M., Mürset, G., Schellong, G.: Acta paediat. scand. 58, 385 (1969).
Turner, S.J.: Obstet. and Gynec. 24, 218 (1964).
Tyler, E.T., Levin, M., Elliot, J., Dolman, H.: Fertil. and Steril. 21, 469
    (1970).
Vorys, N., Ullery, J.C., Stevens, V.: Amer. J. Obstet. Gynec. 93, 641 (1965).
Whitty, C.W.M., Hockeday, J.M., Whitty, M.M.: Lancet 1966 I, 856.
Zañartu, J., Pupkin, M., Rosenberg, D., Davansens, A., Guerrero, R.,
    Rodriguez-Bravo, R., Garcia-Hudobro, M.: Fertil. and Steril. 21, 525
    (1970).

Symp. Dtsch. Ges. Endokrin. _17_, 131-146 (1971)
© by Springer-Verlag

# Podiumsgespräch: Nebenwirkungen der Gestagentherapie

## Panel Discussion: Side-Effects of Gestagen Therapy

<u>Moderator</u>:  F. Neumann

<u>Teilnehmer</u>: G. Bettendorf, K. Dämmrich, J. Hammerstein, F. Neumann, G.A. Overbeek, H.-D. Taubert, K.D. Voigt, J. Zander

Mit 6 Abbildungen[+]

<u>Neumann</u>:  Meine Damen und Herren, ich eröffne das Rundtischgespräch über Nebenwirkungen der Gestagentherapie. Diese Thematik ist deshalb nicht ganz einfach abzuhandeln, weil in den meisten Untersuchungen und Statistiken die Nebenwirkungen von Kombinationspräparaten diskutiert werden. Es gibt nur relativ wenige Untersuchungen über Nebenwirkungen der reinen Gestagentherapie. Heute enthält diese Thematik eine gewisse Brisanz, bedingt durch die Zurückziehung des Chlormadinonacetats und der Chlormadinonacetathaltigen Präparate sowie einiger anderer Gestagene oder Gestagenpräparate. Als von dem Präsidenten unserer Gesellschaft, Herrn Prof. Tamm, dieses Thema ausgewählt wurde, war diese Entwicklung noch nicht abzusehen. Wir haben uns deshalb entschlossen, auch kurz auf die Befunde einzugehen, die zur Zurückziehung des Chlormadinonacetats geführt haben. Es gab Ende 1970 zwei Tagungen, auf denen über diese Thematik diskutiert wurde:
1. ein Symposium in Berlin über "Nebenwirkungen kontrazeptiver Steroide", das übrigens vom Bundesgesundheitsministerium finanziert wurde. Schirmherren dieser Veranstaltung waren Prof. Herken und Prof. Kewitz;
2. eine Tagung in Bonn. Das Thema dieser Tagung lautete: "Ist die zur Zeit geübte Methodik der Steroidtoxikologie für Forschung und klinische Anwendung der Steroide vertretbar?" Moderator dieser Tagung war Prof. Plotz.
Die Vorträge und Diskussionen beider Veranstaltungen werden übrigens publiziert. Aber nicht zuletzt weil der Teilnehmerkreis auf beiden Tagungen sehr klein war, hielten wir es für sinnvoll, ja sogar für notwendig, auch heute noch einmal kurz auf die - ich bitte das in Anführungsstrichen zu lesen - "Chlormadinonaffäre" und vielleicht kurz auf die tumorinduzierende Wirkung von Sexualhormonen überhaupt einzugehen. Es erscheint mir dabei nicht so sehr sinnvoll, einzelne Substanzen oder Präparate isoliert zu betrachten.

Ich möchte Ihnen nun die "Hauptprogrammpunkte" nennen und schlage vor, daß wir die einzelnen Punkte auch in der angegebenen Reihenfolge diskutieren:
1. Versuch einer Definition des Begriffes "Gestagen" vom Standpunkt des Experimentalendokrinologen und des Klinikers.

---

[+] Halbtonbilder s. Anhang S. 193

2. Toxikologische Befunde nach Langzeitbehandlung mit Gestagenen und eventuell auch mit Östrogenen. Zum Verständnis des
   Ganzen müssen wir auch kurz auf die Östrogene eingehen.
3. Klinische Nebenwirkungen der reinen Gestagentherapie und Nebenwirkungen der oralen Kontrazeptiva bzw. der Kombinationspräparate.
Nun zum ersten Programmpunkt: "Versuch einer Definition des Begriffes 'Gestagen'". Wir haben heute schon von Herrn Overbeek
gehört, wie schwierig das werden dürfte. Ich möchte darum vielleicht zuerst Herrn Overbeek fragen.

Overbeek: Ja, Herr Vorsitzender, ich habe eigentlich keine besonderen
Schwierigkeiten mit dem Begriff "gestagen", und zwar deshalb
nicht, weil man heutzutage eigentlich nur von "Gestagenen"
spricht. Wenn man alles gestagen nennt, dann kann man sagen,
wie ich es auch schon in meinem Vortrag getan habe, daß der
Tierversuch, insbesondere der Clauberg-Versuch, entscheidend
ist und daß man sagen kann, wenn eine Substanz im Clauberg-
Versuch aktiv ist, ist sie ein Gestagen. Ich möchte hinzufügen,
daß ich eigentlich bedauere, daß das Wort "gestagen" jetzt immer verwendet wird, denn das möchte ich eigentlich lieber re-
servieren für Substanzen, die wirklich die Schwangerschaft erhalten. Obwohl ich mir sehr wohl bewußt bin, daß sich hieraus
eine neue Schwierigkeit ergibt, z. B. daß eine Substanz gestagen sein kann in einer Tierart, aber nicht in einer anderen.
Unsere Gruppe hat versucht, zwischen Prägestagenen, welche nur
die Schwangerschaft vorbereiten, und Gestagenen, welche die
Schwangerschaft erhalten, zu unterscheiden. Diese Nomenklatur
wurde jedoch nicht anerkannt. Vielleicht ist es auch besser,
nicht weiter darauf zurückzukommen.

Neumann: Ich stimme dem im Prinzip zu, nur gilt dies immer nur für eine
Spezies. Es gibt nämlich Substanzen (Norethinodrel), die bei
der Ratte die Schwangerschaft nicht erhalten. Man würde diese
Verbindung als Prägestagen einstufen. Beim Hamster erhält aber
die gleiche Substanz die Schwangerschaft; für den Hamster gilt
also diese Klassifizierung nicht. Es muß hier vielleicht überhaupt klargestellt werden, daß das Wirkungsspektrum eines Gestagens für jede Spezies anders aussieht.

Bettendorf: Diese Definition ist für den Menschen gar nicht möglich, da
sich die schwangerschaftserhaltende Wirkung nicht beweisen läßt.
Wir müssen also etwas anderes suchen, und da bietet sich das an,
was wir am einfachsten beurteilen können, nämlich die Reaktion
am Endometrium. Ich meine, daß man klinisch ein Gestagen als
eine Substanz definieren sollte, die eine sekretorische Umwandlung des östrogenvorbehandelten Endometriums bewirkt.

Zander: Ja, ich würde auch sagen, entscheidend ist die Wirkung am Endometrium. Man sollte bei dieser spezifischen Definition ausgehen
von dem natürlichen Progesteron. Im übrigen fällt mir bei dem,
was jetzt als gestagen bezeichnet wird, nicht ein, wie man das
gemeinsam definieren könnte; ich weiß auch gar nicht, ob das so
wichtig ist. Sie heißen nun mal Gestagene und sind so eingefah-
ren in der Welt, und wir wissen, was damit gemeint ist. Dabei
sollte man es vielleicht lassen.

Neumann: Das war vielleicht auch der Sinn dieser Frage. Man muß sich nur
im klaren darüber sein, daß die Benutzung des Begriffes "Gestagen" nur eine Hilfskonstruktion darstellt. Übrigens hat ein Re-
ferent, der eine unserer Arbeiten in der "Acta Endocrinologica"
zu beurteilen hatte, vorgeschlagen, den Passus "Wirkungen verschiedener Gestagene" zu ersetzen durch "Wirkungen verschiede-

ner Steroide mit bekannter Gestagenwirkung"; zu Recht, wie wir
meinen. Noch eine Frage tauchte hier auf, die von Herrn Taubert
aufgeworfen worden ist.

Taubert: In Bonn hatte Herr Breuer darauf hingewiesen, daß man aufgrund
neuerer Untersuchungen nicht mehr annimmt, daß gewisse Gestagene, z. B. Norethisteronacetat, im Organismus in Östrogene umgewandelt werden können. Muß man damit den Begriff der "östro-
genogenen" Gestagene aufgeben?

Hammerstein: Sicherlich nicht. Daß manche Gestagene östrogene Partialwirkungen bei Mensch und Tier entfalten, läßt sich schwerlich abstrei-
ten. Hierfür nur 2 Beispiele. Im Gegensatz zu fast allen anderen Gestagenpräparaten führt Norethinodrel nicht zu den entsprechenden Veränderungen des Cervixschleims, wie wir sie zur Zeit
der Gestagendominanz in der Gelbkörperphase anzutreffen gewohnt
sind. Es finden sich nach dieser Substanz reichlich Spermien
beim Postcoitaltest im Cervixsekret.

Laumas u. Mitarb. haben kürzlich gefunden, daß die Mutter-
milch von Frauen nach Verabfolgung von 19-Norandrostan-Deriva-
ten biologisch nachweisbare Östrogenaktivitäten enthält, nicht
dagegen nach Gabe von Megestrolacetat. Es geht also bei der von
Herrn Neumann aufgeworfenen Frage nicht so sehr darum, ob bestimmte Gestagene überhaupt östrogen wirksam sein können, sondern lediglich um die richtige Interpretation. Nachdem sich
durch die zitierten Untersuchungen von H. Breuer herausgestellt
hat, daß die seit 10 Jahren gültigen Vorstellungen der partiel-
len Metabolisierung der 19-Nor-Androstanderivate in Östrogene
nicht mehr länger haltbar ist, rücken zwei auch schon früher
diskutierte Deutungsmöglichkeiten wieder in den Vordergrund:
1. daß diese Gestagene eine dem Molekül selbst innewohnende
Östrogenaktivität besitzen und 2. daß bei ihrer großtechnischen
Herstellung östrogene Nebenprodukte nicht vollständig entfernt
werden. Letzteres ist z. B. vom Norethynodrel bekannt. Ob das
gleiche auch für andere Gestagene gilt, bleibt fraglich. Insgesamt hat die Vorstellung von einer inhärenten Östrogenpartialwirkung einiger Gestagene mehr für sich.

Neumann: Ich bin der gleichen Meinung. Es ist nämlich gleichgültig, ob
die östrogene Wirkung von Gestagenen über eine Verstoffwechse-
lung in Östrogene zustande kommt oder ob es sich um eine inhärente östrogene Wirkung handelt. Entscheidend ist der biologische Effekt.

Overbeek: Wegen der möglichen Beimischung von östrogenen Substanzen, also
einer Verunreinigung, habe ich oft unsere Chemiker gefragt, ob
das heutzutage noch möglich sei, und sie haben mir immer gesagt,
dies sei bestimmt nicht der Fall. Ich glaube also, man muß immer annehmen, daß es sich doch um eine inhärente Aktivität der
Gestagene handelt.

Ich bin nicht ganz mit Herrn Neumann einverstanden, daß es
nicht wichtig sein würde zu wissen, ob die aktive Substanz ein
Metabolit ist oder nicht. Wenn es ein Metabolit wäre, dann muß
man auch damit rechnen, daß es unter bestimmten Umständen, z.B.
wenn die Leber nicht richtig funktioniert, zu einer Beeinträchtigung des Metabolismus kommen kann.

Neumann: Ich möchte noch eine Frage aufwerfen: Wir sprechen teilweise
von - sagen wir - androgenen oder östrogenen Nebenwirkungen der
Gestagene. Ist es eigentlich gerechtfertigt, dabei von Nebenwirkungen zu sprechen? Kann ein bestimmtes Wirkungsspektrum mit
östrogenen und/oder androgenen Partialwirkungen nicht sogar erwünscht sein? Diese Frage möchte ich an einen Kliniker weiter-
leiten.

Taubert:      Ich möchte Ihre Frage bejahen, denn gerade im Fall des Chlorma-
              dinonacetats war die antiandrogene Nebenwirkung auf Akne,
              Seborrhoe und Hirsutismus ja durchaus erwünscht. Auch die ande-
              ren hier anwesenden Kliniker haben sicher mit der Verabreichung
              von Chlormadinonacetat in Kombination mit Mestranol Besserungen
              der Beschwerden bei Mastopathia fibrocystica erzielt und das
              Wachstum der unterentwickelten Brust günstig beeinflussen kön-
              nen.

Zander:       Vielleicht ganz kurz zur Definition des Begriffes "Nebenwir-
              kung". Herr Tausk hat das ja sehr schön in seiner Monographie
              dargelegt. Wir haben es mit Substanzen zu tun, die verschieden-
              artige Wirkungen haben, und wir benutzen nun eine dieser Wirkun-
              gen, die uns für den Effekt, den wir haben möchten, gerade er-
              wünscht ist, und alles andere bezeichnen wir dann als Nebenwir-
              kung. Wenn wir einen anderen Effekt wünschen und dafür die Sub-
              stanz einsetzen, so kann es sein, daß wir das, was wir vorher
              als Hauptwirkung bezeichneten, als Nebenwirkung bezeichnen. Das
              ist eigentlich das Problem der Geschichte. Man sollte deswegen
              vielleicht gar nicht so sehr von Nebenwirkungen, sondern einfach
              von anderen Wirkungen sprechen; dann ist von vornherein klar,
              daß diese Substanzen eben viele Wirkungen haben.

Hammerstein:  Ein klassisches Beispiel hierfür ist das Cyproteronacetat. Die-
              ses Steroid gelangte 1962 wegen seiner starken gestagenen
              Eigenschaften in die klinische Prüfung, ohne daß man zunächst
              wußte, daß es auch über antiandrogene Potenzen verfügt. Heute
              konzentriert sich das Interesse an dieser Substanz auf diese
              antiandrogenen Eigenschaften, während die gestagene Partial-
              komponente mehr oder weniger als Nebeneffekt in Kauf genommen
              werden muß.

Neumann:      Wir sollten jetzt zu unserem zweiten Punkt übergehen, "Toxiko-
              logische Befunde nach chronischer Behandlung mit Gestagenen und
              Östrogenen".
                  Wir können hier die Östrogene nicht ausklammern, schon des-
              halb nicht, weil in Langzeitstudien Östrogene bei Ratten und
              Mäusen Hypophysen- und Mammatumoren induziert haben und weil
              verschiedene Gestagene auch östrogene Partialwirkungen besitzen.
              Herr Overbeek wollte dazu etwas sagen.

Overbeek:     Ja, ich möchte einmal die rohen Ergebnisse eines Versuches zei-
              gen, der bei uns durchgeführt wurde im Auftrag des Dunlop
              Committee in England, das die Maus immer gern empfiehlt für
              carcinogene Teste (Tab. 1.). In einem solchen Versuch wurden

Tab. 1. Tumorbefunde bei männlichen und weiblichen Mäusen.

Behandlungsdauer: 80 Wochen, oral.

   I Plazebo

   II Lyndiol   2,5 :   12,5  x  Humandosierung

  III Lyndiol   2,5 :   50    x        "

  IV Lyndiol   2,5 :   300   x        "

|              | ♂ | | | | ♀ | | | |
|              | I | II | III | IV | I | II | III | IV |
|--------------|---|----|-----|----|---|----|-----|----|
| Anzahl Tiere | 49 | 36 | 30 | 44 | 47 | 38 | 37 | 48 |
| Anzahl Tumoren | 14 | 3 | 0 | 12 | 15 | 4 | 8 | 12 |

die Mäuse 80 Wochen lang behandelt. Sie sehen im oberen Teil
der Tab. 1 die Dosierungen. Es wurde die 12,5 bis 300fache
Humandosis, umgerechnet auf kg Körpergewicht, verabfolgt. Nach
18 Wochen haben wir uns die Anzahl der Tumoren angeschaut. Wie
Sie sehen können, wird eine beträchtliche Anzahl Tumoren sowohl
bei den männlichen als bei den weiblichen Kontrollen angetroffen. Vielleicht ist es am auffälligsten, daß bei den niedrigen
Dosierungen die Anzahl der Tumoren am niedrigsten war, viel
niedriger eigentlich als man erwarten würde aufgrund des Vorkommens bei den Kontrollen. Die Tumoren sind von sehr verschiedener Art. Ich glaube übrigens, daß man sagen muß, daß die Maus,
die in England so sehr empfohlen wird, eigentlich ein sehr ungeeignetes Versuchstier ist; denn jedermann weiß schon seit
den Versuchen von Korenchevsky, daß es bei der Maus sehr
leicht ist, mit Östrogenen Tumoren, insbesondere Mammatumoren,
zu erzeugen. Dies ist in diesem Versuch eigentlich kaum der
Fall. Das bringt uns natürlich auf die Schwierigkeit, die, wie
auch Herr Neumann schon gesagt hat, so oft diskutiert wurde vor
einigen Monaten, nämlich ob der Hund ein geeignetes Versuchstier ist. Zum Hund als Versuchstier in der Toxikologie können
andere sicher mehr sagen. Ich möchte feststellen, daß ich persönlich auch die Maus nicht als ein geeignetes Versuchstier für
Karzinogenteste betrachte.

Neumann: Ich meine, man sollte hier in aller Deutlichkeit sagen, daß jedes Östrogen, an Ratten und Mäusen über längere Zeit verabfolgt,
Hypophysen- und Milchdrüsentumoren induziert. Diese Erkenntnis
ist nicht neu. Die tumorinduzierende Wirkung von Östrogenen und
östrogenhaltigen Präparaten ist auf einen Reglermechanismus bei
Ratten und Mäusen zurückzuführen, der möglicherweise beim Menschen gar nicht existiert oder jedenfalls nicht in diesem Maße
ausgeprägt ist. Östrogene wirken nämlich bei Ratten und Mäusen
nicht direkt auf die Milchdrüse, sondern letztlich via Stimulierung der Prolaktinsekretion. Auch dies ist seit langem bekannt (F. Neumann u. W. Elger: Vortrag auf einer Arbeitstagung
in Bonn (Dezember 1970) über "Ist die zur Zeit geübte Methodik

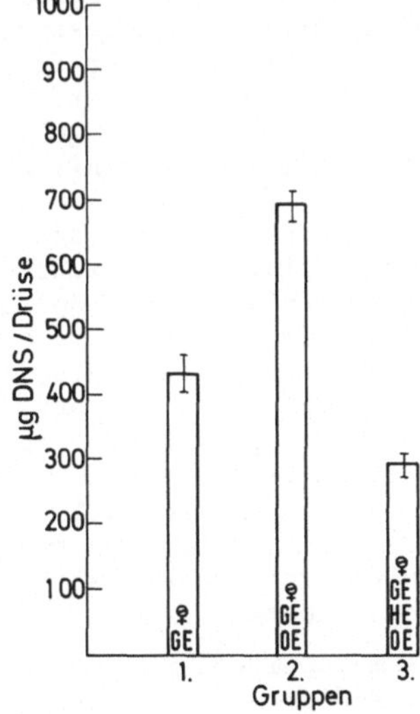



Abb. 1. Einfluß von Östradiol auf das Milchdrüsenwachstum kastrierter bzw. kastrierter und hypophysektomierter Ratten. DNS-Gehalt der Brustdrüsen. K=Unbehandelte Kontrollen. OE 10 µg Östradiol 21 Tage lang sc. HE = Hypophysektomie.

der Steroidtoxikologie für Forschung und klinische Anwendung
der Steroide vertretbar?" (Georg Thieme Verlag, im Druck). Dennoch möchte ich dazu einige Bilder zeigen. Abb. 1 zeigt das
Milchdrüsenwachstum kastrierter und hypophysektomierter weiblicher Ratten nach 21tägiger Behandlung mit täglich 10 µg Östradiol subkutan. Als Maß des Wachstums wurde hier der DNS-Gehalt
der Milchdrüse herangezogen. Man erkennt aus dieser Abbildung,
daß Östrogene nur bei den Tieren mit intakter Hypophyse die
Milchdrüse stimulieren. Noch deutlicher zeigt sich dies in den
Milchdrüsenganzdarstellungen ('whole mount'). Abb. 2a und 2b
sind Übersichtsbilder. Bei den hypophysektomierten Tieren
(Abb. 2a) ist unter Östrogenen allenfalls ein Gangwachstum zu
beobachten; bei den nicht-hypophysektomierten Tieren fällt ein
ausgeprägtes Alveolarwachstum auf (Abb. 2b). Diese Unterschiede
werden noch deutlicher in Ausschnittsvergrößerungen (vgl.
Abb. 3a und 3b). Ich sollte noch sagen, daß Milchdrüsentumoren
bei Ratten und Mäusen nicht spontan entstehen, sondern daß dem
immer eine enorme Stimulierung des Organs vorausgeht. Letztlich
ist also das Auftreten von Hypophysentumoren bei Ratten und
Mäusen nach Gabe von Östrogenen oder östrogenhaltigen Präparaten auf den bei dieser Spezies stark ausgeprägten positiven
Feedback-Mechanismus zwischen Östrogenen und Prolaktin zurückzuführen. Es ergeben sich zwei wichtige Fragen, die ich an die
Kliniker weiterreichen möchte.
1. Welche physiologische Rolle hat Prolaktin beim Menschen,
   speziell bei der Frau?
2. Gibt es Hinweise dafür, daß es beim Menschen einen solchen
   positiven Feedback-Mechanismus zwischen Östrogenen und Prolaktin gibt?
Ich möchte mich jetzt auch an das Auditorium wenden und halte
diese beiden Fragen für sehr wesentlich, wenn wir die Übertragbarkeit dieser tierexperimentellen Daten diskutieren.

Voigt:

Ich glaube, daß die Fragen, die Herr Neumann hier aufwirft,
eigentlich noch nicht beantwortbar sind. Sie sind nicht beantwortbar, weil der vorhandene biologische Nachweis für Prolaktin,
der sog. Taubenkropftest, nicht empfindlich genug ist, um physiologische Prolaktinspiegel zu messen, und weil bis jetzt noch
kein Radioimmunoassay besteht, der irgendeine Spezifität für
das Prolaktin des Menschen aufweist. Von daher gesehen muß man
also Ihre zweite Frage nach der Regulation völlig offen lassen,
wenn es auch nicht sehr wahrscheinlich ist, daß hier Analogien
zum Tier bestehen.
   Zur Frage 1 kann man ergänzen, daß es in der menschlichen
Pathologie Syndrome mit einer vermehrten Milchbildung gibt. Es
sind dies Galaktorrhoe-Fälle, die bei Männern und Frauen auftreten können. Hier wird ziemlich regelmäßig im Taubenkropftest
nachweisbare Prolaktinaktivität gefunden, etwa in der Größenordnung zwischen 40-120 Einheiten pro 100 ml Blut.
   Herr Berle und Herr Apostolakis haben bei uns Untersuchungen
dazu durchgeführt. Es ist aber nach wie vor offen, ob es sich
wirklich ausschließlich um Prolaktin handelt oder ob diese Substanzen nicht auch noch Wachstumshormonaktivitäten aufweisen.
Ich hoffe, daß wir diese Frage vielleicht in Zusammenarbeit mit
Herrn Jungblut beantworten können.

Runnebaum:

Ich darf in diesem Zusammenhang Untersuchungen von Fried u.
Rakoff (J. clin. Endocr. 12, 321 (1952)) erwähnen, die Frauen
mit regelmäßigem menstruellen Zyklus in der Lutealphase über
eine Zeit von 14 - 18 Tagen täglich 2000 IE HCG verabreichten
und keine signifikante Verlängerung der Gelbkörperphase feststellen konnten. Wenn sie jedoch in der Gelbkörperphase jeden

2. Tag 2000 IE HCG und zusätzlich 200 E Prolaktin gaben, dann konnte dadurch die hypertherme Phase des Zyklus um 13 Tage verlängert werden im Vergleich zu der Lutealphase bei Frauen ohne Behandlung. Ich stelle diese Befunde zur Diskussion.

Dhom: Ich möchte auf eine kuriose Beobachtung hinweisen, die vielleicht hierher gehört. Es handelt sich um eine laktierende Mamma bei einem Mann, der wegen Prostatakarzinoms mit Östrogenen behandelt worden ist. Der Fall wurde uns auf einem Mammaschnittseminar demonstriert. Das histologische Bild wurde von allen Pathologen, die an dem Schnittseminar teilgenommen haben, als voll laktierende Mamma diagnostiziert.

Voigt: Herr Runnebaum, Ihr Zitat hebt ab auf eine der für das Prolaktin geforderten biologischen Aktivitäten, nämlich die Verlängerung der Corpus luteum-Phase. Leider wissen wir noch nicht, ob das für den Menschen ebenfalls zutrifft. Ich würde fragen wollen, was für ein Prolaktin verabfolgt worden ist. Gerade hinsichtlich dieser biologischen·Aktivitäten, nämlich der Prolaktin- und der Wachstumshormonaktivität, bestehen außerordentliche Unterschiede bei den einzelnen Spezies. Herr Overbeek hat darauf schon hingewiesen. Ich darf es noch einmal kurz zusammenfassen. Es gibt eine große Zahl von Befunden, die eigentlich mehr dafür sprechen, daß beim Menschen Prolaktin-, oder exakter, Taubenkropfaktivitäten und Wachstumshormonaktivität etwa am selben Molekül sitzen, während es ganz sicher ist, daß bei Säugern, insbesondere auch Nagern, zwei unterschiedliche Entitäten bestehen, die die Wachstumshormon- oder die Prolaktinaktivität tragen. Ganz kurz zu dem Fall von Galaktorrhoe, den Herr Dhom genannt hat. Ich möchte bestätigen, daß auch die Patienten, die Herr Berle gesehen hat, keinen Hypophysentumor hatten. Aber wir sind etwas unsicherer geworden in der Beantwortung der Frage, ob das, was die Hypophysen vermehrt produzierten, ausschließlich Prolaktin war.

Hachmeister: Zur Frage des Prolaktins beim Menschen möchte ich erwähnen, daß nach Untersuchungen von Frantz u. Mitarb. (Science 170, 745 (1970)) Prolaktinaktivitäten im Serum vom Menschen gemessen werden können, welche durch Zusatz von Antikörpern gegen Human-STH zum Testansatz deutlich von der laktotropen Wirkung des Wachstumshormons differenziert werden. Es konnte z. B. unter Therapie mit Chlorpromazin ein Anstieg des Prolaktinserumspiegels bei unveränderten Wachstumshormonspiegeln festgestellt werden. In Fällen mit aktiver Akromegalie ist der Wachstumshormonspiegel erhöht, die Prolaktinspiegel sind normal. Klinische Zeichen einer laktotropen Wirkung bei der Akromegalie, wie sie gelegentlich in der Literatur angemerkt werden, müssen im Regelfall auf eine laktotrope Partialfunktion des Wachstumshormons zurückgeführt werden. Im Gegensatz dazu finden sich beim Galaktorrhoe-Amenorrhoe-Syndrom normale STH-Spiegel. Bei diesem Krankheitsbild werden häufig Hypophysenvorderlappentumoren gefunden, welche von den klassischen eosinophilen Adenomen der Akromegalie auch ultrastrukturell eindeutig abzugrenzen sind (G.T. Peake et al.: J. clin. Endocr. 29, 1383 (1969)). Es ist daraus zu folgern, daß auch beim Menschen ein vom Wachstumshormon zu unterscheidendes Prolaktin existiert; die laktotrope Partialfunktion des Wachstumshormons erschwert die Unterscheidung.

Neumann: Vielen Dank, Herr Hachmeister, aber ich glaube, wir sollten die Problematik der Trennung bzw. Identifizierung von Wachstumshormon und Prolaktin nicht weiter diskutieren. Eines ist vielleicht noch interessant. Herr Elger und Herr von Berswordt-Wallrabe haben in diesem Jahr an einer Workshop-Konferenz über Prolaktin

in Washington teilgenommen. Dort wurde von Frantz berichtet,
daß unter der Einnahme oraler Kontrazeptiva kein Anstieg des
Serumprolaktins zu beobachten war. Herr Elger, können Sie etwas
dazu sagen?

Elger: Frantz hat in Seren von Frauen, die orale Kontrazeptiva einnah-
men, keine gesteigerte Prolaktinaktivität gemessen, sondern
erst dann eine gesteigerte Aktivität gefunden, wenn die Frauen
die oralen Kontrazeptiva abgesetzt hatten (Workshop Conference
on Prolactin, Washington 1971).

Voigt: Entschuldigung, ich muß ganz kurz auf die Bemerkung von Herrn
Hachmeister eingehen. Herr Hachmeister, ich muß nach wie vor
festhalten, daß es keinen spezifischen Radioimmunoassay für
Prolaktin beim Menschen gibt. Sie wissen, daß wir früher auch
sehr dafür gefochten haben, daß es zwei unterschiedliche Mole-
küle sind. Ich hob eben auf die Zusammenarbeit mit Herrn Jungblut zu diesem Punkt ab und darf eine kurze Bemerkung machen.
Wir erhielten von Herrn Jungblut ein Präparat zur Testung. Wir
nahmen an, daß es eine Aufarbeitung unserer Prolaktinchargen
wäre. Wir haben den Tibia-Test für die Wachstumshormonaktivität
und den Taubenkropftest in der Modifikation von Berswordt-
Wallrabe zur Messung der Prolaktinaktivität durchgeführt. Nun
ganz trocken das Resultat: 40 Einheiten Taubenkropfaktivitäten
pro mg, damit das beste auf der Welt überhaupt vorhandene Prolaktin; und dann der Wachstumshormontest, ca. 3 Einheiten
Wachstumshormon pro mg, d. h. eine Wachstumshormonaktivität,
die auch als ausgesprochen gut angesprochen werden muß. Und
dann hinterher von Herrn Jungblut die Nachricht: was ihr getestet habt, war eine besondere Wachstumshormonaufbereitung.

Neumann: Ich glaube, diesen Punkt müssen wir abschließen. Es ist er-
schreckend, wie wenig man über die physiologische Bedeutung des
Prolaktins beim Menschen weiß. Wenn es einen ausgeprägten positiven Rückkopplungsmechanismus zwischen Östrogenen und Prolaktin beim Menschen gäbe, hätten Pathologen doch im Sektionsmate-
rial - etwa bei Männern, die wegen eines Prostatakarzinoms über
längere Zeit mit hohen Östrogendosen behandelt wurden - gelegentlich Hypophysentumoren beobachten müssen. Bei Ratten treten
nämlich immer und ganz regelmäßig solche Tumoren auf, quasi als
Substrat einer übermäßig gesteigerten Prolaktinsekretion.
Um Ihnen einen Eindruck zu vermitteln, wie stark Östrogene
die Prolaktinsekretion bzw. das Wachstum der Hypophyse stimulieren, möchte ich Ihnen einen Situs zeigen, der aus unserer
Abteilung für Toxikologie stammt (Abb. 4). Sie sehen an der
Schädelbasis gelegen die enorm vergrößerte Hypophyse. Eine normale Rattenhypophyse wiegt etwa 5 - 10 mg, ein solcher
östrogeninduzierter Hypophysentumor wiegt einige Gramm.
Die Tiere sterben irgendwann, weil wesentliche Gehirnteile
druckatrophisch zugrunde gehen.
In histologischen Präparaten von solchen Tumoren tritt nur
noch ein artgranulierter Zelltyp in Erscheinung, nämlich jene
Zellen, die Prolaktin bilden (Abb. 5). Etwas ähnliches ist meines Wissens beim Menschen unter einer chronischen Östrogentherapie nie beobachtet worden.
Wir sollten jetzt die Ratte verlassen und noch kurz auf die
toxikologischen Befunde beim Hund eingehen, denn diese Befunde
zu diskutieren erscheint mir noch wesentlicher als jene bei
Ratte und Maus. Der Hund stellt insofern mit Sicherheit eine
Ausnahme dar, als er neben dem Frettchen die einzige Spezies
ist, bei der mit Progesteron bzw. Gestagenen allein auch ein
erhebliches tubulo-alveoläres Milchdrüsenwachstum ausgelöst
werden kann (S.J. Folley: in: A.S. Parkes (ed.) "Marshall's

Physiology of Reproduction", 3rd Edition, vol. II, p. 525.
Longmans, Green Co., London 1952; S.J. Folley: Brit. med. Bull.
11, 145 (1955); C. Huggins u. P.V. Moulder: J. exp. Med. 80, 441
(1944); C.W. Turner u. E.T. Gomez: Mo. Agr. Exp. Sta. Res. Bull.
207 (1934)). Nun, es ist unwahrscheinlich, daß beim Hund Progesteron allein einen vollständigen Aufbau der Milchdrüsen indu-
zieren soll. Bei allen anderen bisher untersuchten Spezies
kommt immer ein hypophysärer oder plazentarer Faktor hinzu. Wir
haben Ende letzten Jahres Hypophysen von Hunden untersucht, die
mit einem sog. reinen Gestagen - ähnlich dem Chlormadinonacetat
- über einen längeren Zeitraum behandelt wurden. Tatsächlich
zeigten sich in der Hypophyse nach Gestagenbehandlung Veränderungen, wie sie für andere Spezies nach Gabe "reiner" Gestagene
unseres Wissens bisher nicht beschrieben und bisher auch nicht
beobachtet wurden.

Es lag z.T. ein völliger Umbau des Organs vor. In extremen
Fällen wurde das Bild von acidophilen Zellen beherrscht. Ich
möchte hier auf Abb. 6 verweisen, die die Hypophysenveränderun-
gen gut wiedergibt.

Man weiß, daß bei anderen Spezies eosinophile Zellen auch
der Entstehungsort des Prolaktins sind. Dennoch können wir natürlich nicht mit Sicherheit sagen, ob unter der Behandlung mit
einem "reinen" Gestagen beim Hund die Prolaktinsekretion stimuliert worden ist oder nicht. Das muß noch geklärt werden. Immerhin deutet aber dieser erste Befund daraufhin, daß bei der
Spezies Hund Progesteron oder Gestagene überhaupt hinsichtlich
des Milchdrüsenaufbaues die Rolle spielen, die bei anderen Spezies - wie etwa Ratten und Mäusen - den Östrogenen zuzuschreiben ist. Daß diese Überlegung nicht völlig absurd ist, beweist
die Tatsache, daß es nach Langzeitbehandlung mit Gestagenen
bzw. Östrogen-/Gestagen-Kombinationen beim Hund zu einer enormen
Entwicklung des Gesäuges kommt; es läßt sich dann sogar bei
männlichen Tieren Milchsekret gewinnen.

Das gleiche findet man auch bei der recht häufig beobachteten Pseudoschwangerschaft der Hündin.

Wieso in einer solchermaßen über längere Zeit stimulierten
Milchdrüse Knötchen und auch Tumoren entstehen können, kann uns
vielleicht Herr Dämmrich erklären.

Dämmrich: Wenn man sich den Ausführungen von Herrn Neumann anschließt,
dann ist zunächst festzustellen, daß der Hund eine Besonderheit
darstellt. Einmal in bezug darauf, daß es möglich ist, beim
Hund mit Gestagenen allein die Entwicklung des Gesäuges bis zur
Laktation voranzutreiben. Zum anderen liegen beim Hund besondere physiologische Gegebenheiten vor. Der Hund ist ein monöstri-
sches Tier und zeigt, auch wenn keine Konzeption zustande gekommen ist, nach dem Östrus über eine Dauer von etwa 50 Tagen
voll funktionsfähige Gelbkörper. Diese funktionsfähigen Gelbkörper können bei Hündinnen proliferative Vorgänge auslösen,
bzw. das Progesteron wirkt beim Hund als Proliferationshormon,
zu dessen Wirkungsspektrum es gehört, daß die Milchdrüse auch
ohne Trächtigkeit bis zur Laktation entwickelt werden kann. Anschließend an den Vortrag von Herrn Overbeek, in dem er das
Wirkungsspektrum der Gestagene beim Menschen mit 3 Punkten charakterisiert hatte, ist für den Hund als 4. Punkt die Entwicklung des Gesäuges anzuführen.

Die Entwicklung des Gesäuges bis zur Laktation durch Progesteron beim Hund, die spontan auch häufig ohne Konzeption erfolgt, ist sicherlich der Grund dafür, daß beim Hund in außerordentlich großer Zahl - man schätzt etwa bei 60% der Hündinnen - Neoplasien der Mamma spontan auftreten. Vielleicht kann

hier auch gleich ergänzt werden, daß der Beagle dabei keine
Ausnahme bildet.

Werden Hunde mit Gestagenen über lange Zeit behandelt, entwickelt sich das Gesäuge in gleicher Weise bis zur Laktation.
Diese beiden Grundlagen, das spontan entwickelte Gesäuge auch
ohne Konzeption und die durch Gestagene experimentell induzier-
te Laktation, führen dazu, daß das Gesäuge vorbereitet wird für
das Entstehen von Neoplasien. Die beim Hund vorkommenden Mammatumoren sind - auch das ist ein Charakteristikum beim Hund - in
etwa 85% maligner Art und überwiegend epitheliale Neubildungen,
die ein außerordentlich vielgestaltiges Bild zeigen.

In der Frage der Wirksamkeit der Gestagene in Bezug auf die
Tumorhäufigkeit im Gesäuge von Hündinnen kann gesagt werden,
daß beim Hund die Gestagene auf dem Umweg über die Proliferation des Milchdrüsengewebes sicherlich die Tumorrate erhöhen,
daß aber die Ergebnisse nicht auf den Menschen übertragbar
sind, weil beim Menschen die Mammaentwicklung - sowohl die
Milchgangs- als auch Alveolenentwicklung - nicht allein durch
Progesteron ausgelöst werden kann.

|  |  |
|---|---|
| Neumann: | Ich meine, die entscheidende Frage an die Kliniker ist doch jetzt: wie wirkt sich eine Gestagentherapie auf die Brustent-wicklung (Milchdrüsenentwicklung) bei der Frau aus, und zwar die Gabe eines Gestagens allein, ohne Östrogenzusatz? |
| Hammerstein: | Herr Dämmrich hat eben darauf hingewiesen, daß Gestagene beim Hund einen proliferationsfördernden Effekt an der Mamma ausüben. Beim Menschen ist das offenbar genau umgekehrt. Zur Belegung dieser Feststellung gibt es allerdings nur wenig Literaturhin-weise. Greenblatt hat kürzlich über eine Patientin mit ausge-prägter Mammahyperplasie berichtet, bei der eine bemerkenswerte Rückbildung des Brustdrüsenkörpers unter dem Einfluß von Gesta-genen mammographisch dokumentiert worden ist. Wir selbst ver-wenden Gestagene, und zwar solche ohne östrogene Partialwirkun-gen, bei der rezidivierenden Mastopathie und haben bisher unter dieser Therapie keine weiteren Rezidive mehr gesehen. Unsere Zahlen sind allerdings für eine statistische Sicherung zu klein. |
| Neumann: | Möchte sich noch jemand dazu äußern? |
| Zander: | Ja, ich möchte ganz kurz noch etwas dazu sagen aus der "persön-lichen" Erfahrung. Man sieht ja zweifellos Patienten, bei denen das so ist, wie Sie eben beschrieben, man sieht aber auch gele-gentlich  das Umgekehrte. |
| Taubert: | Noch eine abschließende Frage an Herrn Neumann. Sie haben ge-zeigt, daß langdauernde Gestagenbehandlung auch beim Hund im HVL zu histologischen Veränderungen führt, die durchaus den Bildern bei der Ratte nach Langzeitbehandlung mit Östrogenen entsprechen.<br><br>Wäre es nicht angebracht, hypophysektomierte Tiere zu verwen-den, wenn man den Hund als Versuchstier zur Überprüfung des Gestageneffektes auf Mammagewebe beibehalten will? |
| Neumann: | Ja, das haben wir vor. |
| Bettendorf: | Ein ähnlicher Effekt könnte beim Menschen möglich sein, und zwar sieht man ja gelegentlich unter hoch dosierter Therapie mit Gestagenen, daß es zu einer Laktation kommt. Wir haben es bei Cyproteronacetat und auch bei 17α-Medroxyprogesteronacetat beobachtet. |
| Overbeek: | Es ist nun einmal heutzutage üblich, daß man für neue Substan-zen Toxizitätsversuche macht. Zu den Toxizitätsversuchen gehört ohne jeden Zweifel ein Karzinogentest. Nur, wie heute gesagt |

wurde, uns fehlt ein zuverlässiger Karzinogentest. Ich glaube,
daß man aus den verschiedenen Diskussionen schließen muß, daß
jedenfalls für Gestagene der Hund ein ungeeignetes Tier ist,
wie für Östrogene die Maus ein ungeeignetes Tier ist. Das bedeutet natürlich nicht - und hier möchte ich doch unsere Beagle
noch einmal verteidigen - daß, obwohl ungeeignet für diese Art
von Karzinogentesten, es dennoch ideale Tiere für Toxizitäts-
versuche sind. Man sagt, die Maus ist nicht brauchbar, der Hund
ist nicht brauchbar, andere Tiere sind vielleicht auch nicht
brauchbar, was ist die Alternative? Die Alternative muß doch
das Sammeln von Daten an Menschen sein, die über lange Zeit behandelt wurden. Wir haben so einen Versuch veranlaßt. Ich möchte Ihnen die Ergebnisse zeigen. Dr. Rice-Wray in Mexico City
sammelt schon seit vielen Jahren Daten über mit O.C. Präparaten
behandelte Frauen, die regelmäßig kontrolliert werden und gerade auch auf das Vorkommen von Mamma- und Uterustumoren. Tab. 2
zeigt Ergebnisse, die das Lyndiol 2.5 betreffen und die aus diesen Untersuchungen stammen.

Tab. 2. <u>Langzeitstudie über Mamma- und Uterustumoren nach
Lyndiol 2,5 bei der Frau (E. Rice-Wray)</u>

| | | |
|---|---|---|
| 76 Frauen | | > 2 Jahre |
| 72 " | | > 3 Jahre |
| 60 " | | > 4 Jahre |
| 38 " | | > 5 Jahre |
| 17 " | | > 6 Jahre |
| 5 " | | > 7 Jahre |
| Unter Behandlung | 67 Frauen | |
| Tumoren | 0 | |

(Untersucher E. Rice-Wray)


Es ist natürlich noch nicht sehr eindrucksvoll, denn Sie sehen, es sind jetzt 76 Frauen, die länger als zwei Jahre behandelt wurden. Es sind nur 5, die mehr als 7 Jahre behandelt wurden, aber noch immer sind 67 Frauen in Behandlung und in der Zukunft werden derartige Untersuchungen lehren, ob die Substanzen
wirklich Gefahren in dieser Hinsicht beinhalten. Vorläufig sind
keine Tumoren oder etwas tumorenähnliches beobachtet worden,
aber ich glaube, daß es sehr wichtig ist, daß derartige klinische Versuche weitergeführt werden.

<u>Neumann</u>: Sie haben völlig recht, Herr Overbeek, man muß natürlich auch
weiterhin Toxizitätsstudien machen, und man wird auch nicht auf
Ratten, Mäuse und Hunde verzichten, nur muß man vielleicht die
Befunde etwas vorsichtiger oder etwas anders interpretieren.
Vielleicht noch Herr Hammerstein zu der Bemerkung von Herrn
Bettendorf, bevor wir dann weitergehen.

<u>Hammerstein</u>: Eine ganz kurze Bemerkung zur Laktation: Auch wir haben unter
Cyproteronacetat 2 Patientinnen beobachtet, bei denen eine
Milchsekretion in Gang gekommen ist. Man muß in diesem Zusammenhang aber wohl zwei Dinge auseinanderhalten: Nämlich den direkten, mit relativ niedrigen Dosen zu erzielenden Effekt der Gestagene am Endorgan, den wir bei der Therapie der Mastopathie

ausnützen, und den zentralen Effekt, der möglicherweise über das
Prolaktin oder den Prolaktin-Inhibiting-Faktor zur Wirkung ge-
langt. Ein Antagonismus dieser beiden Gestageneffekte ist denk-
bar.

<u>Neumann</u>: Eine weitere Diskussion zum Prolaktin muß ich aus Zeitgründen
abbrechen. Wir kommen nun zum dritten Punkt: Klinische Nebenwirkungen der reinen Gestagentherapie, einschließlich der sog.
Minipille[+].

Ich habe vergessen zu sagen, daß wir die Beeinflussung von
Gerinnungsfaktoren heute deshalb nicht diskutieren werden, weil
im Mai - wie mir berichtet wurde - eine spezielle Tagung darüber in Hamburg stattfindet.

Eine möglicherweise sehr wichtige Nebenwirkung der Gestagene
ist die Beeinflussung des Kohlenhydrat- und Fettstoffwechsels.
Da die Zeit fortgeschritten ist, möchte ich gleich an Herrn
Voigt übergeben.

<u>Voigt</u>: Die metabolischen Effekte der Kontrazeptiva, d. h. bisher
eigentlich immer der Kombinationspräparate, sind von einer großen Anzahl von Forschern außerordentlich sorgfältig untersucht
worden. Im Hintergrund standen immer zwei bzw. drei Überlegungen, wenn ich die praktische Seite zuerst betonen darf. Erstens
die Frage, ob ein erhöhtes Risiko für arteriosklerotische Prozesse durch die Einnahme von oralen Kontrazeptiva befürchtet
werden muß. Die zweite Frage, die Herr Neumann ausklammern
möchte, war die, ob solche Veränderungen evtl. ein höheres
Thromboserisiko für die Patientinnen, die hormonale Kontrazeptiva nehmen, bedingen. Die dritte Frage war: Sind solche Substanzen geeignet, evtl. auch ein höheres Diabetesrisiko hervorzurufen? Wie ich eingangs sagte, sind solche Untersuchungen in
größerer Zahl und mit sorgfältiger Technik bisher eigentlich
nur mit Kombinationspräparaten durchgeführt worden. Die entscheidende Frage, ob es der Gestagen- oder Östrogenanteil in
solchen Präparaten ist, der die Hauptwirkungen entfacht, muß
weitgehend offen bleiben. Eine Literaturzusammenstellung etwa
von Herrn Mauvais-Jarvis hat gezeigt, daß man für beides Argumente sammeln kann. Ich möchte darauf im einzelnen nicht eingehen. Reine Gestagene sind in dieser Beziehung nach meiner vielleicht mangelhaften Kenntnis noch nicht sorgfältig untersucht
worden. Was bewirken nun hormonale Kontrazeptiva? Im Prinzip
kann man zwei Arten von Effekten messen. Einmal finden sich
Veränderungen des Kohlehydratstoffwechsels, etwa ein Anstieg
von Pyruvat und Laktat und eine schlechtere Glucose-Utilisation.
Zum anderen läßt sich ein Anstieg der freien Fettsäuren, des
freien Glycerins und ab und zu auch eine eindeutige Zunahme des
Triglyceringehaltes im Blut nachweisen. Die bisherigen Daten
zeigen aber ziemlich regelmäßig, daß die Veränderungen zwar
statistisch signifikant sind, wenn man die Patienten mit sich
selbst vergleicht, daß sie aber nicht in pathologische Berei-
che hereingehen. Diese Feststellung erscheint mir wichtig.
Vielleicht darf ich noch  etwas anschließen. Interessant ist,
daß ein Vergleich dieser Veränderungen mit Befunden, wie man
sie mit denselben Methoden in der Schwangerschaft erheben kann,
zeigt, daß sie unter hormonaler Kontrazeption viel weniger aus-

---

[+] Aus Zeitgründen war es nicht möglich, die klinischen Nebenwirkungen der
Gestagentherapie auszudiskutieren. Der Leser möge entschuldigen, daß auf
einige Nebenwirkungen nur hingewiesen wird. Vielleicht hat das Rundtischgespräch aber schon einen Zweck erfüllt, wenn in Zukunft den hier z. T. nur
kurz erwähnten Nebenwirkungen mehr Beachtung geschenkt wird.

geprägt sind als in der Spätschwangerschaft. Wie meine Mitar-
beiterin, Fräulein Reinke, gezeigt hat, kann man auch während
des normalen Zyklus und gerade in der Corpus luteum-Phase eine
signifikante Vermehrung der freien Fettsäuren im Blut nachweisen. Ich sage das, um auf den Punkt von Herrn Zander zurückzukommen. Die Frage der Nebenwirkungen wird so gesehen immer
problematischer.

Hammerstein: Was die Frage der Nebenwirkungen reiner Gestagenpräparate anlangt, so ist es außerordentlich mühsam, aus der weit gestreu-
ten Literatur überhaupt greifbare Anhaltspunkte zu gewinnen. In
diesem Zusammenhang wären z. B. Camanni u. Mitarb. (1963) zu
zitieren, die zeigen konnten, daß Medroxyprogesteronacetat in
der Lage ist, bei adrenalektomierten Patienten Cortisol zu ersetzen. Dieses Gestagen kann ferner bei diabetischen Patienten
zur Verschlechterung des Kohlenhydratstoffwechsels führen. Das
mag als Hinweis darauf dienen, daß bei der Verabfolgung mancher
Gestagene bezüglich des Kohlenhydrathaushaltes Vorsicht am
Platze ist. Auch beim Norethisteronönanthat, über das heute
Nachmittag berichtet werden wird, sind Stoffwechselveränderun-
gen bekannt geworden.

Taubert: Als kurze Ergänzung zu den Ausführungen von Herrn Voigt möchte
ich auf zwei Veröffentlichungen der Arbeitsgruppen Larson-Cohn
(Acta endocr. (Kbh.) 62, 242 (1969)) und Dittmar u. Waidl
(Geburtsh. Frauenheilk. 29, 1026 (1969)) hinweisen, die bei Verabreichung der sog. Minipille, d.h. 0,5 mg Chlormadinonacetat/die,
keine Abweichungen des k-Wertes von der Norm nach intravenöser
Glukosebelastung sahen. Spellacy et al. (Fertil. Steril 21,
457 (1970)) beobachteten dagegen bei Anwendung des von Ihnen
erwähnten Depot-Medroxyprogesteronacetats eindeutig pathologische Werte.

Vermeulen: In letzter Zeit gibt es auch mehrere Publikationen über den
Effekt von reinen Progesteronen auf den Kohlenhydratstoffwech-
sel. Man hat mit Progesteron in einer Dosierung von 200 - 300 ng
pro Tag schon nach 6 - 10 Tagen Veränderungen der K-Werte des
Glukose-Stoffwechsels und Veränderungen im Insulinspiegel gesehen. Wir selber haben bei Patienten, die mit Chlormadinon
(0.5 ng/Tag über 1 Jahr) behandelt wurden, die Insulinspiegel
und die K-Werte bei intravenös verabreichter Glukose gemessen
und wir fanden, daß der K-Wert signifikant erniedrigt war und
daß die Insulinspiegel signifikant höher waren als bei den
Kontrollen.
 Dies ist in Kontradiktion mit den Werten von Larsson-Cohn.
Beck hat bei Affen ähnliche Resultate gesehen, nicht jedoch bei
Menschen.

Neumann: Haben Sie vielen Dank, Herr Vermeulen. Ich darf vielleicht auch
als Nichtkliniker etwas dazu sagen. Bisher ist es doch häufig
nur mehr oder weniger ein Sammeln von Fakten. Wie man die Befunde zu interpretieren hat, weiß man eigentlich noch nicht so
genau.

Voigt: Ich habe bewußt auf Theorien zur Erklärung dieser Dinge verzichtet. Ich sollte jetzt ganz kurz das Pyruvat-Dehydrogenase-
System erwähnen, das als Schlüsselenzym bei diesen Prozessen
eine bedeutsame Rolle spielt. Ich habe Progesteron- bzw. Gestagen-
arbeiten darum nicht zitiert, weil die Befunde sehr unterschiedlich sind, Herr Vermeulen. Es gibt Beobachtungen, die mit
reinen Gestagenen einen Einfluß nachweisen, wie auch solche,
die nichts finden. Wichtig ist mir zu betonen, daß im Zyklus

nur in der Corpus luteum-Phase - wir haben gleichzeitig Progesteron und Oestrogene bestimmt - ein signifikanter Anstieg der
freien Fettsäuren eintritt.

Neumann: Eine wichtige Nebenwirkung ist der Einfluß von Gestagenen auf
die Leberfunktion und vielleicht auch auf die Nebennierenfunk-
tion. Ich glaube, Herr Hammerstein und Herr Taubert sollten sich
dazu äußern.

Hammerstein: Auch hier gibt es nur wenige Untersuchungen, die den Einfluß
der Gestagene auf die Leberfunktion ohne Östrogenzusatz zum
Ziele haben. Dabei hat sich herausgestellt, daß Unterschiede in
dieser Hinsicht zwischen den 17-Acetoxy-Progesteronderivaten
und den 19-Nor-Androstanabkömmlingen bestehen. Nur die letzteren scheinen in ganz geringem Maße die Leberfunktion zu beeinflussen; dies könnte auf der bereits mehrfach erwähnten
Östrogenpartialwirkung dieser Gestagengruppe beruhen.

Taubert: Der bisherige Verlauf der Diskussion hat zweifellos ergeben,
daß die Zahl der Nebenwirkungen bei reiner Gestagentherapie,
ganz im Gegensatz zu Kombinationspräparaten, relativ gering ist.
Über die Beeinflussung der Leberfunktion durch reine Gestagenpräparate sind bisher nur wenige Studien durchgeführt worden,
die auch keine wesentlichen Befunde ergaben. Im Gegensatz hierzu sind zwei vor kurzem in Acta endocrinologica (65, 193 u. 207
(1970)) erschienene Arbeiten als durchaus besorgniserregend
anzusehen. Martinez-Manautou et al. führten sowohl in der
Schwangerschaft als auch bei Patientinnen, die mit Kombinationspräparaten oder 0.5 mg Chlormadinonacetat behandelt worden waren, Leberbiopsien durch. Die elektronenoptische Auswertung des
Materials ergab bei Patientinnen im letzten Schwangerschafts-
drittel eine deutliche Dilatation und Vesikulierung des glatten
als auch des rauhen endoplasmatischen Retikulums, Verlängerung
und Vergrößerung der Mitochondrien, Bildung von Cristae und
multiple, kristalloide Einschlüsse. Die gleichen morphologischen
Veränderungen an den Mitochondrien wurden bei Patientinnen, die
0,5 mg Chlormadinonacetat eingenommen hatten, noch 5 Monate
nach dem Absetzen des Präparates gefunden. Nach Martinez-
Manautou et al. erklären sich die Änderungen der Feinstruktur
der Leberzellen aus den erhöhten Anforderungen an die Stoff-
wechselleistung der Leber während der Schwangerschaft bei
gleichzeitig größerer Belastung durch Steroide. Nach der
Schwangerschaft ist der Leber meist ausreichend Zeit zur Normalisierung der Zellorganellen gegeben. Man muß sich jedoch fragen, ob der Leber bei Langzeitbehandlung mit Gestagenen, auch
in kleinsten Dosen, diese Chance gegeben wird.

Neumann: Es tut mir leid, daß wir die einzelnen Aspekte nicht bis zu
Ende diskutieren können, wahrscheinlich wäre jeder Punkt für
sich ein Rundtischgepräch wert.

Voigt: Um Ihnen zu zeigen, Herr Taubert, wie kompliziert das Problem
ist, darf ich auf Arbeiten verweisen, die zeigen, daß der Inter-
mediärstoffwechsel sich nach einem Jahr unter der Einnahme der
"Pille" wieder normalisiert.

Neumann: Eine Frage erscheint mir besonders wichtig, sie wurde von Herrn
Bettendorf aufgeworfen. Nimmt im Anschluß an eine Behandlung
mit oralen Kontrazeptiva die Zahl der Frauen mit Sterilität und
anovulatorischen Zyklen bzw. Amenorrhoe zu?

Bettendorf: Ich habe die Frage aufgeworfen, kann aber selbst nicht allzu-
viel dazu sagen, und zwar ist uns aufgefallen, daß zunehmend

Frauen mit einer Sterilität kommen, bei denen sich aus der
Anamnese ergibt, daß diese Sterilität nach Einnahme von Kontrazeptiva aufgetreten ist. Dies ist eine rein klinische Beobachtung, die ich noch nicht in Zahlen belegen kann. Wir haben den
Eindruck, daß es sich vor allem um junge Frauen handelt und um
solche, die vor der Medikation schon Zyklusstörungen hatten.

Neumann:      Wurden ähnliche Beobachtungen auch an anderen Kliniken gemacht?

Hammerstein:  Wir haben diese Beobachtungen nicht gemacht. Wenn Patientinnen
nach Beendigung der hormonalen Kontrazeption mit Zyklusstörungen zu uns kommen, dann ergibt die Anamnese meistens, daß Zyk-
klusanomalien, anovulatorische Störungen usw. auch schon vor
Beginn der Pilleneinnahme bestanden haben. Oft treten diese
Störungen allerdings nach Absetzen der Pille noch ausgeprägter
zutage; deshalb vertreten wir die Ansicht, daß man bei Patientinnen mit gröberen Zyklusstörungen in der Anamnese mit der
Verabfolgung von kontrazeptiven Kombinationspillen sehr zurück-
haltend sein sollte. Auch unsere Zahlen sind zu klein, als daß
man von mehr als nur einem klinischen Eindruck sprechen könnte.

Neumann:      Nun, die Frage liegt auf dem Tisch. Ich glaube, man muß es eben
weiter verfolgen.
    Eine weitere Frage: Spielt die Gestagendosis, unabhängig von
der gestagenen Potenz eines Präparates, bei Nebenwirkungen,
z. B. auf den Leberstoffwechsel, eine Rolle?

Overbeek:    Ich glaube, daß dies überhaupt keine Rolle spielt. Was man mit
diesem Effekt auf die Leber mißt, ist wahrscheinlich das, was
jedes 17-Alkyl-Präparat tut, und das ist ganz unabhängig von
der hormonalen Aktivität. Es gibt hormonal vollständig inaktive
Substanzen, die die gleichen Wirkungen auf die Leber haben. Wir
haben früher einmal eine Anzahl hormonal aktiver und inaktiver
Steroide beim Kaninchen getestet. Die Kaninchenleber ist sehr
empfindlich für verschiedene Steroideffekte. Es wurde gefunden,
daß die hormonal inaktiven Steroide genau so aktiv waren wie
andere, soweit es die Wirkung auf die Leberfunktion betraf.

Neumann:      Vielleicht eine allerletzte Frage: Liegen Erfahrungen über
allergische Reaktionen unter einer Therapie mit Gestagenen vor?

Taubert:     Ich hatte die Frage eigentlich nicht gestellt, da ich darüber
viel weiß, sondern da ich gern mehr darüber erfahren würde.
Meines Wissens nach liegen nur 2 - 3 Beobachtungen in der Literatur über prämenstruell auftretende Exazerbationen einer vesi-
kulobullösen Dermatitis herpetiformis vor. Diese Eruptionen
konnten durch Verabreichung von Norethisteron und Progesteron
provoziert, durch Ovulationshemmung mit Äthinylöstradiol verhindert werden.

Hammerstein:  In Deutschland ist 1966 ein ähnlicher Fall von Hoischen u.
Steigleder mitgeteilt worden. Hier exazerbierte ein Pruritus
simplex regelmäßig während der Gelbkörperphase. Aus verschiedenen Laborergebnissen wurde gefolgert, daß es sich dabei um eine
Gestagenallergie gehandelt haben könnte. Ferner wurde im vorigen Jahr von Malgre ein Fall publiziert, wo eine Konzeption
trotz hormonaler Kontrazeption mit Antikörpern gegen das zuge-
führte Gestagen zu erklären versucht wurde. Viel mehr ist, so
glaube ich, zu diesem Punkt im Moment nicht zu sagen.

Zander:      Ich möchte etwas allgemeines zu den Nebenwirkungen der Gestagene sagen. Wenn man das Gespräch jetzt gehört hat, so hat man
an sich den Eindruck, es gibt kaum wesentliche Nebenwirkungen,

die irgendwie ins Gewicht fallen. Ich persönlich glaube auch,
daß die größten Nebenwirkungen eigentlich die sind, die wir
nicht kennen. Es ist so, daß sich doch ganz zweifellos ein grö-
ßeres Unbehagen ausbreitet. Das beobachtet man hier in unserem
Land einfach deswegen, weil man das Gefühl hat – und das ist
gar nicht irgendwie von der Ratio her so genau zu begründen –,
daß hier Substanzen über riesige Zeiträume ständig und kontinuierlich gegeben werden, die ein Ereignis verhindern sollen,
das nur in wenigen Stunden abläuft. Ohne es durch Zahlen belegen zu können, sieht es so aus, als ob die orale Kontrazeption
mit der "Pille" zurückgeht, genauso wie das Intrauterinpessar,
und daß demgegenüber die operative Methode deutlich in einem
Anstieg begriffen ist. Ich glaube, wir sollten uns darüber im
klaren sein, daß wir im Augenblick immer noch beim Menschen
mitten in einem Experiment sind und daß wir über unendlich
viele Dinge – das haben wir bei diesem Gespräch gesehen, z. B.
bei der Diskussion über Mamma und Hypophyse usw. – unendlich
wenig wissen und daß bisher wenige Beobachtungen in dieser
Richtung vorliegen.

<u>Neumann:</u>     Ich glaube, diesen Ausführungen von Herrn Zander können wir uns
alle nur anschließen. Auf viele Fragen wissen wir heute noch
keine Antwort.

     Ich darf mich bei allen Teilnehmern des Rundtischgespräches
und auch bei dem Auditorium herzlich bedanken.


## Anschriften der Teilnehmer des Podiumgespräches und der Diskussionsredner

Prof. Dr. G. Bettendorf, Universitäts-Frauenklinik, 2 Hamburg 20,
Martinistr. 52

Prof. Dr. K. Dämmrich, Institut für Veterinärpathologie der Freien Universität Berlin, 1 Berlin 33, Harnackstr. 3

Prof. Dr. G. Dhom, Pathologisches Institut der Universitätskliniken im
Landeskrankenhaus, 6650 Homburg/Saar

Dr. W. Elger, Schering AG, Abt. Endokrinpharmakologie, 1 Berlin 65,
Müllerstr. 170-172

Dr. U. Hachmeister, Pathologisches Institut der Universität, 63 Gießen,
Klinikstr. 32g

Prof. Dr. J. Hammerstein, Klinikum Steglitz der Freien Universität Berlin,
1 Berlin 54, Hindenburgdamm 30

Priv.-Doz. Dr. F. Neumann, Schering AG, Abt. Endokrinpharmakologie,
1 Berlin 65, Müllerstr. 170-172

Dr. G. A. Overbeek, N.V. Organon, Oss/Holland

Priv.-Doz. Dr. B. Runnebaum, I. Frauenklinik und Hebammenschule der Universität München, 8 München 15, Maistr. 11

Prof. Dr. H.-D. Taubert, Universitäts-Frauenklinik, Abteilung für Gynäkologische Endokrinologie, 6 Frankfurt 70, Ludwig-Rehn-Str. 14

Prof. Dr. A. Vermeulen, Department of Endocrinology, Akademisch Ziekenhuis,
De Pintelaan 115, Ghent/Belgien

Prof. Dr. K. D. Voigt, Universitäts-Krankenhaus Eppendorf, 2 Hamburg 20,
Martinistr. 52

Prof. Dr. J. Zander, I. Frauenklinik und Hebammenschule der Universität
München, 8 München 15, Maistr. 11

Symp. Dtsch. Ges. Endokrin. _17_, 147–156 (1971)
© by Springer-Verlag

# Recent Developments in the Study of Hormone Effects and Metabolism in Prostate Tissue[1]

P. OFNER

Steroid Biochemistry Laboratory of the Medical Services, Lemuel Shattuck Hospital, Department of Public Health of the Commonwealth of Massachusetts, Departments of Pharmacology of the Harvard School of Dental Medicine and Harvard Medical School and Department of Medicine, Tufts University School of Medicine, Boston, Massachusetts, USA.

With 3 Figures

I should like to express my thanks to the German Endocrine Society for the invitation to participate in this Symposium and my appreciation of the timely choice of topics for this morning's discussion. Recent developments in the study of hormone effects and metabolism in prostate tissue hold the promise of bridging the large gap in biochemical knowledge between the developmental, histologic and clinical contributions of Lacassagne (1933 a, b), Huggins (1941, 1948), Moore (1947, 1952), Price (1957, 1963) and Woodruff (1960, 1962) and current concepts of the molecular mode of hormone action so admirably stated by Karlson (1963).

The action of a hormone involves responsiveness of the target tissue, specificity and amplification of the initiating signal. Our understanding of these components of androgen action has been greatly expanded by a stream of recent information derived from studies of nuclear RNA and cytoplasmic protein synthesis, ATP metabolism, effects of cyclic adenosine 3',5'-monophosphate (c-AMP) linked to the concept of a second messenger, $C_{19}$-steroid transformations and primary hormone receptors in male accessory sex organs. Since I intend to tell you of recent work done in my laboratory, I shall only have time to touch long enough on each of these important aspects to convey the impression of new complexities and controversies and to raise some points for subsequent round-table discussion. By beginning with the later events of amplification of male hormone action, we shall arrive at consideration of intracellular androgen metabolism and retention, where attempts are now being made to apply the new knowledge gained in recent years to the understanding of prostatic disease.

Androgen replacement in castrated mammals induces rapid prostatic formation of many new cytoplasmic ribosomes and restores the network of granular endoplasmic reticulum which had regressed on orchiectomy (Williams-Ashman et al., 1964). Liao and co-workers (1966 a, b, 1967) gathered persuasive evidence from studies with isolated prostatic nuclei that the normal function of the prostatic mucleolus plays a central role in androgenic control of RNA and protein synthesis. These investigations revealed that the increased RNA polymerase activity of rat ventral prostate induced by testosterone administration was much more sensitive to inhibition by low levels of concomitantly-injected actinomycin D than was the base-line activity of the castrate. In mammalian cells low concentrations of actinomycin D preferentially inhibit synthesis of ribosomal RNA precursors in the nucleolus with little or no effect on formation of nonribosomal RNA at extranucleolar re-

---

[1] Supported by United States Public Health Service Grant CA-10343 from the National Cancer Institute and by appropriations of the Commonwealth of Massachusetts, Item 2026-50-02.

gions of the chromatin. Moreover, nearest neighbor base frequency analysis showed that the nucleotides of actinomycin D-sensitive $^{32}$P-labeled RNA formed by prostatic nuclei of testosterone-treated castrates had the characteristically high guanine and cytosine composition of prostatic ribosomal RNA.

These findings led Liao and Fang (1969) to divide the chromatin into four regions, the relative proportion of each being determined by its ability to bind actinomycin D and to support RNA synthesis by endogenous nuclear, or additional exogenous bacterial, polymerase. A masked region (M) accounted for 70-80% of the chromatin DNA und was unavailable to actinomycin D or for transcription. A repressed region (R) did not seem to participate in RNA synthesis in intact nuclei, since it could transcribe only in the presence of exogenous bacterial polymerase. Only a small region (1%) of active chromatin (Ch) was actively engaged in RNA synthesis in prostatic nuclei of castrates. As we have seen, another small region (1%) of nucleolar chromatin (No), highly sensitive to actinomycin D, is activated if testosterone is given to castrated animals. Liao and Stumpf (1968) also provided autoradiographic support for a selective increase by testosterone of nucleolar RNA synthesis in the ventral prostate of the rat. While the evidence thus points to a selective enhancement of nucleolar rRNA synthesis as an early consequence of androgen action, tRNA and mRNA synthesis at extranucleolar, and possibly even nucleolar, sites of the chromatin appear to be affected also, but the coupling mechanisms are at present largely unknown (Liao and Fang, 1969; Williams-Ashman, 1970). The function of rRNA remains a major mystery (Watson, 1970).

Villee and Fujii (1968, 1969) reported that the testosterone-induced increase in seminal vesicle weight and protein synthesis can also be initiated by instilling RNA extracted from this organ of a testosterone-treated immature rat into the lumen of the seminal vesicle of an immature rat. The same effect could be obtained with RNA extracted from the liver, kidney, prostate or seminal vesicle of adult, intact, untreated male rats, but not with RNA preparations of the thymus of adult rats nor with RNA from any tissue of untreated three-week-old rats. The 18S RNA fraction from livers of adult male rats had the greatest biological activity of any liver fraction tested. It remains to be seen whether the extracted, biologically-active RNA has mRNA or rRNA function *per se* or serves to stimulate endogenous activity.

The point made by Voigt (1968) during the Tremsbüttel Workshop discussions that it would be useful to study androgen regulation of the synthesis of a single protein rather than overall stimulation of general protein synthesis, remains as a challenge to investigators.

ATP-levels of the ventral prostate decline precipitously as early as 30 minutes after intramuscular injection of testosterone into the castrate rat, reach a nadir after one hour and then return to the normal value which is not significantly different from that of the castrate. This observation made by Ritter (1966) and subsequently confirmed by Coffey et al. (1968) may have far-reaching implications and has a parallel in a similar swift fall of uterine ATP in response to estrogens (Aaronson et al., 1965). It may reflect the synthesis of AA $\sim$ tRNA compounds or of precursors of new RNA and account for the testosterone-mediated restoration of cation-dependent ATPase which declines drastically on orchiectomy (Ahmed and Williams-Ashman, 1969). Farnsworth (1970) published data which implicate the K$^+$-dependent dephosphorylation, and not the Na$^+$-dependent phosphorylation, as the testosterone-sensitive component of the system.

Another possible pathway for the transient utilization of ATP is the formation of c-AMP as a second messenger of hormone action. Recent evidence that c-AMP

1. has an "andromimetic action" in the presence of theophylline as demonstrated by its actinomycin D- and cycloheximide-sensitive induction of hexokinase, phosphofructokinase, pyruvate kinase and glucose-6-phosphate dehydrogenase in the seminal vesicles of both orchiectomized and immature rats (Singhal et al., 1970);
2. tissue levels in rat ventral prostate and seminal vesicles decrease markedly after castration and increase following a single injection of tes-

tosterone propionate to castrates (Singhal, 1971) - on the other hand,
Rosenfeld and O'Malley (1970) were unable to elicit a stimulation of adenyl cyclase activity by testosterone in the ventral prostate of hypophysectomized rats *in vivo or in vitro*;
3. is synthesized in significant amounts from ATP by prostatic cell nuclei
   in the presence of KF and $Mg^{++}$ or $Mn^{++}$, the adenyl cyclase activity of
   the purified nuclear fraction amounting to 25 to 35% of that of cytoplasmic particulate fractions and, on nuclear extraction, associating tightly
   with the remaining chromatin sediments (Liao et al., 1971)
4. and protein kinase may play an important role in the regulation of RNA
   synthesis through a phosphorylation mechanism (Martelo et al., 1970)
tempts speculation that prostatic androgen receptors may be part of regulatory adenyl cyclase subunits.

Circulating testosterone and androstenedione appear to exert their biological activities in man as precursors of estrogens (Ryan, 1959; McDonald et
al., 1970) and of "active 5α-reduced metabolites". Tracing the fate of labeled
testosterone, administered by various routes, in accessory sex organs of
the rat has proved very helpful in characterizing the early events of uptake
(Butenandt et al., 1960; Tveter and Attramadal, 1969; Tveter, 1970) and retention which must precede the triggering mechanism of hormone action. Published work on uptake and androgen metabolism up to 1967 has been reviewed
by Ofner (1968). In contrast to the exclusive uterine retention of unmeta-
bolized estradiol-17β as cytoplasmic and nuclear protein complexes (Jensen
et al., 1968, 1970), radioactivity administered as tagged testosterone has
been recovered from analogous subcellular fractions of rat ventral prostate
predominantly as 5α-dihydrotestosterone (Bruchovsky and Wilson, 1968 a;
Mainwaring, 1969; Unhjem et al., 1969; Fang et al., 1969; Baulieu and Jung,
1970). We may cite the following lines of evidence for formation of "active"
intracellular transformation products of circulating androgens in target
tissues:
1. the fraction of administered labeled testosterone metabolized by the 17β-
   hydroxysteroid pathway (Baulieu and Mauvais-Jarvis, 1964) to urinary 5α-
   androstane-3a,17β-diol glucuronide increases after puberty and provides
   a measure of male sexual differentiation in patients with gonadal dys-
   function syndromes (Mauvais-Jarvis et al., 1968, 1970)
2. the cytosol of rat ventral prostate contains a protein fraction which exhibits an extremely high specificity toward 5α-dihydrotestosterone and
   will not bind tritiated testosterone, estradiol-17β, progesterone or cortisol at concentrations between 0.01 and 1 nM (Fang and Liao, 1971). Moreover, the retention of 5α-dihydrotestosterone by prostate-cell nuclei *in
   vivo*, in minced tissue preparations and in cell-free systems is a highly
   steroid- and tissue-specific phenomenon (Bruchovsky and Wilson, 1968 b;
   Fang et al., 1969; Fang and Liao, 1971)
3. work in this laboratory to be described later has shown that administration of a pharmacologic dose of estrogen, a standard practice in the
   treatment of hormone-sensitive prostatic cancer in man, has the effect of
   inhibiting $\Delta^4$-3-keto-$C_{19}$-steroid 5α-reductase, and thereby 5α-dihydro-
   testosterone formation, in the canine prostate with squamous metaplasia
   (Leav et al., 1971)
4. we have also noted a less marked, but substantial decrease in the *in vitro*
   conversion of testosterone to 5α-dihydrotestosterone by prostate tissue
   of the dog castrated one month before assay (Leav et al., 1971). This
   finding is in agreement with the reported decrease of 5α-reduction in the
   rat ventral prostate on castration, and restoration of the activity to
   the control level on testosterone replacement (Shimazaki et al., 1969)
5. antiandrogens like cyproterone acetate which reduce the increase in the
   weight of accessory sex organs induced by testosterone in the hypophysectomized animal (Neumann and von Berswordt-Wallrabe, 1966) have been
   shown to antagonize the formation of cytoplasmic and nuclear 5α-dihydro-
   testosterone complexes in the ventral prostate (Fang and Liao, 1969,

1971; Baulieu and Jung, 1970; Walsh and Korenman, 1971) and seminal
vesicles (Stern and Eisenfeld, 1969)

6. prostatic 5α-dihydrotestosterone derived from administered testosterone
is bound tightly to the protein moiety of nuclear chromatin (Wilson et
al., 1969; Fang et al., 1969) and may therefore form part of a regulatory
system for RNA transcription
7. 5α-dihydrotestosterone has a higher androgenic potency than testosterone
when assayed by the weight increase of the ventral prostate of the castrate weanling rat (Hilgar and Hummel, 1964) and is also more active than
the male hormone in promoting cell division and inducing hyperplasia in
rat prostate epithelium grown in organ culture (Baulieu et al., 1968)
8. the prostatic testosterone metabolite 5α-androstane-3β,17β-diol exhibits
very weak androgenic activity when administered systemically (Hilgar and
Hummel, 1964), yet is more effective than testosterone in stimulating
cell growth and secretion in the rat organ culture test (Baulieu et al.,
1968). The diol and its 17-ketosteroid analogue, *iso*androsterone, may be
responsible for a part of the effect at the target site previously attrib-
uted to the circulating androgens (Ofner et al., 1970).

It is apparent from these considerations that transformation of circulating
$C_{19}$-steroids in accessory sex organs must play a significant role in androgen action and be intimately involved in the mechanism of hormone responsiveness of these glands.

Our own major concern has been the study of comparative $C_{19}$-steroid metabolism in the canine and human prostate, motivated by the consideration that
the dog is the only species which shares with man an age-dependent tendency
to develop both benign prostatic hypertrophy (BPH) and adenocarcinoma of the
gland (Ofner, 1968; Leav and Ling, 1968). I sympathize with the urologist
(Valensin, 1965) who blames the inhibitions imposed by urban society that
prevent man and dog from urinating at will.

We have reported the fate of testosterone-4-[14]C and androstenedione-4-[14]C
in the prostate of six mature dogs and of two patients with BPH and previously untreated adenocarcinoma after infusion of physiological amounts of the
labeled steroids into the arterial supply of the glands (Morfin et al.,
1970, 1971; Ofner et al., 1970). The infusion procedures are depicted in
Figs. 1 and 2 and Table 1 compares the resulting prostatic radiometabolite

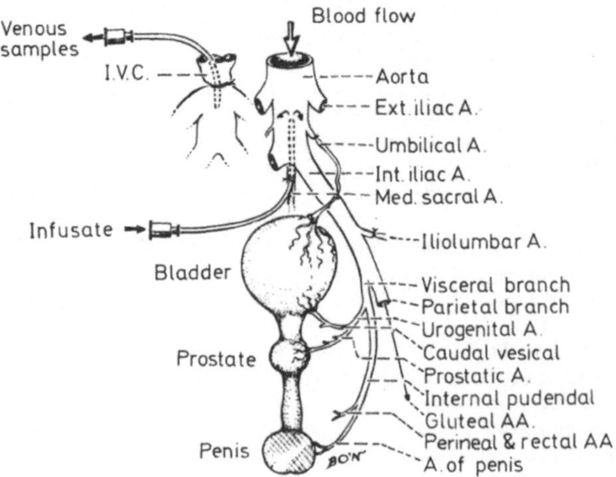


Fig. 1. A ventral dorsal representation of the prostatic infusion procedure
in the mature dog. Testosterone-4-[14]C or androstenedione-4-[14]C in
physiological saline were infused into the median sacral artery and
carried by blood flow to the prostate and urinary bladder (Morfin
et al., 1970; courtesy of Lippincott Co.).

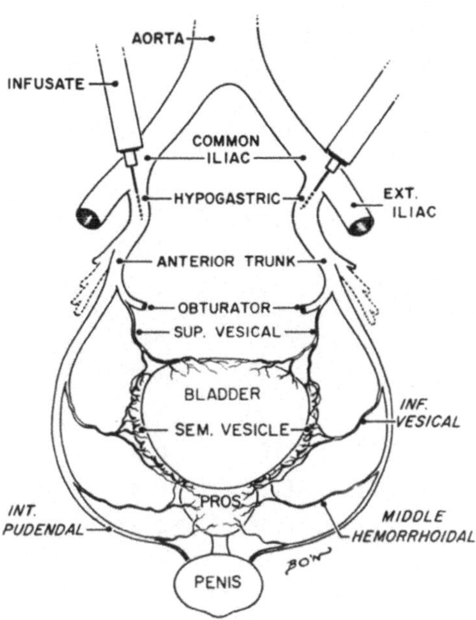



Fig. 2. An anterior representation of the prostatic infusion procedure in a patient with benign hypertrophy of the gland. Testosterone-4-$^{14}$C in physiological saline containing 2.5% ethanol was infused into each hypogastric artery and carried by blood flow to the prostate (Morfin et al., 1970; courtesy of Excerpta Medica Foundation, Amsterdam).

patterns. 5α-dihydrotestosterone was the major identified transformation product of testosterone-4-$^{14}$C both in the canine and human hypertrophic prostate, accounting for at least 35% and 63% of the radioactivity found in the respective glands. Whereas there was a several-fold predominance of 5α-reduced 17β-hydroxysteroids over 17-ketosteroids when the labeled male hormone was infused, canine prostatic metabolism of androstenedione-4-$^{14}$C *in vivo* resulted in almost exclusive conversion into 5α-reduced ketosteroids. 5α-androstane-3,17-dione and *iso*androsterone accounted equally for 65% of the prostatic radioactivity, testosterone and 5α-dihydrotestosterone for less than 1% and 5%, respectively; this finding may explain the low androgenicity of androstenedione. 3β-hydroxyradiometabolites predominated over the 3α-hydroxy epimers in the canine prostate, whereas the 3α-hydroxysteroids were preponderant in the human gland.

Our infusion experiments and incubation studies with minced and homogenized tissue preparations and subcellular fractions (Chamberlain et al., 1966; Ofner et al., 1971; Morfin, 1970; Morfin et al., 1970, 1971) have demonstrated that while all accessory sex organs assayed by us contain the four $C_{19}O_2$-steroid-metabolizing enzymes, $\Delta^4$-3-keto-$C_{19}$-steroid 5α-reductase, 17β-hydroxy-$C_{19}$-steroid oxido-reductase and 3α (and 3β)-hydroxysteroid oxido-reductase, the relative abundance of these activities may vary considerably with age, between the different lobes of the rat prostate, from one accessory sex organ to another, and between species. We have found these four enzymes in human epithelial-like cells (Fraley et al., 1970) harvested from explants of BPH tissue and subsequently grown in monolayer subcultures as uniform cell populations. This tissue-culture system may prove a useful model in which to study steroid binding and enzyme induction.

By a happy coincidence, important complementary results of determinations of prostatic $C_{19}$-steroid contents were published from Dr. Wilson's labora-

Table1. <u>Prostatic radiometabolite patterns after infusion of testosterone-4-[14]C and androstenedione-4-[14]C</u>

| Infused [14]C-Steroid | Canine | | Human |
| --- | --- | --- | --- |
| | TST | $A^4$-3,17-one | TST |
| Polar fraction | 3.7 | 4.5 | 4.0 |
| αA-3α,17β-ol | 1.9[a] | | 12.6[a] |
| αA-3β,17β-ol | 11.7[a] | 1.9 | 8.9[a] |
| Testosterone | 19.4[a] | 1.1 | 3.7 |
| Androsterone | 2.0[a] | 6.5 | 2.3 |
| Isoandrosterone | 6.7[a] | 32.7[a] | - |
| αA-17β-ol-3-one | 38.0[a] | 5.1 | 62.7[a] |
| Unknown | 7.8 | 0.9 | - |
| $A^4$-3,17-one | 1.2 | 5.4 | 2.8 |
| αA-3,17-one | 7.6[a] | 32.6[a] | 2.8 |
| Unknown | - | 9.3 | 0.2 |
| $\frac{17β\text{-ol Metabs}}{17\text{-one Metabs}}$ | 2.95 | 0.11 | 10.7 |

[a] Identified by crystallization to constant SA.

Steroid nomenclature according to Bush (1961); TST = testosterone.

The values are expressed as percentage contributions of the listed steroids to the total radioactivity found in the glands at sacrifice, based on elution of homogenous chromatographic fractions isopolar with submicrogram quantities of authentic reference radiosteroids.
(Morfin et al., 1971; courtesy of Excerpta Medica Foundation, Amsterdam)


tory (Siiteri and Wilson, 1970; Gloyna et al., 1970) simultaneously with our infusion data. Table 2 gives an abbreviated composite version of these results obtained by the remarkable $^3$H-exchange-labeling procedure for keto-

Table 2. <u>$C_{19}$-steroid content of human and canine prostate</u> (μg/100 g)

Procedure: Siiteri et al. (1968) Endocrinology <u>82</u>, 837

| | No. | Weight | T | α-DT | $A^4$-3,17-one |
| --- | --- | --- | --- | --- | --- |
| Human | | | | | |
| Normal prostate[a] | 15 | 18 ± 2 | 0.09 ± 0.03 | 0.13 ± 0.05 | 0.09 ± 0.03 |
| BPH[b] | 10 | 46 ± 10 | 0.09 ± 0.03 | 0.60 ± 0.10 | 0.04 ± 0.03 |
| Canine | | | | | |
| Immature | 7 | 2.3 ± 0.5 | 0.03 | 0.23 | |
| Mature | 9 | 10.2 ± 0.8 | 0.03 | 0.33 | |
| BPH | 8 | 27.7 ± 2.7 | 0.10 | 1.60 | |

[a] Autopsy Material
[b] Surgical Specimen
Siiteri and Wilson (1970); Gloyna et al. (1970) (courtesy of J. Clin. Invest.)

152

steroids of Siiteri et al. (1968). The presence of testosterone, 5α-dihydro-
testosterone and androstenedione in the human organ is thus established.
There is a five-fold increase in the concentration of 5α-dihydrotestosterone
in the hypertrophic as compared to the normal glands. Moreover, a regional
study of autopsy specimens of human hypertrophied prostate demonstrated that
"the dihydrotestosterone content was two and three-fold greater in the peri-
urethral area where prostatic hypertrophy usually commences than in the
outer regions of the gland". These results are in accord with the biological
properties of 5α-dihydrotestosterone in the organ-culture system mentioned
previously and implicate the "active" androgen in the pathogenesis of benign
prostatic hypertrophy.

I should now like to describe a series of experiments in which the effects
of estrogen administration and gonadectomy on canine prostatic fine structure
and $C_{19}$-steroid metabolism *in vitro* were investigated in tissue specimens
taken from the same gland before and 30 days after treatment (Leav et al.,
1971). Squamous metaplasia and atrophy of prostatic epithelium were used as
characteristic histologic markers of the respective procedures. Estrogen
treatment and castration produced similar nuclear and cytoplasmic alter-
ations in glandular epithelium, notably change from nuclear euchromatin to
heterochromatin patterns. Only the presence of numerous lysosomes and more
severe effects on granular endoplasmic reticulum served to distinguish the
prostatic fine structure of castrates from that of estrogen-treated dogs.
As shown in Table 3, estrogen treatment caused substantial increase in testosterone and 5α-dihydrotestosterone transformation by minced tissued preparations over levels found in biopsy specimens and a shift from reductive to

Table 3. <u>Proportion of normal values of canine prostatic $C_{19}$-steroid metabolism following estrogen administration or castration</u>

| | Testosterone | | | 5α-Dihydrotestosterone | | |
|---|---|---|---|---|---|---|
| | Substrate Metabolized | 5α-Reduction | 17β-ol Oxid. | Substrate Metabolized | 5α-3-Ketone Reduction | 17β-ol Oxid. |
| E | 2.16 | 0.23 | 6.86 | 1.96 | 0.33 | 8.80 |
| C | 0.80 | 0.46 | 1.57 | 0.89[a] | 0.79[a] | 1.60[a] |

[a] NADPH-Supplementation

Incubation Conditions:  610 pmoles, 0.48 μM substrate, 100 mg unsupplemented
tissue mince, 1 hr.

oxidative metabolism. The shift in the metabolism of testosterone from 5α-
reduced 17β-hydroxysteroids to androstenedione occurred to a lesser extent
also in the gland of the castrate, but reduction of 5α-dihydrotestosterone
to the 5α-androstanediols remained substantial. The relative contributions
of the reductive and oxidative pathways of prostatic $C_{19}$-steroid metabolism
may reflect the androgen-estrogen balance of the endocrine milieu to which
the gland is exposed.

In the foregoing discussion only the epithelium of the prostate has been
considered. There is great need for precise biochemical information about
factors which maintain the functioning epithelial cell-fibromuscular complex
(Franks et al., 1964, 1970). However, Tveter's autoradiographic study (1970)
has shown that the radioactivity of administered [3]H-testosterone is preferentially localized in the epithelial cells of the rat ventral prostate,
with significantly less labeling in the stroma and none in the secretions of
the glandular lumina.

In summary, if response of the prostatic epithelium to circulating androgen is mediated by 5α-dihydrotestosterone and its metabolites, the separate
enzyme and receptor proteins (Mainwaring, 1970; Unhjem, 1970) concerned with
its intracellular metabolism and translocation (Fig. 3) must play a regulatory role. The reported stimulation of cytidine incorporation into RNA of

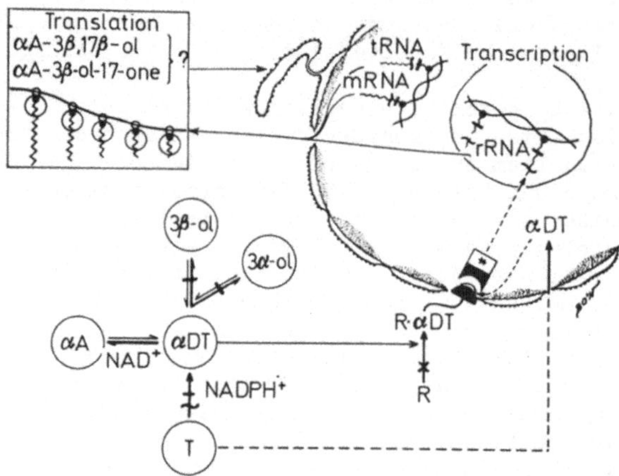


Fig. 3. The effect of estrogen, castration and anti-androgen on intracellular
        prostatic formation of "active" testosterone metabolites and on
        transport of 5α-dihydrotestosterone to sites of RNA transcription
        upon nucleolar DNA template. Abbreviations and symbols: T, testos-
        terone; αDT, 5α-dihydrotestosterone; 3α-ol, 3β-ol, 5α-androstane-3α
        (and 3β),17β-diols; R, 5α-dihydrotestosterone-specific cytoplasmic
        receptor protein; -, ~,x, adversely affected by estrogen, castration
        and anti-androgen, respectively; +, possibly adenyl cyclase with
        regulatory subunit. Use of the dashed line signifies lack of pub-
        lished evidence.

prostatic nuclei from immature rats in the absence of cytoplasmic protein
(Bashirelahi and Villee, 1970) by added 5α-dihydrotestosterone, but not by
testosterone, provides a rationale for the presence of nuclear 5α-reductase.
This finding does not agree with the view that 5α-dihydrotestosterone-
specific cytoplasmic receptor protein fulfills a function other than that
of transport vehicle.

We have seen different components of this postulated regulatory system
respond to a variety of anti-androgenic measures. It seems not unreasonable
to hope that biochemical characterization of these catalytic and receptor
proteins and further study of the effects of hormone treatment and with-
drawal on prostatic androgen metabolism and cellular transport will yield
new understanding and remedies of prostatic disease.

*Acknowledgments:* The author gratefully acknowledges the artistic contributions of Mr. Barry T. O'Neil and thanks Mr. Richard L. Vena for his invalu-
able technical support.

References

Aaronson, S.A., Nator, Y., Tarver, H.: Proc. Soc. exp. Biol. (N.Y.) 120, 9
  (1965).
Ahmed, K., Williams-Ashman, H.G.: Biochem. J. 113, 829 (1969).
Bashirelahi, N., Villee, C.A.: Biochim. biophys. Acta (Amst.) 202, 192 (1970).
Baulieu, E.E., Jung, I.: Biochem. biophys. Res. Commun. 38, 599 (1970).
- Lasnitzki, I., Robel, P.: In: "Advances in the Biosciences 3", p. 169
  (G. Raspé, ed.) New York: Pergamon Press, Braunschweig, Berlin: Vieweg,
  1968.
- Mauvais-Jarvis, P.: J. biol. Chem. 239, 1569 (1964).
Bruchovsky, N., Wilson, J.D.: J. biol. Chem. 243, 5953 (1968 a).
- - J. biol. Chem. 243, 2012 (1968 b).
Bush, I.E.: In: "The Chromatography of Steroids", p. xviii. New York:
  Pergamon Press, 1961.
Butenandt, A., Gunther, H., Turba, F.: Z. physiol. Chem. 322, 28 (1960).
Chamberlain, J., Jagarinec, N., Ofner, P.: Biochem. J. 99, 610 (1966).
Coffey, D.S, Ichinose, R.R., Shimazaki, J., Williams-Ashman, H.G.: Molec.
  Pharmacol. 4, 580 (1968).
Fang, S., Anderson, K.M., Liao, Sh.: J. biol. Chem. 244, 6584 (1969).
- Liao, Sh.: Molec. Pharmacol. 5, 428 (1969).
- - J. biol. Chem. 246, 16 (1971).
Farnsworth, W.E.: Biol. Reprod. 3, 218 (1970).
Fraley, E.E., Ecker, S., Vincent, M.M.: Science 170, 540 (1970).
Franks, L.M., O'Shea, J.D., Thomson, A.E.R.: Cancer (Philad.) 17, 983 (1964).
- Riddle, P.N., Carbonell, A.W., Gey, G.O.: J. Path. Bact. 100, 113 (1970).
Gloyna, R.E., Siiteri, P.K., Wilson, J.D.: J. clin. Invest. 49, 1746 (1970).
Hilgar, A.G., Hummel, D.J.: "Androgenic and Myogenic Endocrine Bioassay Data",
  Issue 1, pp. 1-243. Cancer Chemotherapy Nat. Serv. Center, Nat. Inst.
  Health, Bethesda, Md., 1964.
Huggins, C., Hodges, C.V.: Cancer Res. 1, 293 (1941).
- Webster, W.O.: J. Urol. (Baltimore) 59, 258 (1948).
Jensen, E.V.: In: "Some Aspects of Aetiology and Biochemistry of Prostate
  Cancer" (K. Griffiths and C.G. Pierrepoint, eds.), Third Tenovus Workshop,
  Cardiff, Wales, 1970, p. 151.
- Suzuki, T., Kawashima, T., Stumpf, W.E., Jungblut, P.W., DeSombre, E.R.:
  Proc. nat. Acad. Sci. (Wash.) 59, 632 (1968).
Karlson, P.: Perspect. Biol. Med. 6, 203 (1963).
Lacassagne, A.: C. R. Soc. Biol. (Paris) 113, 590 (1933 a).
- Villela, E.: C. R. Soc. Biol. (Paris) 114, 870 (1933 b).
Leav, I., Ling, G.V.: Cancer (Philad.) 22, 1329 (1968).
- Morfin, R.F., Ofner, P., Cavazos, L.F., Leeds, E.B.: Endocrinology 89,
  465 (1971).
Liao, Sh., Barton, R.W., Lin, A.H.: Proc. nat. Acad. Sci. (Wash.) 55, 1593
  (1966 a).
- Fang, S.: Vitam. and Horm. 27, 17 (1969).
- Lin, A.H.: Proc. nat. Acad. Sci. (Wash.) 57, 379 (1967).
- - Barton, R.W.: J. biol. Chem. 241, 3869 (1966 b).
- - Tymoczko, J.L.: Biochim. biophys. Acta. (Amst.) (1971, in press).
- Stumpf, W.E.: Endocrinology 83, 629 (1968).
- Mainwaring, W.I.P.: J. Endocr. 45, 531 (1969).
- Biochem. biophys. Res. Commun. 40, 192 (1970).
Martelo, O.J., Woo, S.L.C., Reimann, E.M., Davie, E.W.: Biochemistry (Wash.)
  9, 4807 (1970).
Mauvais-Jarvis, P., Bercovici, J.-P., Crépy, O., Gauthier, F.: J. clin.
  Invest. 49, 31 (1970).
- Floch, H.H., Bercovici, J.-P.: J. clin. Endocr. 28, 460 (1968).
Moore, R.A.: In: "Endocrinology of Neoplastic Diseases" (G.H. Twombly and
  G.T. Pack, eds.) Chapter 8. London and New York: Oxford Univ. Press,
  1947.

Moore, R.A.: In: "Cowdry's Problems of Ageing" (A.I. Lansing, ed.), 3rd ed.
    Baltimore, Md.: Williams and Wilkins, 1952.
Morfin, R.F.: "Métabolisme des Androgènes au Niveau des Prostates Humaines
    et Canines". Thesis for the Doctorat d'Etat of the Faculty of Sciences
    of the University of Paris (1970).
-   Aliapoulios, M.A., Bennett, A.H., Harrison, J.H., Ofner, P.: Proc. 3rd
    Intern. Congr. Hormonal Steroids, Hamburg, 1970, Excerpta med. Intern.
    Congr. Ser. Excerpta Med. Found., Amsterdam (in press 1971).
-   - Chamberlain, J., Ofner, P.: Endocrinology 87, 394 (1970).
McDonald, P., Beyer, C., Newton, F., Brien, B., Baker, R., Tan, H.S.,
    Sampson, C., Kitching, P., Greenhill, R., Pritchard, D.: Nature 227,
    964 (1970).
Neumann, F., Berswordt-Wallrabe, R. von: J. Endocr. 35, 363 (1966).
Ofner, P.: Vitam. and Horm. 26, 237 (1968).
-   Morfin, R.E., Vena, R.L., Aliapoulios, M.A.: In: "Some Aspects of Aetio-
    logy and Biochemistry of Prostate Cancer" (K. Griffiths and C.G. Pierre-
    point, eds.) Third Tenovus Workshop, Cardiff, Wales, 1970, p. 551.
Price, D.: In: "The Biology of the Prostate and Related Tissues" (E.P.
    Vollmer and G. Kauffmann, eds.), Nat. Cancer Inst. Monograph 12 (1963).
-   Ingle, D.J.: Rev. suisse Zool. 64, 743 (1957).
Ritter, C.: Molec. Pharmacol. 2, 125 (1966).
Rosenfeld, M.G., O'Malley, B.W.: Science 168, 253 (1970).
Ryan, K.J.: J. biol. Chem. 234, 268 (1959).
Shimazaki, J. Matsushita, I., Furuya, N., Yamanaka, H., Shida, K.: Endocr.
    jap. 16, 453 (1969).
Siiteri, P.K., Tippit, P., Yates, C., jr., Porter, J.C.: Endocrinology 82,
    837 (1968).
-   Wilson, J.D.: J. clin. Invest. 49, 1737 (1970).
Singhal, R.L.: Personal Communication (1971).
-   Vijayvargiya, R., Ling, G.M.: Science 168, 261 (1970).
Stern, J.M., Eisenfeld, A.J.: Science 166, 233 (1969).
Tveter, K.J.: Acta Endocr. (Kbh.) 63, 489 (1970).
-   Attramadal, A.: Endocrinology 85, 350 (1969).
Unhjem, O.: Acta Endocr. (Kbh.) 65, 525 (1970).
-   Tveter, K.J., Aakvaag, A.: Acta Endocr. (Kbh.) 62, 153 (1969).
Valensin, G.: In: "La prostate", p. 38. La Jeune Parque, Utrecht: Bosch,
    1965.
Villee, C.A., Fujii, T.: In: "Testosterone". (J. Tamm, ed.). Proc. Workshop
    Conf. Tremsbüttel, Germany, 1967, p. 96. Stuttgart: Thieme, 1968.
-   - Proc. nat. Acad. Sci. (Wash.) 62, 836 (1969).
Voigt, K.D.: In: "Testosterone" (J. Tamm, ed.). Proc. Workshop Conf. Trems-
    büttel, Germany, 1967, p. 101. Stuttgart: Thieme, 1968.
Walsh, P.C., Korenman, S.G.: J. Urol. 105, 850 (1971).
Watson, J.D.: "In: "Molecular Biology of the Gene", 2nd ed. p. 372. New
    York: Benjamin, Inc., 1970.
Williams-Ashman, H.G.: In: "The Androgens of the Testis", (K.B. Eik-Nes,
    eds.), p. 117. New York: Marcel Dekker, Inc., 1970.
-   Liao, Sh., Hancock, R.L., Jurkowitz, L., Silverman, D.A.: Recent Progr.
    Hormone Res. 20, 247 (1964).
Wilson, J.D., Bruchovsky, N., Chatfield, J.N.: Proc. 3rd Intern. Congr. Ser.
    No. 184, p. 17. Excerpta Med. Found., Amsterdam (1969).
Woodruff, M.W., Perez-Mesa, C.: J. Urol. 88, 273 (1962).
-   Umiker, W.O.: J. Urol. 84, 162 (1960).

Symp. Dtsch. Ges. Endokrin. _17_, 157–164 (1971)
© by Springer-Verlag

# Neue Gesichtspunkte der Therapie von benignen und malignen Neoplasien der Prostata

## New Aspects in the Treatment of Benign and Malignant Neoplasias of the Prostate

H. KLOSTERHALFEN

Urologische Universitätsklinik Hamburg

Mit 3 Abbildungen

Der Formulierung des Themas könnte man entnehmen, daß es neue Gesichtspunkte der Therapie bedeutenderer Art tatsächlich gibt, eine Vermutung, die nur in beschränktem Ausmaß zutrifft und wenn sie zutrifft, den konservativ endokrinologischen Sektor ebenso wie den instrumentellen bzw. operativen Sektor betrifft.

Die Forderung nach der Frühoperation der Prostatiker ist eigentlich keine Novität mehr, neu daran ist allerdings, daß sich die Frühoperation auch tatsächlich in der Praxis durchzusetzen beginnt. Unter Frühoperation verstehen wir die Operation in einer Phase der Erkrankung, in der erst geringe Restharnmengen (50 ml) oder auch noch gar kein Restharn, aber der Größe der "Prostata" entsprechende Beschwerden mit Dysurie und Nykturie vorhanden sind. In diesen Fällen kann man den Zeitpunkt absehen, an dem die konservative Behandlung nichts mehr nützt, und hier stellen wir die Indikation zur Frühoperation, für die es überzeugende Gründe gibt:

Es ist eine inzwischen bewährte Erfahrungstatsache, daß verhältnismäßig junge Prostatiker der Operationsbelastung besser gewachsen sind, eine geringere Komplikationsrate aufweisen und kürzere Rehabilitationszeiten haben. Tatsächlich ist es in der Praxis aber noch vielfach so, daß dem Prostatiker mit bereits entwickeltem Adenom aber noch geringen Restharnmengen zwar angedeutet wird, später sei eine Operation erforderlich, aber dies erst in einigen Jahren, weil die geringe Höhe des Restharns den Eingriff zur Zeit noch nicht rechtfertige. Man behält den Kranken "in Beobachtung", und das heißt mit anderen Worten: Therapeutisch geschieht nichts Entscheidendes, bis der Zeitpunkt kommt, an dem die Blasenentleerung bedenklich oder unmöglich wird.

Diese allgemein noch verbreitete Art der Beratung bedeutet nichts anderes als die Verschiebung schließlich doch unumgänglicher therapeutischer Maßnahmen auf einen späteren, und das kann nur heißen, auf einen ungünstigeren Zeitpunkt. Die Kranken so lange der konservativen Behandlung zuzuführen, bis sich – man möchte fast sagen wie erwartet – ihre Unzulänglichkeit bestätigt hat, ist eine unärztliche Einstellung, zumal die bei längerer abwartender Behandlung eintretende Nieren- und Kreislaufinsuffizienz die schon bei Beginn der Behandlung als letzten Endes unvermeidlich angesehene Operation zu einem Risikoeingriff macht.

Im Gegensatz zu dieser leider noch viel geübten Praxis steht die Forderung nach der Frühoperation des Prostatikers. Sie basiert auf der längst bewährten Erfahrung, daß die operative Behandlung, mit der heutigen Technik und zum richtigen Zeitpunkt durchgeführt, kein Risikoeingriff mehr ist (Mortalität 2%), und daß es für eine leistungsfähige Klinik nur noch wenige inoperable Prostatiker gibt. In diesem Zusammenhang gilt im allgemeinen, daß ein Patient, der einen Katheter oder eine suprapubische Blasenfistel auf die Dauer verkraftet, auch eine Prostatektomie überstanden hätte.

Für Patienten mit tatsächlich hohem Operationsrisiko (Herz- oder Lungen-
insuffizienz) oder für Patienten, die effektiv als inoperabel bezeichnet
werden müssen, gibt es ein neues Behandlungsverfahren, die *Kryochirurgie*.
Kryochirurgie heißt lokale Anwendung extremer Kälte. Das Prinzip der Methode
besteht darin, daß über eine transurethral eingeführte Sonde flüssiger
Stickstoff mit Temperaturen um -190° an die Prostata gebracht wird, die dann
nach wenigen Minuten regelrecht vereist ist. Durch den Gefriervorgang zer-
reißen die intrazellulären Strukturen und es kommt zum Zelltod. Je länger
der Vereisungsvorgang dauert, umso größer wird die Gefrierzone und umso größer wird auch die Gefahr der Schädigung von Nachbarorganen, wobei vor allem
das Rektum gefährdet ist. Nach Entfernung der Kältesonde wird lediglich ein
Dauerkatheter eingelegt, über den die nekrotischen Anteile der Prostata her-
ausgespült werden. Dieser Abstoßungsvorgang kann wochenlang dauern und
stellt einen unangenehmen Nachteil der ansonsten mit relativ geringem Risiko
belasteten Methode dar.

Dieses neue Behandlungsverfahren kommt nicht nur für die gutartigen Neubildungen der Prostata sondern auch  für das Carcinom in Frage und hier gibt
es noch einen weiteren interessanten Gesichtspunkt: In der Literatur der
letzten Monate gibt es mehrere Mitteilungen darüber, daß es nach mehrfachem
Vereisen der Prostata zu röntgenologisch gesicherten Konsilidierungsvorgän-
gen an Knochenmetastasen gekommen ist. Eine stichhaltige Erklärung für diese klinische Beobachtung gibt es noch nicht, dagegen sind Überlegungen in
der Richtung angestellt worden, ob es aufgrund der mehrfachen Vereisungsvor-
gänge in den Randgebieten der Vereisungszone, die der Nekrose nicht komplett
anheimfallen, zu gewissen zellulären Reaktionsabläufen kommt, die zu einer
Art Antikörperbildung gegen die Carcinomzellen führen.

Soviel zu den instrumentellen bzw. operativen neuen Aspekten der Therapie,
und damit komme ich zu den *endokrinologischen Gesichtspunkten* der Behandlung.

Die Sexualhormone haben in der konservativen Behandlung der benignen und
malignen Neoplasien der Prostata seit Jahren ihren festen, wenn auch nicht
immer wissenschaftlich fundierten Platz. Die Auswahl der applizierten Hormone unterlag bei der Therapie der benignen sogenannten Prostata-Hypertrophie
nicht immer objektiven Kriterien und wechselte in den letzten 40 Jahren
mehrfach:

Von der Androgen-Therapie kam man um 1945 zur Östrogen-Behandlung und
propagierte dann die kombinierte Androgen-Östrogen-Therapie, und wie Sie
wissen, wird zur Zeit mit erheblichem Werbeaufwand die Gestagen-Behandlung
propagiert. In Anbetracht dieser therapeutischen Sprünge erscheint es angebracht, zunächst nach der Abhängigkeit der Prostata vom endokrinen System zu
fragen, und dazu ist folgendes zu sagen: Frühkastraten erkranken weder an
einer benignen noch an einer malignen Veränderung der Prostata, während die
Kastration nach der Pubertät zur Atrophie der Vorsteherdrüse führt. Diese
Atrophie tritt normalerweise auch mit zunehmendem Lebensalter, das heißt mit
nachlassender männlicher Hormonproduktion auf. Fast gesetzmäßig kommt es
gleichlaufend zu einer Wucherung des periurethralen Gewebes zwischen Colli-
culus seminalis und Blasenhals, und daraus entsteht das klinische Bild des
periurethralen Adenoms oder, wie man es auch nennt, des Blasenhals-Adenoms,
das wir fälschlicherweise als Prostata-Adenom oder Prostata-Hypertrophie be-
zeichnen. Dieser Terminus hat sich inzwischen so eingebürgert, daß es unmöglich ist, ihn zu korrigieren und ihn durch einen passenderen zu ersetzen.
Formalgenetisch handelt es sich immer um ein Adenomyofibrom des periurethralen Gewebes, das nichts mit der Vorsteherdrüse des Mannes zu tun hat. Die
Bezeichnung "Prostata-Hypertrophie" ist im morphologischen und pathogeneti-
schen Sinne also falsch. Das, was man mit Prostata-Hypertrophie meint, ist
eine echte Gewebsneoplasie in der Pars prostatica der Harnröhre. Mit der
Prostata hat diese Wucherung nur insofern etwas zu tun, als der gutartig
expansiv wachsende Tumor das Prostatagewebe vom Harnröhrenvolumen her zur
Druckatrophie bringt.

Tierexperimentelle Untersuchungen haben Hinweise für eine Öströgen-Abhän-
gigkeit des Adenomwachstums ergeben. In bestimmten Teilen der Prostata-Anlage
lassen sich durch Östrogene Wucherungen erzielen, die denen der Situation

beim Menschen vergleichbar sind. Diese östrogenbedingten Wucherungen können
durch Androgen-Applikation verhindert werden.

Überprüft man nun die endokrine Situation des Mannes in verschiedenen
Stadien des Lebensalters, so findet man folgende Situation:

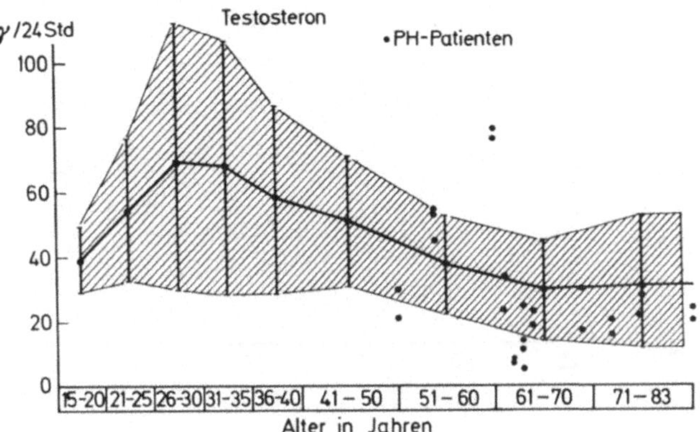


Abb. 1. Testosteron-Tagesausscheidung bei gesunden Männern.

Das Maximum der Testosteron-Ausscheidung, hier gemessen an der Tagesausscheidung, liegt um das 30. Lebensjahr. Dem folgt ein kontinuierlicher Abfall bis zum 65. Lebensjahr. Die Testosteronausscheidungswerte sinken bei
den Normalpersonen auf 41% des Maximalwertes.

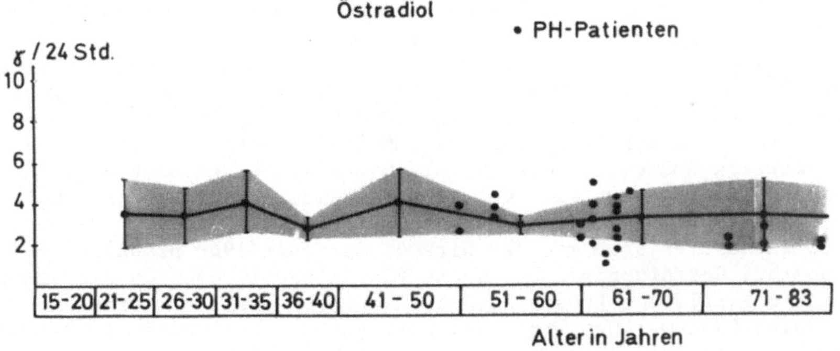


Abb. 2. Östradiol-Ausscheidung bei gesunden Männern.

Die *Östrogene*, ebenfalls an der Tagesausscheidung gemessen, zeigen zwischen dem 4. und 6. Lebensjahrzehnt dagegen weder eine Altersabhängigkeit
noch einen Unterschied zwischen den behandlungsbedürftigen Prostatikern und
den Normalpersonen.

Die kontinuierliche Testosteronverminderung bei gleichbleibenden Östrogen-
werten kommt einer relativen Östrogenvermehrung im alternden männlichen
Organismus gleich. Das Verhältnis Testosteron zu Östradiol verschiebt sich

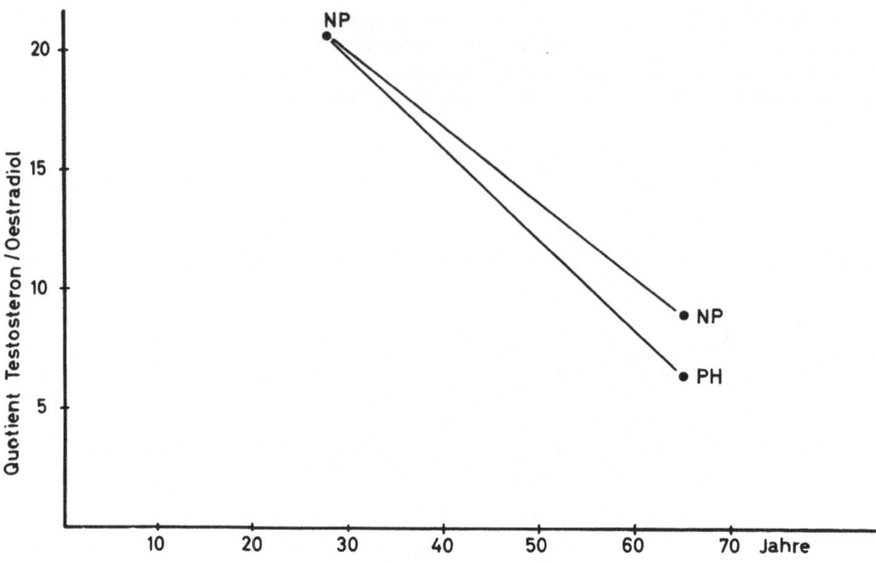


Abb. 3. Quotient Testosteron/Östradiol.

bei den behandlungsbedürftigen Prostatikern stärker zugunsten der Östrogene als bei den nicht behandlungsbedürftigen Männern.

Nun zur klinischen Bedeutung dieser Befunde: Wenn dem so ist und alle Anzeichen sprechen dafür, daß es so ist, dann wäre es nur konsequent, das Testosteron-Defizit durch eine entsprechende *Testosteron*-Substitution zu korrigieren. Das ist auch bis etwa 1945 geschehen, ohne daß man derart exakte Angaben über die endokrine Funktion im männlichen Organismus zur Verfügung hatte. Danach entdeckte man die Androgenabhängigkeit des Prostata-Carcinoms und dessen gute Ansprechbarkeit auf die Behandlung mit weiblichen Hormonen. Und unter dem Eindruck der klinischen Erfolge dieser gegengeschlechtlichen Hormontherapie beim Carcinom begann man nun auch, das gutartige Adenom mit *weiblichen Hormonen* zu behandeln.

Den enthusiastischen ersten Berichten über Therapieerfolge, die in dem Schlagwort "Hormon statt Messer" gipfelten, folgten bald kritischere Stimmen, insbesondere nachdem man erfahren mußte, daß eine meßbare Größenverminderung einer einmal bestehenden Adenombildung nicht eintrat. Blasendruckmessungen zeigten später, daß das klinisch bessere Miktionsvermögen der hormonell behandelten Adenomträger auf eine Tonisierung der Blasenmuskulatur zurückzuführen war.

Ähnlich verhält es sich mit der Wirkung der seit 1966 propagierten *Gestagene* (Depostat, Gestafortin). Gestagene führen zur Involution der Samenblasen und der eigentlichen Prostatadrüse, nicht aber zur Verkleinerung des periurethralen Adenoms. Dem Wirkungsmechanismus entsprechend klagen gestagenbehandelte Patienten über Potenzminderung bis zur kompletten Impotenz und an dieser Stelle sei der Hinweis gestattet, daß die Zahl derjenigen Männer im "Prostatiker-Alter", die sexuell noch aktiv sind und die es auch bleiben wollen, überraschend hoch ist, und daß dieser Gesichtspunkt zum Beispiel auch bei der Wahl des jeweiligen Operationsverfahrens berücksichtigt werden muß.

Auch durch die Verabreichung von *Androgenen* ist keine Rückbildung eines bereits vorhandenen Adenoms zu erzielen. Wir sind aber berechtigt anzunehmen, daß durch eine frühzeitige Substitution des Testosterondefizits die Entwicklung eines Prostata-Adenoms zeitlich verzögert bzw. in seinem Wachstum gebremst werden kann. (Dosierung 3 Tabletten Mesterolon = Proviron).

Mesterolon ist deshalb zu empfehlen, weil im Gegensatz zur Testosteron-
Applikation die Gonadotropinsekretion der Hypophyse nicht gehemmt wird.

Im Zusammenhang mit der Androgenapplikation taucht natürlich die Frage
nach der möglichen Aktivierung eines klinisch okkulten Prostata-Carcinoms
auf. Diesbezügliche Untersuchungen sind 1955 an einem größeren Kollektiv
durchgeführt worden. Obwohl diese Untersuchung keine signifikanten Hinweise
auf die mögliche Provokation eines latenten Carcinoms ergab, andererseits
die Prostata-Carcinom-Zelle in ihrem Wachstum androgenabhängig ist, muß man
als Voraussetzung für die Behandlung mit männlichen Hormonen regelmäßige und
gewissenhafte rektale Kontrollen des Tastbefundes an der Prostata fordern.

Dieses Kapitel *zusammenfassend* ist abschließend festzustellen:

Es gibt zur Zeit kein Medikament und kein Hormon, das eine Verkleinerung
oder eine rückläufige Entwicklung eines bereits existenten periurethralen
Adenoms bewirken könnte. Das gilt sowohl für die Behandlung mit weiblichen
und männlichen Hormonen als auch für die Gestagene. Die Wirkung der Hormone
erschöpft sich im Symptomatischen. Bestenfalls kommt es zu einer vorüberge-
henden Besserung der Miktion auf der Basis der Tonusänderung der Blasenmuskulatur. Heilung ist nur durch operative Entfernung der gutartigen Neoplasie
zu erzielen, und für jeden Einzelfall ist hier die jeweilig beste Methode in
Form der offenen Prostatektomie, der transurethralen Elektroresektion oder
der Vereisung der Prostata heranzuziehen. Ist bei einem Kranken bereits Restharn vorhanden, dann darf der günstigste Zeitpunkt für den operativen Eingriff nicht durch konservative Behandlungsversuche hinausgeschoben werden,
die sich im übrigen im Funktionellen erschöpfen und deren Effekt meist
passagerer Natur ist.

Wenn man aus im einzelnen hier nicht zu erörternden Gründen gezwungen
ist, konservative Behandlungsmaßnahmen einzuleiten, dann sind nicht Östrogene sondern - unter der Voraussetzung regelmäßiger rektaler Palpationskon-
trollen der Prostata - Androgene zu empfehlen. Beide Hormone wirken über
eine Tonusverbesserung der Blasenmuskulatur. Während die Androgene jedoch
unter Umständen einen Stillstand, eine Wachstumsverzögerung der Adenombildung bewirken können, liefern die Östrogene aufgrund der pathogenetisch bekannten Faktoren optimale Wachstumsbedingungen für die Adenombildung.


## Prostata-Carcinom

Neue Aspekte der Therapie beim Prostata-Carcinom gibt es sowohl auf dem operativen bzw. instrumentellen Sektor als auch auf konservativem, das heißt
endokrinologischem Gebiet. Die *Kryochirurgie* als neue instrumentelle Behand-
lungsmethode wurde schon erwähnt, und in diesem Zusammenhang möchte ich noch
einmal auf die interessante Beobachtung der Metastasenveränderung nach mehr-
maligem Vereisen eines carcinomatösen Prostata-Primärherdes hinweisen.

Was die *Radikaloperation* angeht, so kann man konstatieren, daß die bisher
zu beobachtende Zurückhaltung in der Indikationsstellung vor allem darauf
zurückzuführen war, daß man wegen der Symptomenarmut nur ganz selten einmal
einen Fall im ersten oder zweiten, also noch dem radikal operablen Stadium,
zu Gesicht bekam. Mit den jetzt anlaufenden Vorsorgeuntersuchungen des Mannes wird sich diese Situation sicher ändern, denn dann wird man auch die
bisher im Krankengut fehlenden Frühstadien erfassen. Nach der Erfahrung amerikanischer Operateure läßt sich mit der Radikaloperation immerhin eine
5-Jahres-Überlebenszeit bei 70% der Patienten erreichen, und damit dürfte
die Bedeutung der Präventivuntersuchung für den Mann und hier insbesondere
für den häufigsten Organkrebs des über 50jährigen Mannes klargestellt sein.
Ebenso wichtig ist es zu wissen, daß nur die Radikaloperation eine echte
Heilung verbürgt, während alle konservativen Behandlungsmethoden trotz oft
verblüffender Anfangserfolge zeitbegrenzte Wirkung haben und letzten Endes
versagen. Der Objektivität halber muß man jedoch darauf hinweisen, daß auch
fortgeschrittene Prostata-Carcinome mit einer durchaus erlebenswerten Le-
bensverlängerung zu rechnen haben, wenn man das Gesamtregister der verschiedenen konservativen Behandlungsmöglichkeiten kennt und es auf den Einzelfall
abgestimmt einzusetzen weiß.

Grundlage der konservativen Behandlung und jeder Variation dieser konservativen Behandlung ist die vor 30 Jahren von Huggins gewonnene Erkenntnis, daß Androgene auf das Wachstum der Carcinomzelle fördernd, Östrogene dagegen hemmend einwirken. Von dieser heute unverändert gültigen Konzeption sind alle weiteren Behandlungsverfahren ausgegangen. Behandlungsziel ist also die möglichst rigorose Ausschaltung der vom Organismus in verschiedenen Produktionsstätten sezernierten androgenen Substanzen. Über den optimalen Weg, Hoden und Nebennieren funktionell oder organisch auszuschalten, gibt es eine weltweite Kontroverse, die sich vor allem auf die Frage Orchidektomie oder Behandlung mit östrogenen Substanzen oder einer Kombination der beiden zugespitzt hat. Für beide Ansichten gibt es gute Gründe und auch Gegenargumente, die teilweise subjektiv gefärbt sind, die aber aus der psychologisch bedingten Situation der Patienten und der behandelnden Ärzte zu verstehen sind: Immerhin handelt es sich, gleichgültig, ob Orchidektomie oder Östrogen-Behandlung, um einen schwerwiegenden Eingriff in das gesamte hormonelle Gefüge und damit in die Gesamtpersönlichkeit eines Menschen.

Zur Zeit sieht es nicht so aus, als ob man in absehbarer Zeit zu einer gemeinsamen Meinungsbildung in dieser Frage kommen würde und deshalb möchte ich Ihnen unsere Ansicht zu diesem Fragenkomplex vortragen.

90% der Androgene werden in den Hoden gebildet und es ist deshalb konsequent, diese Hauptproduktionsstätte durch den einfachen und nicht belastenden Eingriff der Orchidektomie auszuschalten. Der vor allem von amerikanischer Seite geäußerte Einwand, die Orchidektomie stelle für den Patienten eine zu große psychische Belastung dar, trifft für unsere Verhältnisse sicher nicht in diesem Ausmaß zu. Wir kennen keinen Patienten, der die Operation abgelehnt hat, insbesondere wenn man ihm die Dringlichkeit mit für ihn einsehbaren und verständlichen Argumenten erklärt. Hier spielt natürlich auch die Art der Formulierung eine Rolle: Die Reaktion wird eine andere sein, wenn man anstelle der Kastration die subkapsuläre Entfernung des Hodenparenchyms in Vorschlag bringt, wenn man also nur das Keimepithel entfernt und die Hodenhüllen wie auch die Nebenhoden als Scrotalinhalt beläßt. Wir sind also für die Orchidektomie und gegen eine alleinige Östrogen-Behandlung.

Die *Östrogen-Behandlung* wird nunmehr seit 30 Jahren geübt. Die Erfolge sprechen für sich, sie sind besonders zu Anfang der Behandlung eindrucksvoll und sie sind bequem zu erzielen. Trotzdem steht das Prostata-Carcinom in der Häufigkeit der Krebstodessterblichkeit der Männer unverändert an 3. Stelle. Während also an der Huggins'schen Konzeption keinerlei Zweifel bestehen, ist neuerdings über die Frage der lebenslänglichen Hormonzufuhr, die schon als entschieden galt, eine gewisse Diskussion in Gang gekommen. Es sind sozusagen die Nebenwirkungen der Östrogene, die die Argumente gegen ihre lebenslängliche Applikation liefern und denen unter dem Eindruck der imposanten Anfangserfolge wenig Beachtung geschenkt wurde. Unter den Nebenwirkungen einer ausweichenden Östrogen-Behandlung sind die Gynäkomastie, die allgemeine Feminisierung, Verlust von Libido und Potenz die bekanntesten, vom krebstherapeutischen Standpunkt betrachtet aber auch die nebensächlichsten Nebenwirkungen.

Für die Kliniker wird es interessant sein, daß man die für viele Patienten so lästige und auch schmerzhafte therapiebedingte Gynäkomastie durch eine prophylaktische Röntgenbestrahlung der Mammille verhindern kann (800 r Herddosis).

Wesentlich wichtiger als die eben genannten Nebenwirkungen in Bezug auf den Verlauf der Carcinom-Erkrankung ist die auf eine fortlaufende Östrogen-Zufuhr folgende Hyperplasie der Nebennierenrinde, die nach dem Ausfall der Androgenproduktion in den Hoden vikariierend ihre Androgenproduktion steigert. Damit wird ein gewisser Teil der das Krebswachstum blockierenden Maßnahme der Östrogenbehandlung bzw. der Orchidektomie wieder aufgehoben. In dem Zusammenhang ist ein weiterer Gesichtspunkt von Bedeutung: Unter ständiger Zufuhr östrogenwirksamer Substanzen wird die hypophysäre *Prolaktin*-sekretion gesteigert und dies wiederum führt zu einer Potenzierung der verbliebenen Testosteronwirkung, also auf längere Sicht gesehen zu einer Aktivierung des Carcinom-Wachstums.

Eine weitere wichtige Nebenwirkung der fortlaufenden Östrogen-Applikation
ist die *Natrium-Retention* mit allgemeiner Ödemneigung und verstärkter Anfäl-
ligkeit des männlichen Organismus zu Kreislauferkrankungen. In diesem Zusammenhang muß auf die Mellinger-Studie hingewiesen werden, deren Schlußfolgerungen zwar umstritten sind, deren Ergebnis jedoch so gewichtig ist, daß man
für die Therapie Konsequenzen ziehen sollte. Die Mellinger-Studie besagt,
daß die Quote der Kreislauftoten unter fortlaufender Östrogen-Behandlung die
Zahl der durch die Östrogen-Behandlung überlebenden Carcinom-Patienten nahezu aufwiegt. An dieser Stelle sei mir die Bemerkung gestattet, daß wir
selbst bereits 1961, also lange vor der Mellinger-Studie, auf diesen gravie-
renden Nebeneffekt der Östrogen-Behandlung beim Mann hingewiesen und aus
diesem Grunde die gleichzeitige Verabreichung von Natriuretica empfohlen haben. Die Bedenken hinsichtlich der Nebenwirkungen gelten für die natürlichen
Hormone und die synthetischen Östrogene gleichermaßen, am wenigsten jedoch
für die östrogen-wirksamen Substanzen in der Form des Diäthyl-Dioxy-Stilben
(Honvan).

Um das Für und Wider der Östrogen-Behandlung beim Prostata-Carcinom abzu-
runden und abzuschließen, noch folgende Feststellung:

Wenn man das antiandrogene Behandlungsprinzip anerkennt, dann ist es logischer, anstelle der Östrogene *Cortison* einzusetzen. Die nach einer Orchidektomie im Urin noch meßbare nebennierenbedingte Testosteronausscheidung
kann man, wie wir das gemeinsam mit Voigt u. Tamm 1965 gezeigt haben, komplett zum Verschwinden bringen. Auch die Anhänger einer ausschließlichen
Östrogen-Behandlung müssen das antiandrogene Behandlungsprinzip notgedrungen
anerkennen, weil es bis heute keinen exakten Beweis dafür gibt, ob Östrogene
lokal an der Prostata überhaupt wirksam werden können, d. h. im antimitoti-
schen Sinne wirksam werden können.

Wie Sie wissen, gibt es inzwischen Substanzen, die ohne den Umweg über
die Hypophyse die Wirkung der Androgene unmittelbar am Erfolgsorgan blockie-
ren können, die sogenannten *Antiandrogene* (Cyproteron). Es lag natürlich nahe, diese neue Substanz gegen das Prostata-Carcinom einzusetzen. Die bisherigen Behandlungsergebnisse lassen sich jedoch noch nicht auf einen Nenner
bringen, und dies ist unter Umständen eine Frage der Dosierung. Wir selbst
betreiben seit über 2 Jahren eine Prospektivstudie mit 5 verschiedenen Be-
handlungsgruppen. Von diesen 5 Gruppen scheint es der Antiandrogen-Gruppe
(25 Patienten) neben der Cortison-Gruppe besonders gut zu gehen, ein subjek-
tiver Eindruck, der von Arbeitskreisen in Essen und München geteilt wird. Es
wäre sicherlich verfrüht, mehr als diesen vorläufigen Eindruck mitzuteilen.
Man muß abwarten, wie die Langzeitergebnisse aussehen, und auch die Frage,
welchen Einfluß die Gestagenwirkung des Cyproteron auf den Tumor selbst und
die Blasenentleerung hat, bedarf noch der Beantwortung.


## Sekundärverfahren

Ebenso wie beim Mamma-Carcinom hat die Adrenalektomie den ihr vorübergehend
zuerkannten Behandlungswert inzwischen an die Hypophysektomie verloren. Die
Voraussetzungen der Hypophysektomie sind insofern besser, als sich die Wirkung auf das gesamte funktionstüchtige Nebennierenrindengewebe erstreckt,
also auch auf ektopische Zellverbände. Was die optimale Methode zur Hypophy-
senausschaltung angeht, scheint sich eine neue Entwicklung anzubahnen: Die
mit uns im Rahmen des Sonderforschungsbereiches Endokrinologie kooperierende
neurochirurgische Arbeitsgruppe Kautzky hat in jüngster Zeit ausgezeichnete
Erfahrungen mit der transnasal-transsphenoidalen mikrochirurgischen Hypophysektomie gemacht, ein Eingriff, der offenbar auch schwerkranken Patienten zu-
gemutet werden kann. Die bisher vorliegenden Ergebnisse zeigen, daß im Gegensatz zu den bisher üblichen Verfahren, die Radioisotopen-Methode eingeschlos-
sen, die Ausschaltung der Hypophyse als total bezeichnet werden kann. Bei allen Patienten entwickelte sich eine in ihrem Ausmaß erwartete Hypothyreose
und Nebenniereninsuffizienz, deren Parameter mit dem Radiojodstudium und der
Cortisol-Bestimmung prä- und postoperativ kontrolliert wurden. Besonders

eklatant war das Verhalten der Gonadotropinausscheidung, und hier ist der Unterschied zwischen den präoperativ kastrationsbedingten überhöhten Ausgangs-
werten und den postoperativen Messungen besonders deutlich.

Die Indikation zur Hypophysektomie wird gestellt, wenn Therapieresistenz
eintritt und wenn unbeeinflußbare Knochenmetastasen-Schmerzen auftreten. Vor
allem die fast mit dem Tag des operativen Eingriffes einsetzende Schmerz-
freiheit nach einer Phase unerträglicher Metastasenschmerzen ist imponierend,
wobei man allerdings wissen muß, daß radikulär verursachte Schmerzen unbeein-
flußbar bleiben können.

Ebenfalls im Rahmen des Sonderforschungsbereiches Endokrinologie versuchen wir zur Zeit in Zusammenarbeit mit der Radioisotopen-Abteilung mit Hilfe der Strontiumkinetik, objektive Kriterien der Metastasenveränderung nach
Hypophysektomie festzulegen. Bisher ist es noch nicht möglich, Beziehungen
zwischen Strontiumanreicherung oder Strontiumverminderung in Metastasen
einerseits und Schmerzbeeinflussung andererseits aufzufinden. Auch die Frage
der Beeinflussung des Carcinom-Wachstums durch die Ausschaltung des somato-
tropen Hormons nach Hypophysektomie ist noch offen.

Trotz aller Fortschritte und neuer Aspekte in der konservativen Behandlung des Prostata-Carcinoms muß festgehalten werden, daß alle diese Methoden
auch unter Berücksichtigung oft verblüffender Anfangserfolge ausnahmslos
zeitbegrenzte Wirkung haben und letzten Endes versagen. Diese Feststellung
mindert nicht den Wert der konservativen Behandlungsverfahren, die, gezielt
eingesetzt, auch Kranken der fortgeschrittenen Stadien die Möglichkeit geben,
über Jahre hinaus ein durchaus lebenswertes Leben zu führen. Heilung allerdings bietet nur die radikale Prostatektomie.

Symp. Dtsch. Ges. Endokrin. 17, 165–174 (1971)
© by Springer-Verlag

# Der Einfluß des Alterns auf die endokrinen Funktionen des Mannes

## The Influence of Aging on the Endocrine Functions of the Male

H. SCHMIDT

II. Medizinische Universitätsklinik, Hamburg

Mit 4 Abbildungen

Summary

From earlier investigations on the morphology of the human testis and sperm
it can be stated that in older age the spermatogenic activity is not affected to a greater degree whereas the number of hormone-producing cells – the
Leydig cells – decreases with age. Histochemically these cells showed a
decrease of lipid constituents in their cytoplasma. In former years the age-
depending androgen production in the human male was studied by bioassay of
these hormones in the urine. A peak of androgen exretion was found between
the ages of 25 und 30 years, føllowed by a steady decrease; beyond 60 and
70 years of age only a third of the maximum excretion was found. Investigations on the excretion of 17-ketosteroids and their fractions, throughout
life, showed a similar relationship to aging, indicating primarily that
adrenal precursors of urinary androgens decrease with age. Studies on the
function of male secondary sex organs in old age have also been performed:
from the third to the seventh decade of life a steady decrease of fructose
content in the seminal plasma has been found, suggesting a reduced function
of seminal vesicles. In the last years modern methods of steroid analysis
gave more precise information about androgens in the older male. The excretion of testosterone glucuronoside – a good parameter for testosterone production in the male – showed a peak in the life span between 25 and 35
years; thereafter a gradual decrease was observed, reaching a constant level
in the $7^{th}$ decade. At this age and afterwards testosterone excretion was
only one third of the maximum. In addition measurements of blood production
rates of testosterone showed a decrease in older age groups; at the same
time the metabolic clearance rate of testosterone was diminished. In contrast
to urinary testosterone, the plasma concentration remained high until the
$8^{th}$ or $9^{th}$ decade of life, a phenomenon which is partly explained by the
diminished clearance rate of testosterone. However, a normal testosterone
level in plasma cannot guarantee normal androgen action in old age, because
the testosterone binding capacity of plasma increases and free plasma testosterone, which is thought to be the really active fraction, decreases with
age. Furthermore some observations suggested that certain target organs demonstrated a decreased sensibility to testosterone in older age. In addition
testosterone metabolism in the older male seemed to be changed to a more female pattern, especially concerning a greater conversion to 5β-metabolites
and androstanediols.
   In contrast to the decrease of androgen production and excretion, estrogen excretion did not diminish with age; a greater conversion of androgens
to estrogens was theorized. Total urinary gonadotrophins showed a slow and
not very extensive rise in a few aging men beyond 50 years, but in the majority of older men the gonadotrophin content in urine remained within the
normal range.

Schon seit den aufsehenerregenden Selbstversuchen des französischen Phy-
siologen Brown-Séquard im vorigen Jahrhundert und den optimistischen Publikationen Steinachs um 1920, die den Hodenextrakten und -einpflanzungen eine
verjüngende Wirkung nachsagten, ist die Frage der Hodeninvolution als
Schrittmacher männlichen Alterns immer aktuell und populär geblieben.

Auch wenn wir uns heute mit dem Einfluß des Alterns auf die endokrinen
Funktionen des Mannes beschäftigen, interessieren uns zunächst Altersveränderungen des Hodens und seiner Androgenbildung, denn die wichtigsten Androgene Testosteron und Androstendion werden hauptsächlich von diesem Organ gebildet und sezerniert. Die Nebenniere steuert einen geringen Anteil an wirksamen Androgenen bei, hauptsächlich durch periphere Konversion des von ihr
sezernierten Dehydroepiandrosterons und Dehydroepiandrosteronsulfats. Beginn-
nen wir mit der Morphologie der Hoden in Abhängigkeit vom Alter. Das Gewicht der Hoden nimmt im Alter nur geringfügig ab, höchstens um 3,5 g bei
einem Normalgewicht von 20 g in jüngerem Lebensalter. Wenn nur Hoden mit
einer intakten Spermatogenese berücksichtigt werden, so finden sich praktisch keine Altersunterschiede (35, 38, 45). Das Hodengewicht ist demnach
hauptsächlich ein Maß für die Erhaltung des Tubulusapparates und der Spermatogenese, der spermatogenetische Apparat des Hodens nimmt volumenmäßig den
Hauptanteil ein und bildet sich im Alter nur wenig zurück (9, 35, 43). Damit
stimmen die Spermatogramm-Befunde überein, die zeigen, daß auch im Alter
meistens eine weitgehend intakte Spermatogenese vorhanden ist (30, 34). Der
eigentliche hormonbildende Anteil der Hoden - die Leydigzellen - macht nur
ca. 12% des Volumens aus und kann bei einer teilweisen Involution gewichts-
mäßig kaum erfaßt werden.

Dagegen läßt sich histologisch eine Abnahme der Leydigzellzahl in Relation zu den Tubuli feststellen. Das umfassendste Material hat Tillinger (45)
hierzu vorgelegt, dessen Befunde in der Abb. 1 wiedergegeben sind. Bei 227

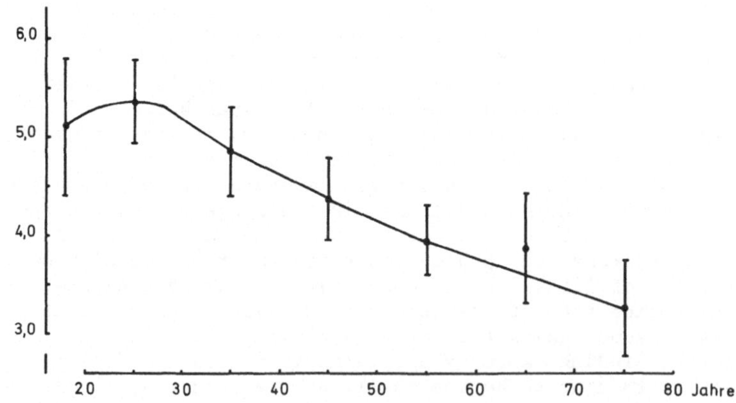


Abb. 1. Abhängigkeit der relativen Leydig-Zellzahl vom Lebensalter bei
227 Männern (Tillinger, 1957).


Männern wurden in Hoden mit intakter Spermatogenese die Leydigzellen pro
Tubulus-Querschnitt ausgezählt. Es findet sich ein Maximum in der Altersgruppe von 20 - 29 Jahren und anschließend ein kontinuierlicher, statistisch
signifikanter Abfall. Ich mache darauf aufmerksam, daß die Ordinate in der
von Tillinger gewählten Darstellung nicht bei 0 beginnt und die gesamte Ab-

nahme der Leydigzellzahl vom Maximum bis zum 8. Lebensjahrzehnt nur wenig
mehr als 1/3 beträgt. Was die Histologie und Histochemie zum Funktionszustand dieser Zellen im Verlaufe des Alterungsvorganges aussagt, möchte ich
im Einzelnen nicht referieren, weil die Aussagekraft dieser mehr subjektiven
Beurteilungen in quantitativer Hinsicht nicht sehr groß erscheint und die
Spezifität hinsichtlich des Funktionszustandes zumindest fraglich ist. Immerhin haben Lynch u. Scott (31) an 160 menschlichen Hoden histochemisch
festgestellt, daß der Lipoidgehalt der Leydigzellen in der Altersgruppe zwischen 20 und 35 Jahren am höchsten ist und mit zunehmendem Alter geringer
wird.

Noch ein weiteres histologisches Phänomen am alternden Hoden bedarf einer
Erwähnung. Mit zunehmendem Alter findet sich nämlich eine Verdickung und
Hyalinose der Basalmembran der Tubuli contorti (9, 35, 43, 44). Aus tierexperimentellen Befunden ergibt sich, daß diese Erscheinung als eine verminderte Androgenaktivität im Sinne einer herabgesetzten androgenen Kontakt-
wirkung auf die Tubuluswände angesehen werden kann (46, 47).

Wie verhält es sich nun mit der Androgenbildung in den Testikeln und Nebennieren in Abhängigkeit vom Alter. Schon vor Jahrzehnten ist man dieser
Frage nachgegangen, indem man mit den damals verfügbaren Methoden die Androgenausscheidung im Urin zu erfassen suchte. In den Jahren zwischen 1937 und
1954 haben eine ganze Reihe von namhaften Autoren mit biologischen Methoden
die Androgenausscheidung bei Männern bestimmt (8, 17, 29, 36, 37, 41). Sie
fanden mit weitgehender Übereinstimmung ein Maximum der biologischen Androgenaktivität um das 30. Lebensjahr und anschließend einen mehr oder weniger-
kontinuierlichen Abfall, dessen Ausmaß von der jeweils angewandten Methode
abhing. Im Durchschnitt war aber im Alter nach dem 60. Lebensjahr nur noch
etwa 1/3 des Maximums festzustellen. Die inter-individuelle Streuung der mit
diesen Methoden festgestellten Werte ist sehr groß, umso erstaunlicher ist
die weitgehende Übereinstimmung des altersabhängigen Verlaufes, wenn man die
mit modernen Methoden gewonnenen Ergebnisse vergleicht.

Über die Ausscheidung der 17-Ketosteroide beim Manne in Abhängigkeit vom
Alter gibt es eine Fülle von Befunden (12, 18, 19, 21, 37, 42, 51). Wenn man
die vom Material her umfassendsten Untersuchungen zusammenstellt, so ergibt
sich ein Maximum der 17-Ketosteroidausscheidung in der Altersgruppe zwischen
25 und 30 Jahren mit ca. 15 - 16 mg pro Tag und anschließend ein ziemlich
gleichmäßiger Abfall auf 5 - 6 mg pro Tag in der Altersgruppe zwischen 60
und 70 Jahren. Von einigen Autoren wird auch in noch höherem Lebensalter ein
weiterer Rückgang der Ausscheidung festgestellt. Die chromatographische
Fraktionierung der 17-Ketosteroide in den verschiedenen Altersgruppen führte
zu keinen zusätzlichen Informationen: die Dehydroepiandrosteron-Fraktion
fiel ebenso stark ab wie die Fraktionen Androsteron und Ätiocholanolon (25).
Zur Interpretation dieser 17-Ketosteroid-Ausscheidungskurven ist es wichtig,
daran zu erinnern, daß die 17-Ketosteroide sich nur zum geringen Teil von
dem biologisch hochaktiven Testosteron herleiten. Nach Injektion radioaktiv
markierten Testosterons werden nach Angaben der verschiedenen Autoren nur
zwischen 25 und 45% der verabreichten Radioaktivität in Form von 17-Keto-
steroiden ausgeschieden (1, 2, 3, 5, 23, 33). Übertragen auf die physiologischen Verhältnisse bei jungen Männern mit einer durchschnittlichen Testos-
teron-Blutproduktionsrate von 7 mg/Tag bedeutet dies, daß nur ca. 2 - 3 mg
der täglich ausgeschiedenen 17-Ketosteroide direkt vom Testosteron herstam-
men. Die altersabhängige 17-Ketosteroidausscheidung stellt also keinen Spiegel der Testosteronproduktion dar; ihr Hauptanteil entstammt den aus der Nebenniere sezernierten Steroiden Dehydroepiandrosteron und Dehydroepiandrosteronsulfat. Ein Rückgang der Testosteronproduktion im Alter wird sich also
an der 17-Ketosteroidausscheidung kaum ablesen lassen. Andererseits würde
auch ein völliges Versiegen der Testosteronproduktion im Alter den geschil-
derten starken Abfall der 17-Ketosteroide quantitativ nicht erklären. Wir
müssen daraus schließen, daß die Produktion von Vorläufern der 17-Keto-
steroide in der Nebennierenrinde im Alter abnimmt.

Bei diesem Stand des Wissens haben wir vor vielen Jahren eine indirekte
Methode zur Beurteilung der Androgenaktivität im Alter herangezogen - nämlich

die Bestimmung der Spermaplasma-Fruktose, die ja einen Anhalt für einen Funktionszustand der Bläschendrüsen gibt (34). Wie aus der Abb. 2 ersichtlich,

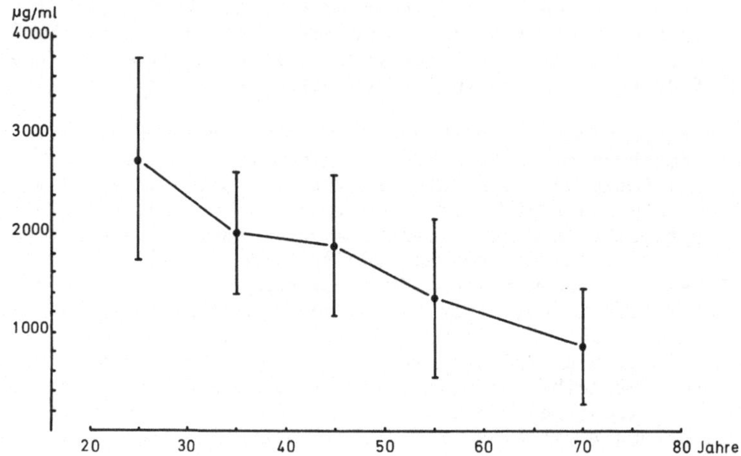


Abb. 2. Fructosekonzentration im Spermaplasma in Abhängigkeit vom Lebensalter (n = 83)

ist eine Altersabhängigkeit eindeutig, statistisch ist sie signifikant. Es ist jedoch bei einer solchen indirekten Methode nicht auszuschließen, daß die Reaktionsfähigkeit des Erfolgsorgans und nicht die Androgenproduktion des Organismus gemessen wird.

Seitdem es möglich ist, Testosteron-Glucuronid im Urin direkt zu bestimmen, haben wir ein besseres Maß für die Testosteronproduktion im männlichen Organismus in der Hand. Zwar werden beim Mann nur ca. 1% des produzierten Testosterons in Form des Testosteronglucuronids ausgeschieden, dieses leitet sich aber ausschließlich vom Testosteron her (3, 5, 7, 23, 33). Die Blutproduktionsrate des Testosterons wird mit dieser Methode nur geringfügig überschätzt.

Die Ausscheidung des Testosteron-Glucuronids ist eindeutig altersabhängig. Die Abb. 3 zeigt die in unserem Laboratorium bei normalen Männern gemessenen

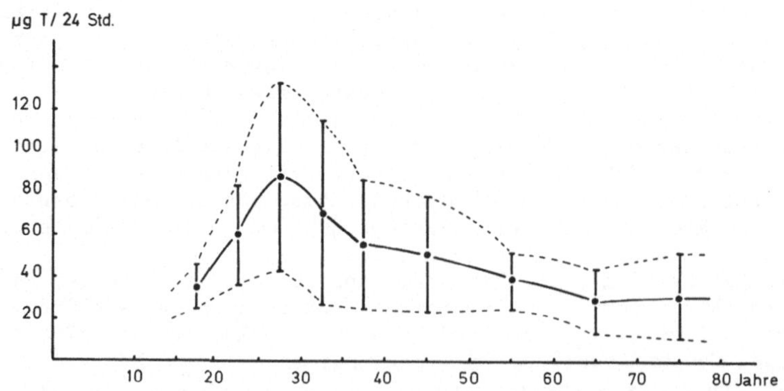


Abb. 3. Testosteronausscheidung bei 109 normalen männlichen Personen.

Werte (40, 53). Auch hier liegt der Gipfel zwischen 25 und 30 Jahren, anschließend findet sich zunächst ein schnellerer, dann mäßiger Rückgang der Testosteron-Ausscheidung bis ins 7. Lebensjahrzehnt. Später ist kein weiterer Abfall mehr zu verzeichnen. Der gesamte Abfall vom Maximum bis zum Alter beträgt etwa 2/3. Damit fallen die Testosteron-Ausscheidungswerte schon in den
Bereich, wie wir ihn bei jüngeren Kastraten gefunden haben (40). Alle Untersucher, die die Altersabhängigkeit der Testosteron-Ausscheidung gemessen haben, finden ebenfalls einen Rückgang mit zunehmendem Alter (14, 15, 39, 48,
49).

Nach diesen Befunden bestand an einer Altersinvolution der Androgenbildung kaum noch ein Zweifel. Es tauchte aber eine erhebliche Diskrepanz zu
den Testosteronwerten im Plasma auf. Wie die Tab. 1 zeigt, fanden Kent u.
Acone (28) bei einem ausreichend großen Kollektiv keinen eindeutigen Rückgang der Testosteronspiegel im Plasma bis zum 8. Lebensjahrzehnt. Erst nach

Tab. 1. <u>Plasma-Testosteron-Spiegel bei Männern in Beziehung zum Lebensalter.</u>

nach
A) Kent u. Acone (1965)        "Double isotope derivative" Technik
B) Coppage u. Cooner (1965)      "        "          "          "
C) Frick (1969)                "Competitive protein binding" Methode

| A) Alter (Jahre) | N | Mittelwert ng/100 ml | Standardabweichung |
|---|---|---|---|
| 20-29 | 13 | 660 | 190 |
| 30-39 | 11 | 560 | 190 |
| 40-49 | 8 | 470 | 90 |
| 50-59 | 9 | 590 | 200 |
| 60-69 | 12 | 590 | 180 |
| 70-79 | 10 | 580 | 170 |
| 80-93 | 5 | 280 | 130 |
| B) 16-43 | 10 | 650 | 170 |
| 46-92 | 19 | 500 | 180 |
| C) 18-35 | 21 | 740 | 290 |
| > 60 | 68 | 435 | 250 |
| > 60 | 17 | 1149 | 730 (nach 5000 E HCG) |

dem 80. Lebensjahr konnten sie bei 5 Personen einen deutlichen Abfall der
Testosteronwerte im Plasma feststellen. Auch Coppage u. Cooner (10) fanden
bei der Gegenüberstellung je einer Gruppe jüngerer und älterer Männer nur
einen geringen Abfall der Testosteronwerte im Plasma, nämlich von 650 auf
500 ng pro 100 ml. Eine sichere Altersabhängigkeit innerhalb der älteren
Gruppe läßt sich aus den in der Arbeit angegebenen Einzelwerten nicht ablesen. Einen etwas deutlicheren Abfall der Plasma-Testosteronwerte findet
Frick (13), wenn er eine Gruppe von 18 - 35jährigen mit einem großen Kollektiv von über 60jährigen vergleicht. Leider ist in dieser Publikation eine
genauere Altersstaffelung nicht angegeben. Erstaunlich ist die in dieser Untersuchung festgestellte Reaktion des Plasma-Testosterons auf eine hochdo-
sierte ICSH-Gabe bei über 60jährigen Männern. Die auf diesem Symposion von
Vermeulen u. Mitarb. (50) vorgetragenen Befunde sind hier noch ergänzend anzuführen; es wurde in diesen Untersuchungen ebenfalls erst im Alter von über
60 Jahren ein deutlicher Abfall des Plasma-Testosterons gefunden. Das Plasma-
Testosteron scheint also nach diesen Befunden später und weniger stark abzu-
fallen als die Testosteronausscheidung im Urin. Wie ist diese Diskrepanz zu
erklären?

Die Testosteronbildung und -sekretion der Testikel und die gesamte Blutproduktionsrate scheinen im Alter tatsächlich abzunehmen. Axelrod (1 a) hat in
vitro an menschlichen Hoden festgestellt, daß nach Zugabe markierten Pregne-
nolons bestimmte Syntheseschritte in Richtung zum Testosteron im älteren Hoden weniger schnell ablaufen. Allerdings sind diese Befunde hinsichtlich der
Quantität der Testosteronbildung in vivo schwer zu interpretieren. Dagegen
haben Hollander u. Hollander (22) die Testosteronkonzentration in der Vena
spermatica direkt gemessen (Tab. 2). Sie fanden eine erhebliche Altersabhän-

Tab. 2. <u>Testosteron-Konzentrationen in der Vena spermatica bei Männern verschiedenen Alters (n = 24) nach Hollander u. Hollander (1958)</u>.

| Alter (Jahre) | Mittelwert µg/100 ml |
|---|---|
| 30-40 | 97 |
| 40-55 | 66 |
| 56-65 | 41 |
| 66-80 | 23 |

Mittelwerte der Testosteron-Produktionsraten im Blut (BPR) und der metabolischen Clearance-Raten (MCR) bei Männern in Abhängigkeit vom Lebensalter
(nach Kent u. Acone, 1965).

| Alter (Jahre) | N | BPR mg/Tag | MCR Liter/Tag |
|---|---|---|---|
| 21-34 | 5 | 7,6 | 1254 |
| 66-86 | 6 | 4,4 | 798 |

gigkeit, wie sie aus dieser Aufstellung hervorgeht. Es muß aber erwähnt werden, daß aufgrund der relativ geringen Anzahl der Fälle insbesondere in den
jüngeren Altersgruppen keine bindenden Schlüsse bezüglich der quantitativen
Verhältnisse möglich sind. Hinzu kommt, daß in den älteren Gruppen ein Teil
der Fälle an einem Prostata-Carcinom litt, 2 davon waren mit Östrogenen behandelt worden.

Messungen der Testosteronproduktionsraten im Blut sind von Kent u. Acone
(28) durchgeführt worden. Sie fanden, wie in Tab. 2 aufgeführt, einen Rückgang um 1/3 bei Vergleich einer jüngeren Gruppe mit einer von über 66jähri-
gen. Die Streuung innerhalb dieser Gruppen ist groß und deshalb die Fallzahl
zu klein, um eine verläßliche quantitative Angabe machen zu können. Es ist
der Tab. 2 weiter zu entnehmen, daß auch die metabolische Clearance-Rate
des Testosterons in der älteren Gruppe abgenommen hat. Vermeulen u. Mitarb.
(50) kommen nach ihren hier vorgetragenen Befunden zu ähnlichen Ergebnissen.
Wenn auch die quantitativen und zeitlichen Verhältnisse nicht voll übereinstimmen, so ergibt sich aus der mit zunehmendem Alter abnehmenden metabolischen Clearance-Rate des Testosterons doch wenigstens eine teilweise Erklärung für die Tatsache, daß die Testosteronspiegel im Plasma nicht entsprechend den Produktionsraten und den Ausscheidungswerten abfallen.

Diese Erklärung enthebt uns aber nicht der Frage, welche physiologische
Bedeutung dieser Androgenkonstellation im Alter zukommt. Dem ersten Anschei-
ne nach möchte man dem im Alter offensichtlich nicht allzu stark reduzierten
Plasmatestosteron-Spiegel eine wichtigere Bedeutung für die Beurteilung bei-
messen als den Produktions- und Ausscheidungswerten. Denn die Androgen-
versorgung der Erfolgsorgane hängt doch, so sollte man meinen, in erster Linie von der Höhe der Plasmakonzentration des Testosterons ab. Eine so einfa-
che Schlußfolgerung ist jedoch nach den heutigen Kenntnissen über die Bindungsverhältnisse des Testosterons im Plasma und den sich daraus ergebenden
Konsequenzen nicht erlaubt. Wie wir aus den Befunden verschiedener Autoren

wissen, wird Testosteron im Plasma teils relativ lose an Albumine, teils wesentlich stärker an Transcortin, vor allem aber auch relativ fest an ein spezifisches β-Globulin, das sog. TBG (testosterone binding globulin) gebunden. In vitro Versuche mit menschlichem Plasma ließen erkennen, daß Testosteronmengen, wie sie physiologischerweise im Blut junger Männer vorkommen, zu 43% an TBG, zu 26% an Transcortin, und zu 29% an Albumin gebunden wurden. An freiem Testosteron blieben nur 2%, was etwa nur 12 ng entspricht (4). Nach den neueren Befunden von Vermeulen et al. (50) scheint das freie Testosteron auch in vivo in der gleichen Konzentration vorzuliegen. Man ist sich heute über die physiologische Bedeutung der Testosteronbindung im Plasma, insbesondere an die spezifischen Proteine, noch nicht völlig im Klaren, kennt auch noch nicht genau die Voraussetzungen der Dissoziation. Die Meinung geht aber dahin - und einige Befunde sprechen auch dafür -, daß nur das freie Testosteron, evtl. auch noch ein Teil des albumingebundenen für die Metabolisierung in der Peripherie und auch für die Wirkung auf die Erfolgsorgane zur Verfügung steht. Aus diesem Grunde ist es für die Beurteilung der Plasma-Testosteronspiegel von entscheidender Bedeutung, etwas über die Bindungsverhältnisse des Testosterons im Plasma in Abhängigkeit vom Alter zu wissen. Vermeulen u. Mitarb. (50) haben auf diesem Symposion hierzu Befunde vorgelegt. Danach scheint das freie Testosteron - ähnlich wie die Testosteron-Ausscheidung - schon vom mittleren Mannesalter an abzunehmen und offensichtlich auch in stärkerem Maße als das gesamte Plasmatestosteron. Gleichzeitig kommt es zu einem Anstieg der Testosteron-Bindungskapazität des Plasmas mit zunehmendem Alter. Diese Befunde sprechen dafür, daß auch bei noch annähernd normalem Gesamttestosteron im Plasma den Erfolgsorganen des alternden Mannes weniger Testosteron verfügbar ist. Sie ergeben gleichzeitig eine mögliche Erklärung für die herabgesetzten metabolischen Clearance-Raten im Alter. Wie die auf diesem Symposion vorgetragenen Befunde von Vermeulen u. Mitarb. (50) zeigen, vollziehen sich im Androgenabbau des alternden Mannes ebenfalls Veränderungen, und zwar in Richtung auf das weibliche Muster des Androgenmetabolismus; insbesondere findet sich eine Bevorzugung der Konversion zu 5β-Metaboliten und zu Androstandiolen. Hier möchte ich aber vor allem noch auf einen Punkt hinweisen, der für die Beurteilung der Androgenwirkung beim alternden Mann von wesentlicher Bedeutung sein dürfte. Wir dürfen das Problem nicht nur von der Seite der Produktion und Verfügbarkeit des Testosterons sehen, sondern müssen auch die Frage nach der Ansprechbarkeit der Zielorgane im Alter stellen. Diese Frage ist nicht neu; klinische Erfahrungen haben seit langem gezeigt, daß die exogene Testosteronzufuhr auf die Spermaplasmafruktose und das Wachstum der Terminalbehaarung mit zunehmendem Lebensalter weniger wirksam ist. Die Frage läßt sich aber heute wesentlich konkreter stellen, denn die Erkenntnisse der letzten Jahre haben gezeigt, daß die Akkumulation und Konversion des Testosterons in bestimmten Erfolgsorganen von ausschlaggebender Bedeutung für die biologische Wirkung ist. Es wäre deshalb wichtig zu wissen, wie sich die Rezeptor- und Konversionskapazität insbesondere zum 5α-Dihydrotestosteron in Relation zum Alter verhält. In unserem Arbeitskreis konnte festgestellt werden, daß nach Verabreichung H$^3$-markierten Testosterons bei älteren Männern mit Prostata-Adenomen die eigentliche atrophische Prostata immer noch etwas mehr Radioaktivität anreichert als z. B. die Skelettmuskulatur und daß außerdem noch eine beachtliche Konversion zu 5α-Dihydrotestosteron stattfindet (6). Wir haben jedoch noch keinen Vergleich zu dem Verhalten der Prostatae jüngerer Männer. Die 5α-Dihydrotestosteron-Bildung aus Testosteron im Präputium scheint, wie aus dem Vortrag von Wilson (52) auf diesem Symposion hervorging, mit zunehmendem Alter abzunehmen - womit sicherlich eine verminderte Androgen-Sensibilität verbunden ist. Weitere gezielte Befunderhebungen zu dieser Frage der Sensibilität androgener Erfolgsorgane in Abhängigkeit vom Alter wären für pathologische Betrachtungen und therapeutische Ansätze von großem Wert.

Bei der Betrachtung der Endokrinologie des Mannes im Alter scheinen außer den Androgenen noch zwei Hormongruppen wichtig zu sein: das Verhalten der

Östrogene und der Gonadotropine. Bei normalen jungen Männern wurden 10 -
36 ng Östron und 2 - 6 ng Östradiol pro 100 ml Plasma (24) und eine Östrogenausscheidung zwischen 10 - 20 µg pro 24 Std. im Urin gefunden. Testes und
Nebennieren sind zwar in der Lage, Östrogene zu bilden, eine signifikante
Sekretion konnte jedoch nicht nachgewiesen werden. Die Östrogene beim Mann
stammen vielmehr zum ganz überwiegenden Teil aus einer Konversion von Androgenen, wie mehrere Untersucher mit Hilfe radioaktiv markierter Androgene
feststellen konnten (1, 5, 11, 32). Plasma-Östrogene im Alter oder Produktionsraten von Östrogenen in Abhängigkeit vom Alter sind mir nicht bekannt.
Dagegen ist die Östrogenausscheidung in Relation zum Alter mehrfach untersucht worden. Wie aus Abb. 4 ersichtlich, fallen weder die biologisch gemessenen Östrogenaktivitäten noch das chemisch bestimmte Östron oder Östradiol

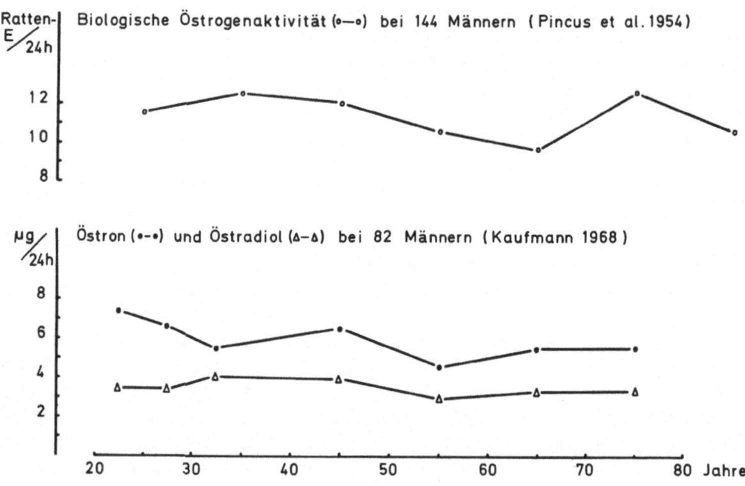


Abb. 4. Östrogenausscheidung im Urin in Bezug zum Lebensalter.

mit zunehmendem Alter ab (27, 37). Dies ist insofern erstaunlich, als ja die
Androgenproduktion zurückgeht und die Östrogene - wie gesagt - sich von den
Androgenen herleiten. Es bleibt eigentlich nur übrig, eine erhöhte Konversionsrate von Androgenen zu Östrogenen im Alter anzunehmen; direkt gemessen
wurde die Konversionsrate bei älteren Männern nicht. Immerhin verschiebt
sicht dadurch im Alter die Östrogen-Androgen-Relation im Urin zugunsten der
Östrogene.
    Bezüglich der Gonadotropine im Alter sind mir lediglich Ausscheidungswer-
te der Gesamt-Gonadotropine bekannt geworden. Übereinstimmend wird von den
Autoren berichtet, daß sie etwa ab 50. Lebensjahr in den Durchschnittswerten
allmählich und nur geringfügig ansteigen (16, 20, 21, 26, 42). Bei der Mehrzahl der alten Männer bleibt die Gonadotropinausscheidung im normalen Bereich. Es ist anzunehmen, daß dieses Verhalten der Gesamtgonadotropine mit
der bis ins hohe Lebensalter relativ gut erhaltenen spermatogenetischen
Funktion der Testikel zusammenhängt.

# Literatur

1.  Ahmad, N., Morse, W.: Canad. J. Biochem. $\underline{43}$, 25 (1965).
1a. Axelrod, L.R.: Biochem. biophys. Acta (Amst.) $\underline{97}$, 551 (1965).
2.  Baulieu, E.E., Mauvais-Jarvis, P.: J. biol. Chem. $\underline{239}$, 1569 (1964).
3.  – – J. biol. Chem. $\underline{239}$, 1578 (1964).
4.  – Raynaud, J.P.: zit. n. Baulieu, E.E. u. Robel, P., in Eik-Nes, K.B. (ed.): The Androgens of the Testis, New York 1970, S. 55.
5.  Bolt, W., Ritzl, F., Bolt, H.M.: Excerpta med. (Amst.), Intern. Congr. Ser. No. 111, Abstr. 526 (1966).
6.  Becker, H., Buric, L., Petersen, C., Voigt, K.D.: 17. Symp. Dtsch. Ges. Endokrin., Acta endocr. (Kbh.) Supp. $\underline{152}$, 28 (1971).
7.  Comacho, A.M., Migeon, C.J.: J. clin. Invest. $\underline{43}$, 1083 (1964).
8.  Dingemanse, E., Borchardt, H., Laquer, E.: Biochem. J. $\underline{31}$, 500 (1937).
9.  Engle, E.T.: In Cowdry's Problems of Aging, Baltimore 1952, S. 708.
10. Coppage, W.S., Cooner, A.E.: New Engl. J. Med. $\underline{273}$, 902 (1965).
11. Epstein, J., Raheja, M.C., Frow, E., Morse, W.I.: Canad. J. Biochem. $\underline{44}$, 971 (1966).
12. Fraser, R., Forbes, A., Albright, F., Sukowitch, H., Reifenstein, E.J.: J. clin. Endocrin. $\underline{1}$, 234 (1941).
13. Frick, J.: Urol. int. (Basel) $\underline{24}$, 481 (1969).
14. Futterweit, W., McNiven, N.L., Guerra-Garcia, R., Gibree, N., Drodowsky, M., Siegel, G.L., Soffer, L.J., Rosenthal, J.M., Dorfman, R.J.: Steroids $\underline{4}$, 137 (1964).
15. – Freeman, R., Siegel, G.L., Grieboff, I.S., Dorfman, R.I., Soffer, L.J.: J. clin. Endocr. $\underline{25}$, 1451 (1965).
16. Goldzieher, M., Goldzieher, J.W.: Geriatrics $\underline{8}$, 1 (1953).
17. Hamburger, C., Halvorsen, K.: Ugeskr. Laeg. $\underline{104}$, 1443 (1942).
18. – Acta endocr. (Kbh.) $\underline{1}$, 19 (1948).
19. Hamilton, H.B., Hamilton, J.B.: J. clin. Endocr. $\underline{8}$, 433 (1948).
20. Heller, C.G., Myers, G.B.: J. Amer. Med. Ass. $\underline{126}$, 472 (1944).
21. Heller, A.L., Shipley, R.A.: J. Geront. $\underline{6}$, Suppl. 3, 101 (1951).
22. Hollander, N., Hollander, V.P.: J. clin. Endocr. $\underline{18}$, 966 (1958).
23. Horton, R., Rosner, J.M., Forsham, P.H.: Proc. Soc. exp. Biol. (N.Y.) $\underline{114}$, 400 (1963).
24. Ichii, S., Forchielli, E., Perloff, W.H., Dorfman, R.I.: Anal. Biochem. $\underline{5}$, 422 (1963).
25. Johnsen, S.G.: Acta endocr. (Kbh.) $\underline{21}$, 146 (1956).
26. – Acta endocr. (Kbh.) $\underline{31}$, 209 (1959).
27. Kaufmann: Z. Urol. $\underline{61}$, 229 (1968).
28. Kent, J.R., Acone, A.B.: 2nd Sympos. on Steroid Hormones, Ghent, Excerpta Medica Found., Amsterdam 1965, S. 31.
29. Kochakian, C.: Endocrinology $\underline{21}$, 60 (1937).
30. Kühnau, J., Nowakowski, H.: Endokrinologie $\underline{40}$, 1 (1960).
31. Lynch, K.M., Scott, W.W.: Fertil. and Steril. $\underline{3}$, 35 (1952).
32. MacDonald, P.C., Rombaut, R.P., Siiteri, P.K.: J. clin. Endocr. $\underline{27}$, 1103 (1967).
33. Mauvais-Jarvis, P., Baulieu, E.E.: J. clin. Endocr. $\underline{25}$, 1167 (1965).
34. Nowakowski, H., Schmidt, H.: Schweiz. Med. Wschr. $\underline{89}$, 1204 (1959).
35. Olesen, H.: Morfologiske Sperma- og Testisundersogelser, Copenhagen 1948.
36. Pedersen-Bjergard, K., Tønnesen, M.: Acta endocr. (Kbh.) $\underline{1}$, 38 (1948).
37. Pincus, G., Romanoff, L.P., Carlo, J.: J. Geront. $\underline{9}$, 113 (1954).
38. Rössle, R., Roulet, F.: Maß und Zahl in der Pathologie. Berlin 1932.
39. Sandberg, D.H., Ahmad, N., Cleveland, W.W., Savard, K.: Steroids $\underline{4}$, 557 (1964).
40. Schmidt, H.: Acta Endocr. (Kbh.) Suppl. 128 (1968).
41. Schou, H.I.: Acta endocr. (Kbh.) $\underline{8}$, 149 (1951).
42. Segal, S.J., Nelson, W.O., Flocks, R.A.: 39th Meet. Amer. Endocr. Soc. (1957).
43. Spangaro, S.: Anatom. Hefte $\underline{18}$, 593 (1902).

44. Stieve, H.: Möllendorfs Handb. der mikrosk. Anatomie des Menschen, VII,
    2 (1930).
45. Tillinger, K.G.: Acta endocr. (Kbh.) Suppl. 30 (1957).
46. Tonutti, E.: 1. Sympos. Dtsch. Ges. Endokrinol. Berlin-Göttingen-Heidel-
    berg: Springer 1954.
47. - Verhdl. Anat. Gesellsch., Jena 1954.
48. Vermeulen, A.: 2nd Sympos. on Steroid Hormones, Ghent. Excerpta Medica
    Found., Amsterdam 1965, S. 71.
49. - 13. Sympos. Dtsch. Ges. Endokrin. Berlin - Heidelberg - New York:
    Springer, 1968.
50. - Rubens, R., Verdonck, L.: 17. Sympos. Dtsch. Ges. Endokrin. Acta
    endocr. (Kbh.) 66, Suppl. 152 (1971), S. 23.
51. Voigt, K.D., Schröder, W., Beckmann, I., Rosenkilde, H.: Dtsch. Arch.
    klin. Med. 202, 1 (1955).
52. Wilson, J.D.: Sympos. Dtsch. Ges. Endokrin., 17 (1971) Berlin - Heidelberg -
    New York: Springer.
53. Morer-Fargas, F., Nowakowski, H.: Acta endocr. (Kbh.) 49, 443 (1965).

Symp. Dtsch. Ges. Endokrin. 17, 175–177 (1971)
© by Springer-Verlag

# Podiumsgespräch „Klimakterium virile"*
## Panel Discussion: „Climacterium virile"

Moderator: P.W. Jungblut

Teilnehmer: H. Klosterhalfen, H. Nowakowski, P. Ofner, H. Schmidt,
A. Vermeulen, J. Wilson

Der Begriff "Klimakterium virile" läßt vermuten, daß es beim alternden
Mann zu einer ähnlichen Abnahme der Testosteronproduktion und entsprechend
drastisch gesenkten Plasmakonzentrationen kommt, wie es für die weiblichen
Sexualhormone bei der klimakterischen Frau bekannt ist. Bisherige Untersuchungen bestätigen diese Vermutung nur bedingt. Jenseits des siebzigsten
Lebensjahres erfolgt zwar ein signifikanter Abfall des (Gesamt-)Testosteron-
plasmaspiegels, jedoch variieren die individuellen Werte beträchtlich und
selbst im neunten Dezennium werden gelegentlich ähnlich hohe Werte wie bei
jungen erwachsenen Männern gemessen. Deutlicher als die Konzentration des
Gesamttestosterons nimmt die Fraktion des freien (nicht proteingebundenen)
Testosterons im Plasma mit zunehmendem Alter ab. Dieser Abnahme entspricht
eine vermehrte Bindung des Testosterons an ein Transportprotein des Blutplas-
mas. Da dieses Protein (sex steroid-binding-globulin) auch Oestrogene bindet
und seine Synthese wahrscheinlich von Testosteron, vor allem aber von Oestrogenen stimuliert wird, stellt sich die Frage einer veränderten Testosteron/
Oestrogen-Relation im "Klimakterium virile". Als Anhaltspunkte dafür liegen
bisher nur die Ausscheidungswerte der beiden Hormongruppen und die ebenfalls
im Urin gemessenen Metabolitenverhältnisse vor, die nach dem 65. Lebensjahr
ähnliche Muster aufweisen, wie sie bei Frauen und in Fällen von Hypogonadismus beobachtet werden. Versuche, diese Befunde mit der Symptomatik des sog.
"Klimakterium virile" ("Wallungen", depressive Symptome, Nachlassen von Ge-
dächtnis und Konzentrationsvermögen, Paraesthesien, Potenz- und Miktionsstö-
rungen, soweit sie nicht zum Komplex der "Prostata"-Störungen gehören, etc.)
zu korrelieren, verliefen negativ. Ein ursächlicher Zusammenhang derartiger
Beschwerden mit dem veränderten Steroidhormonhaushalt ist vorerst nicht zu
sehen.

Für die beiden "typischen" Erkrankungen des alternden Mannes, das Adenom
der paraurethralen Drüsen – bekannt als benigne "Prostatahypertrophie" – und
das Carcinom der Prostata wird die Beteiligung hormonaler Faktoren seit langem vermutet. Aber auch in diesen Fällen sind die auslösenden Mechanismen
noch unbekannt. Paraurethrale Drüsen und Prostata sind Erfolgsorgane in dem
Sinne, daß sie auf Androgene proliferativ reagieren. Der Begriff Erfolgsorgan kann heute auch mit dem Vorkommen spezifischer Hormonrezeptoren in Cytoplasma und Zellkern interpretiert werden. Dem Nachweis von Androgenrezeptoren
in männlichen akzessorischen Geschlechtsorganen ist die Erkenntnis zu ver-

---

<sup></sup>⁺ Das Gespräch wurde durch einen technischen Fehler nicht auf Band genommen
und kann deshalb nur als Zusammenfassung gebracht werden.

danken, daß nicht Testosteron sondern das schon vorher als wirksameres Androgen bekannte 5α-Dihydrotestosteron (Androstan-5α-3 on-17β ol-) im Zellkern
gebunden wird und dort wahrscheinlich die proliferativen Prozesse auslöst.
Testosteron kann daher als zirkulierende (inaktive) Vorstufe des eigentlichen
Hormons angesehen werden, das in der Zelle unter der Wirkung einer 5α-Reduk-
tase entsteht. Das Enzym wird interessanterweise im zentralen Anteil von
Adenomen – den hypertrophierten paraurethralen Drüsen – in 2 bis 3fach höhe-
rer Konzentration gefunden als im peripheren Anteil, der wahrscheinlich
Prostatagewebe repräsentiert. Diese Verteilung unterstützt Spekulationen, wo-
nach eine gesteigerte Synthese von 5α-Reduktase zu einer gegenüber der Norm
vermehrten Retention von 5α-Dihydrotestosteron im Zellkern, gefolgt von er-
höhter Proliferation, zum Adenom führen soll. Als auslösende Faktoren für
die gesteigerte Enzymsynthese werden eine zugunsten der Oestrogene verscho-
bene Relation des Androgen/Oestrogen Plasmaspiegels, eine verringerte adrenale Androgenproduktion und eine Zunahme der Prolaktinsekretion genannt. Beweise für einen ursächlichen Zusammenhang dieser z. T. mehr vermuteten als gefundenen Faktoren mit der genannten Enzymverteilung fehlen. Es ist zudem
fragwürdig, die Konzentration an 5α-Reduktase als zentralen Steuerungsmecha-
nismus anzusehen, da die Bindung von 5α-Dihydrotestosteron im Zellkern die
gleiche Sättigungscharakteristik aufweist, wie sie von anderen Steroidhormonen
und ihren Erfolgsorganzellen bekannt ist. Die Vorstellung einer hormonalen
Genese des Prostatacarcinoms hätte bei der Annahme einer Schlüsselrolle der
5α-Reduktase vom umgekehrten Zustand, nämlich der erniedrigten Enzymkonzen-
tration in der alternden Prostata auszugehen.

Hormonsubstitution zur Behandlung des "Klimakterium virile" muß solange
als zweifelhaft betrachtet werden, als hormonelle Ursachen der Beschwerden
fragwürdig sind. Die früher oft geübte – bestenfalls hinhaltende – "kontra-
hormonale" Therapie des sog. Prostataadenoms sollte zugunsten der wirksame-
ren und technisch hoch entwickelten operativen Verfahren verlassen werden.
Dagegen ist der positive Effekt der Keimdrüsenentfernung und der Gabe pharma-
kologischer Oestrogendosen als (postoperative) Behandlung des Prostatacarcinoms erwiesen. Versuche, die mit unerwünschten Nebenwirkungen verbundene
Oestrogentherapie durch Antiandrogene zu ersetzen, stehen erst am Anfang.
Für den Wirkungsmechanismus beider Substanzgruppen gibt es Anhaltspunkte. Im
Falle der Antiandrogene kommt eine (nachgewiesene) Konkurrenz mit Dihydrotestosteron um die Rezeptorbindung in Betracht. Kürzlich wurde ebenfalls nachgewiesen, daß Oestrogene in hohen (pharmakologischen) Dosen die Bindung von
Dihydrotestosteron an den Androgenrezeptor hemmen. (Die Bedeutung des auch
in männlichen Erfolgsorganen zu findenden cytoplasmatischen Oestrogenrezep-
tors ist ungeklärt.) Da die Rezeptorbindung Voraussetzung der proliferativen
Hormonwirkung ist, kann ihre Verhinderung wachstumshemmend und sogar degene-
rativ wirken, wie morphologische Untersuchungen ergaben. Falls die medikamen-
töse Rezeptorblockade tatsächlich den Therapieerfolg bedingt, so kann aus
der quantitativen Bestimmung des Rezeptors in Operationsmaterial auf die Ansprechbarkeit von nicht entferntem Tumorgewebe geschlossen werden. Derartige
Untersuchungen versprechen in Zukunft zumindest die Möglichkeit einer siche-
reren Prognosestellung.


<u>Anschriften der Teilnehmer</u>

Prof. Dr. P.W. Jungblut, Max-Planck-Institut für Zellbiologie, 2940 Wilhelmshaven, Anton-Dohrn-Weg

Prof. Dr. H. Klosterhalfen, Urologische Universitätsklinik, 2000 Hamburg 20,
Martinistr. 52

Prof. Dr. H. Nowakowski, II. Med. Univ.-Klinik, 2000 Hamburg 20, Martini-
straße 52

Prof. Dr. P. Ofner, Steroid Biochemistry Laboratory, Lemuel Shattuck Hospital, 170 Morton Street, Jamaica Plain, Mass. 02130 (USA)

Doz. Dr. H. Schmidt, II. Med. Univ.-Klinik, 2000 Hamburg 20, Martinistr. 52

Prof. Dr. A. Vermeulen, Rijksuniversiteit, Interne Klinik, De Pintelaan 115, Ghent (Belgien)

Prof. Dr. J.D. Wilson, Department of Internal Medicine, The University of Texas Southwestern Medical School of Dallas, Dallas, Texas (USA)

Symp. Dtsch. Ges. Endokrin. __17__, 179–182 (1971)
© by Springer-Verlag

# Klassifikation der Schilddrüsenkrankheiten

## Classification of Thyroid Diseases

E. KLEIN, H.-L. KRÜSKEMPER, D. REINWEIN, K. SCHWARZ UND P. C. SCRIBA

(Sektion Schilddrüse der Deutschen Gesellschaft für Endokrinologie)

Das Gründungskomitee der Sektion Schilddrüse der Deutschen Gesellschaft
für Endokrinologie hat es in Anbetracht auch einschlägiger ausländischer
Versuche (1., 2.[1]) als erste Aufgabe angesehen, eine Klassifikation der
Schilddrüsenkrankheiten zu erarbeiten, die vornehmlich klinischen Belangen
gerecht wird und praktikabel ist. Es hat sich deshalb entschlossen, Funktion
und Beschaffenheit der Schilddrüse in den Mittelpunkt zu stellen und nicht
nur von der aktuellen peripheren Stoffwechselsituation auszugehen.

Bis auf die Schilddrüsenmalignome, die spätestens mit Therapiebeginn im
Detail zu klassifizieren sind, können Krankheitsformen und -untergruppen
durch die heute mögliche spezielle Schilddrüsen-Diagnostik abgeklärt werden.
Hinsichtlich der Nomenklatur und der sachlich wohlbegründeten Tendenz, Eigen-
namen zu vermeiden, schließt sich die hier vorgelegte Einteilung den bisherigen langjährigen Bemühungen der einzelnen Autoren und denen ausländischer
Arbeitsgruppen weitgehend an. Ihre praktische Anwendung muß erweisen, ob und
wo bei Erweiterung unserer Kenntnisse über Schilddrüsenkrankheiten Ergänzun-
gen oder Korrekturen angebracht sein werden. Unter diesem Aspekt ist hier
die Erörterung der endokrinen Ophthalmopathie bei den Hyperthyreosen zu verstehen, obgleich trotz enger Beziehungen zwischen beiden auch dauerhaft
euthyreotische Verlaufsformen derselben vorkommen.

## 1. Hypothyreosen

| | |
|---|---|
| 1.1. | Angeborene Hypothyreose (sporadischer und endemischer Kretinis-mus) |
| 1.1.1. | Schilddrüsenaplasie (Athyreose, kongenitales Myxödem) |
| 1.1.2. | Schilddrüsendysplasie |
| | 1.1.2.1. ektopisch (z. B. Zungengrundschilddrüse) |
| | 1.1.2.2. an normaler Stelle des Halses |
| 1.1.3. | Struma mit Jodfehlverwertung (Dyshormonogenese); z. Zt. 6 Typen bekannt; Angabe des biochemischen Defektes |
| 1.1.4. | bei endemischer Struma |
| 1.2. | Postnatal erworbene Hypothyreose (höchster Schweregrad: Myxödem) |
| 1.2.1. | Primär (mit oder ohne Struma) |
| | 1.2.1.1. idiopathisch |
| | 1.2.1.2. entzündlich (s. auch unter Schilddrüsenentzündungen) |
| | 1.2.1.3. neoplastisch |
| | 1.2.1.4. postoperativ |
| | 1.2.1.5. nach Strahlenbehandlung (extern oder Radiojod) |
| | 1.2.1.6. medikamentös |
| | 1.2.1.6.1. Jod in hohen Dosen |
| | 1.2.1.6.2. strumigene Medikamente |
| | 1.2.1.7. bei extremem Jodmangel |
| | 1.2.1.8. bei starken Hormonverlusten (renal, intestinal) |

---

[1] 1. Werner, S.C.: J. clin. Endocr. __29__, 860 (1969).
   2. Hedinger, Ch.E.: Thyroid. Cancer. UICC., Vol. XII, Berlin-Heidelberg-
New York: Springer, 1969.

1.2.2     Sekundär (TSH-Mangel bei totaler oder partieller Hypophysen-
          vorderlappeninsuffizienz)

<u>Kommentar</u>

Hypothyreosen sind Krankheitsbilder, die durch einen Mangel an Schilddrüsen-
hormonen in der Körperperipherie gekennzeichnet sind. Das hypothyreotische
Koma stellt kein eigenes Krankheitsbild sondern eine ungewöhnliche und sehr
seltene Verlaufsform der Hypothyreose dar.


*2. Hyperthyreosen*

2.1.      Hyperthyreosen, die mit oder ohne endokrine Ophthalmo- und Dermo-
          pathie einhergehen können.
2.1.1.    Hyperthyreose ohne Struma
2.1.2.    Hyperthyreose mit Struma diffusa
2.1.3.    Hyperthyreose mit Struma nodosa
2.2.      Hyperthyreose ohne endokrine Ophthalmo- und Dermopathie
2.2.1.    Autonomes Adenom mit Hyperthyreose
          2.2.1.1 solitär
          2.2.1.2 multilokulär
2.2.2.    Hyperthyreose durch Adenokarzinom der Schilddrüse
          (Primärtumor oder Metastasen)
2.2.3.    Hyperthyreose bei Thyreoiditis
          (s. auch unter Schilddrüsenentzündungen)
2.3.      Hyperthyreose durch TSH oder TSH-ähnliche Aktivitäten
2.3.1.    Hypophysenvorderlappen-Adenom
2.3.2.    paraneoplastisches Syndrom
2.4.      Hyperthyreosis factitia

<u>Kommentar</u>

Hyperthyreosen sind Krankheitsbilder, die durch einen Überschuß von Schild-
drüsenhormonen in der Körperperipherie gekennzeichnet sind. Die hyperthyre-
otische Krise (s. Koma) stellt keine eigene Hyperthyreoseform sondern den
schwersten Verlauf der Erkrankung dar. Die Bezeichnung "Morbus Basedow"
trifft lediglich auf die Krankheitsform 2.1.2. mit endokriner Ophthalmopathie
zu, der Name "toxisches Adenom" wird ersetzt durch "autonomes Adenom mit
Hyperthyreose" (Krankheitsform 2.2.1.). Der Ausdruck "Jodbasedow" kennzeich-
net nicht die Krankheitsform 2.1.3., sondern nur *eine* Möglichkeit seiner
Pathogenese und wird aufgegeben. Bei der Beschreibung eines autonomen Adenoms ist "kompensiert" nicht mit "euthyreot" und "dekompensiert" nicht mit
"hyperthyreot" gleichzusetzen (s. auch unter blande Strumen). Postoperativ kann
eine Hyperthyreose persistieren oder exacerbieren, beides mit oder ohne Rezidiv-
struma. Die endokrine Ophthalmopathie (und Dermopathie) kann ein- oder beid-
seitig vorkommen, der Manifestation einer Hyperthyreose vorangehen oder nach
deren Remission persistieren. Eine solche euthyreotische Verlaufsform gibt
es auch ohne Entwicklung einer Hyperthyreose.


*3. Blande Strumen*
     (mit oder ohne örtliche Komplikationen)

3.1.      im Halsbereich (ggf. substernal)
3.1.1.    diffus
3.1.2.    einknotig
          3.1.2.1. Zyste, Blutung, hormonell inaktives Gewebe:
                   szintigraphisch kalt
          3.1.2.2. Adenom: szintigraphisch warm
          3.1.2.3. autonomes Adenom ohne Hyperthyreose:
                   szintigraphisch heiß

3.1.3     mehrknotig
          3.1.3.1. Zysten, Blutungen, hormonell inaktives Gewebe:
                   szintigraphisch kalt
          3.1.3.2. Adenome, hormonell aktives Gewebe:
                   szintigraphisch warm
          3.1.3.3. autonome Adenome ohne Hyperthyreose:
                   szintigraphisch heiß
3.2.      dystopisch
3.2.1.    mediastinale oder pulmonale Struma (ggf. Teratom)
3.2.2.    Struma ovarii
3.2.3.    Zungengrundstruma

<u>Kommentar</u>

Die Kennzeichnung "blande" bedeutet, daß diese Strumen nicht entzündlich
und nicht maligne sind und eine euthyreotische Stoffwechselsituation unterhalten. Sie können endemisch oder sporadisch vorkommen, im letzteren Fall
auch durch Jodfehlverwertung, strumigene Substanzen und Medikamente, Trauma,
Haemorrhagie, Amyloidose oder Haemochromatose hervorgerufen sein. Autonome
Adenome (3.1.2.3. und 3.1.3.3.) neigen insbesondere bei Jodzufuhr zu hyper-
thyreotischer Entgleisung.


*4. Schilddrüsenentzündungen*

4.1       akute Thyreoiditis (diffus oder fokal)
4.1.1.    eitrig
4.1.2.    nicht-eitrig (bakteriell, viral, strahlenbedingt)
4.2.      subakute Thyreoiditis (diffus oder fokal)
4.2.1.    infektiös
4.2.2.    parainfektiös
4.3.      chronische Thyreoiditis
4.3.1.    lymphozytär (Autoimmunthyreoiditis)
          4.3.1.1.  ohne Struma
          4.3.1.2.  mit Struma
4.3.2.    fibrös
4.3.3.    spezifisch (Tuberkulose, Lues, Sarkoidose, Lymphogranulomatose,
          Parasiten, Mykosen)

<u>Kommentar</u>

Thyreoiditiden können eine Hyper- oder Hypothyreose nach sich ziehen. Die
de Quervain'sche Thyreoiditis stellt eine durch Riesenzellen gekennzeichnete
Untergruppe der Form 4.2. dar, die Riedel-Struma entspricht der Form 4.3.2.
und die Struma Hashimoto der Form 4.3.1.2.


*5. Schilddrüsenmalignome*

5.1.      Karzinome
5.1.1.    differenziert
          5.1.1.1. follikulär
          5.1.1.2. papillär
          5.1.1.3. Mischformen follikulär-papillär
5.1.2.    medullär
          5.1.2.1. trabekulär
          5.1.2.2. parafollikulär (C-Zellen-Karzinom)
5.1.3.    Anaplastisch
          5.1.3.1. kleinzellig
          5.1.3.2. spindelzellig
          5.1.3.3. riesenzellig

5.2.    Malignes Lymphom
5.3.    Sarkom
5.4.    Haemangioendotheliom
5.5.    Malignes Teratom
5.6.    Metastasen schilddrüsenfremder Tumoren


*Stadien der Tumorausdehnung* (unabhängig vom Tumortyp)

T  (Primärtumor)
   T 0 nicht tastbar
   T 1 kleiner solitärer Tumor, gut verschieblich
   T 2 großer, die Drüse deformierender Tumor oder
       multiple Tumoren in beiden Lappen, gut verschieblich
   T 3 in die Umgebung infiltrierter, fixierter Tumor

N  (Befall regionaler Lymphknoten)
   N 0 nicht nachweisbar
   N 1 homolateral Lymphknoten, gut verschieblich
   N 2 kontralateral oder bilateral Lymphknoten, gut verschieblich
   N 3 verbackene Lymphknoten-Pakete

M  (Fernmetastasen)
   M 0 nicht nachweisbar
   M 1 nachweisbar

## Kommentar

Klassifizierung und Stadieneinteilung beruhen auf den Vorstellungen und Vor-
schlägen des internationalen Krebs-Kongresses der UICC 1968 in Lausanne (s.
Hedinger, 2) und berücksichtigen überdies die Sonderstellung des Calcitonin-
produzierenden C-Zellen-Karzinoms. Da das der internationalen Verständigung
und Vereinfachung dienende TNM-System auf die Wachstumseigenarten der
Schilddrüsenmalignome gut anwendbar ist, wurde die bisherige Einteilung in
4 Stadien zugunsten der 3 Stadien T (Primärtumor), N (Befall von regionalen
Lymphknoten) und M (Fernmetastasen) mit je 3 Untergruppen verlassen.

   Das Tumorstadium ist demnach zu kennzeichnen durch die Angaben 0 - 3 zu
jeweils T, N und 0 - 1 zu M.


## Anschriften der Verfasser

Prof. Dr. E. Klein, Städt. Krankenanstalten, 48 Bielefeld, Oelmühlenstr. 26

Prof. Dr. H.-L. Krüskemper, Medizinische Hochschule, Abt. für Endokrinologie,
3 Hannover, Roderbruchstr. 101

Prof. Dr. D. Reinwein, II. Medizinische Universitätsklinik, 4 Düsseldorf,
Moorenstr. 5

Prof. Dr. K. Schwarz, II. Medizinische Universitätsklinik, 8 München,
Ziemssenstr. 1

Priv.Doz. Dr. P. C. Scriba, II. Medizinische Universitätsklinik, 8 München,
Ziemssenstr. 1

Abbildungsteil

# Hormone und Behaarung (S. 43)

E. Ludwig

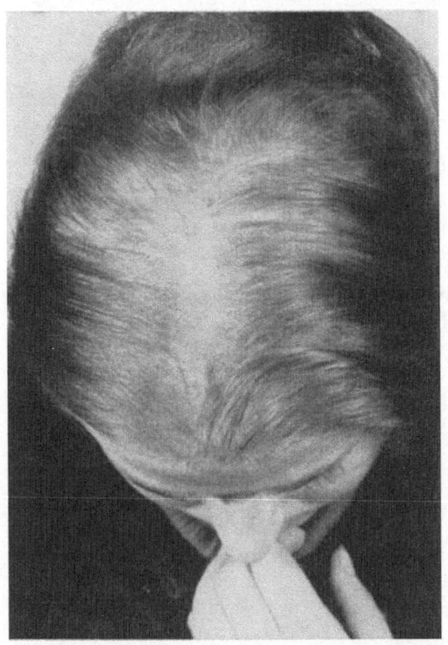

Abb. 1. Leichte Form der andro-genetischen Alopecie bei 20jähr. Patientin. Diffuse Lichtung des (sehr fetten) Haares am Scheitel

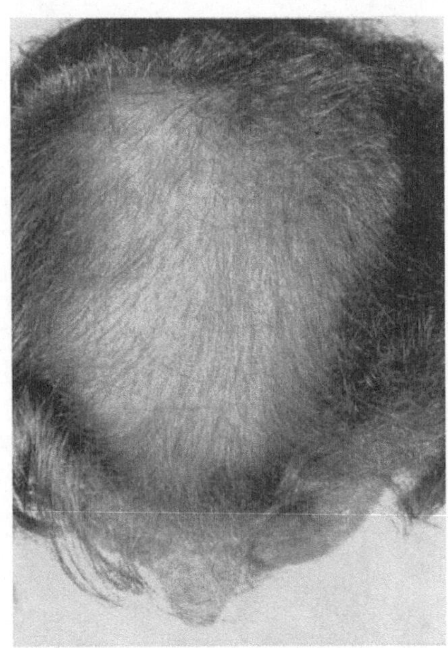

Abb. 2. Schwere Form der andro-genetischen Alopecie bei 26jähr. Patientin. Mittlere Testosteronausscheidung 33,9 $\mu$/24 Std. Urin

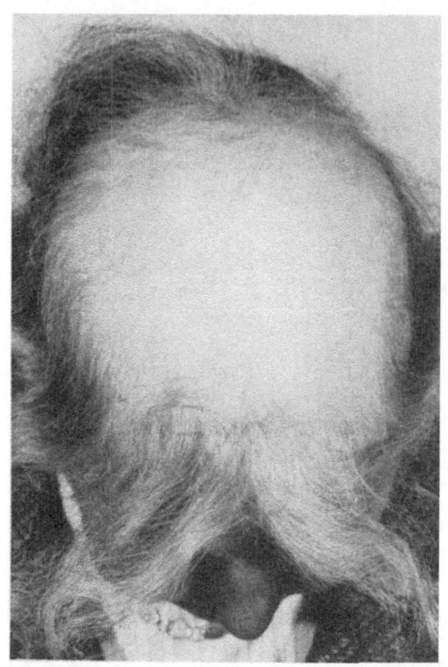

Abb. 3. Scheitel einer 68jähr. Patientin mit hochgradiger andro-genetischer Alopecie. Erhaltener Saum von Haaren entlang der Stirn-Haargrenze

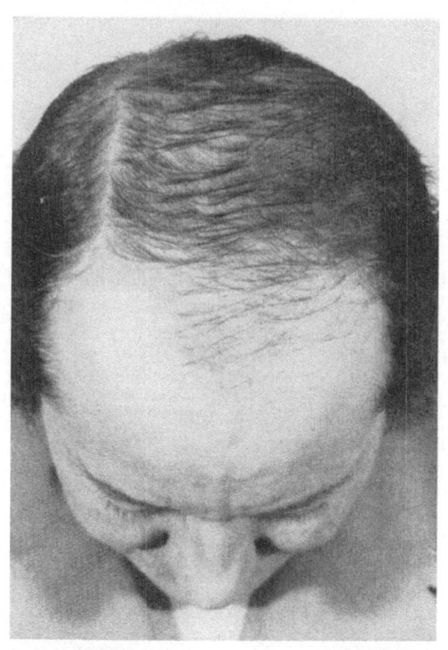

Abb. 4. Männliche Form der andro-genetischen Alopecie bei 46jähr. Patienten mit Hirsutismus und urinärer Testosteronausscheidung von 109 $\mu$/24 Sdt

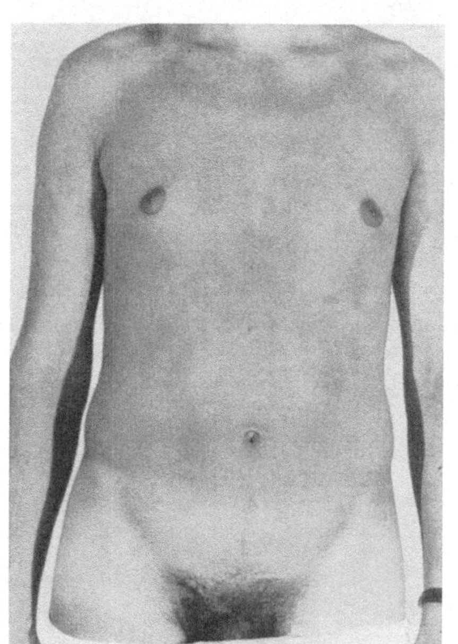

Abb. 5. 22jähr. gesunder Mann mit fehlender Brust- und Bauchbehaarung und sog. weiblicher Schamhaargrenze. Urinäre Testosteronausscheidung an der oberen Grenze der Norm

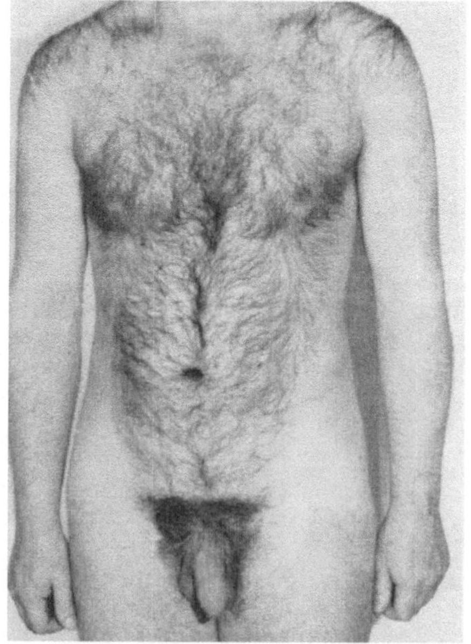

Abb. 6. 24jähr. Mann mit sehr starker Körperbehaarung und urinärer Testosteronausscheidung an der unteren Grenze des altersentsprechenden Normbereiches

Abb. 7. Ausschnitt aus dem Gemälde „la barbuda" von Ribera („lo Spagnoletto"). Museum des Tavera-Hospitals in Toledo

# Physiologie und Pathophysiologie der Corpus Luteum Funktion bei der Frau (S. 57)

J. ZANDER, B. RUNNEBAUM

Abb. 1. LUDWIG FRAENKEL im Alter von 80 Jahren. Reproduktion aus Geburtshilfe und Frauenheilkunde **30**, 397 (1970)

# Stoffwechsel und Wirkung der synthetischen Gestagene (S. 85)

G. A. Overbeek

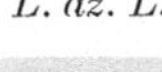



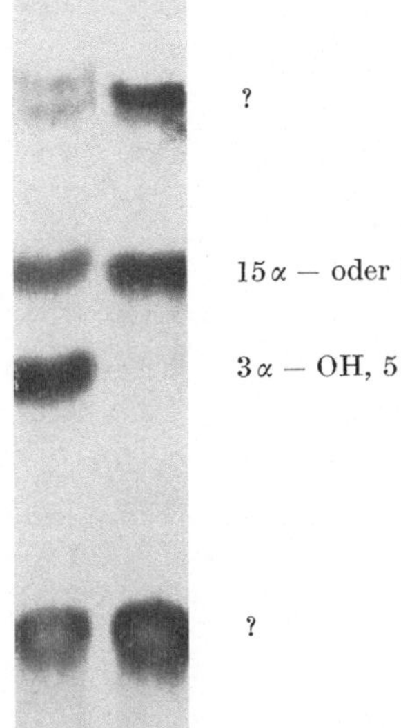

L. az. = Lynestrenolazetat 4-$^{14}$C
L. = Lynestrenol 4-$^{14}$C

Abb. 11

F. Neumann

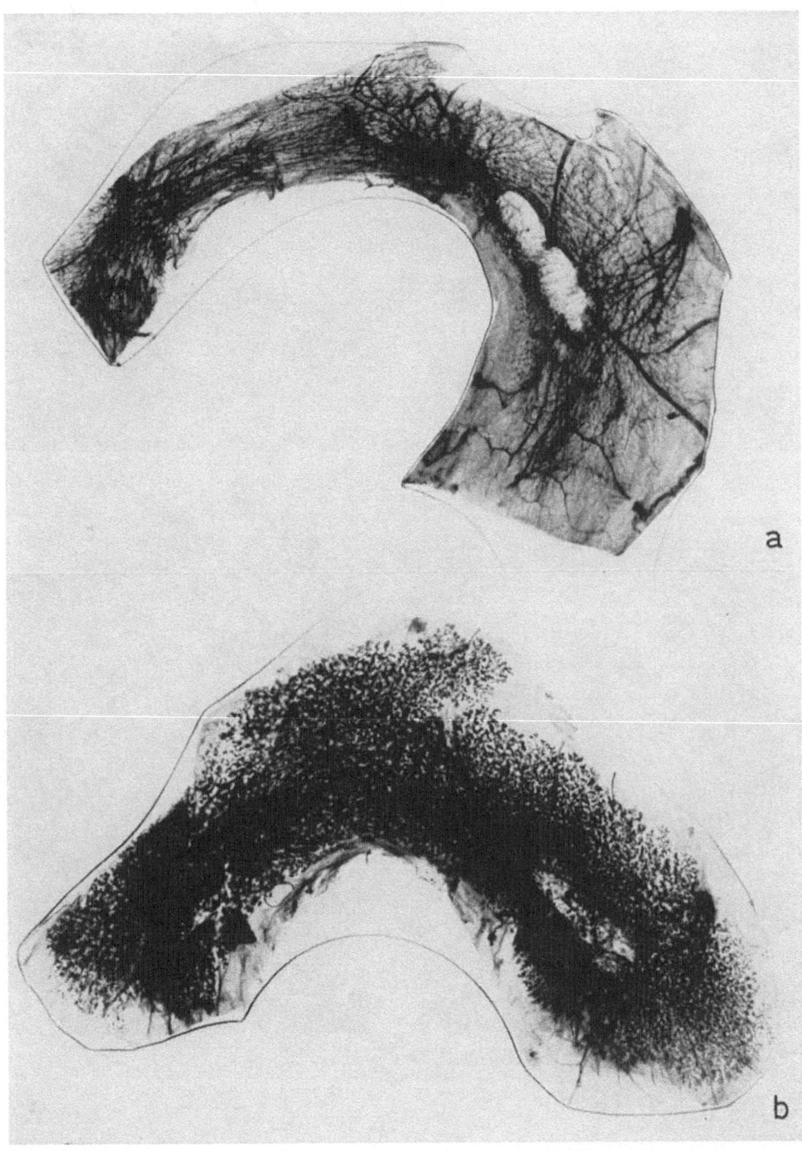

Abb. 2 und 3. Einfluß von Oestradiol auf das Milchdrüsenwachstum kastrierter bzw. kastrierter und hypophysektomierter Ratten (21tägige Behandlung mit täglich 10 µg Oestradiol/Tier subcutan). Milchdrüsenganzdarstellung der drei rechten inguinalen Drüsenkomplexe. Vergr. Abb. 2a und b: 1,5mal; Vergr. Abb. 3a und b: 26mal. Abb. 2a und 3a: kastrierte und hypophysektomierte Tiere; Abb. 2b und 3b: kastrierte Tiere. Bei den hypophysektomierten Tieren ist kein tubulo-alveoläres Wachstum zu beobachten

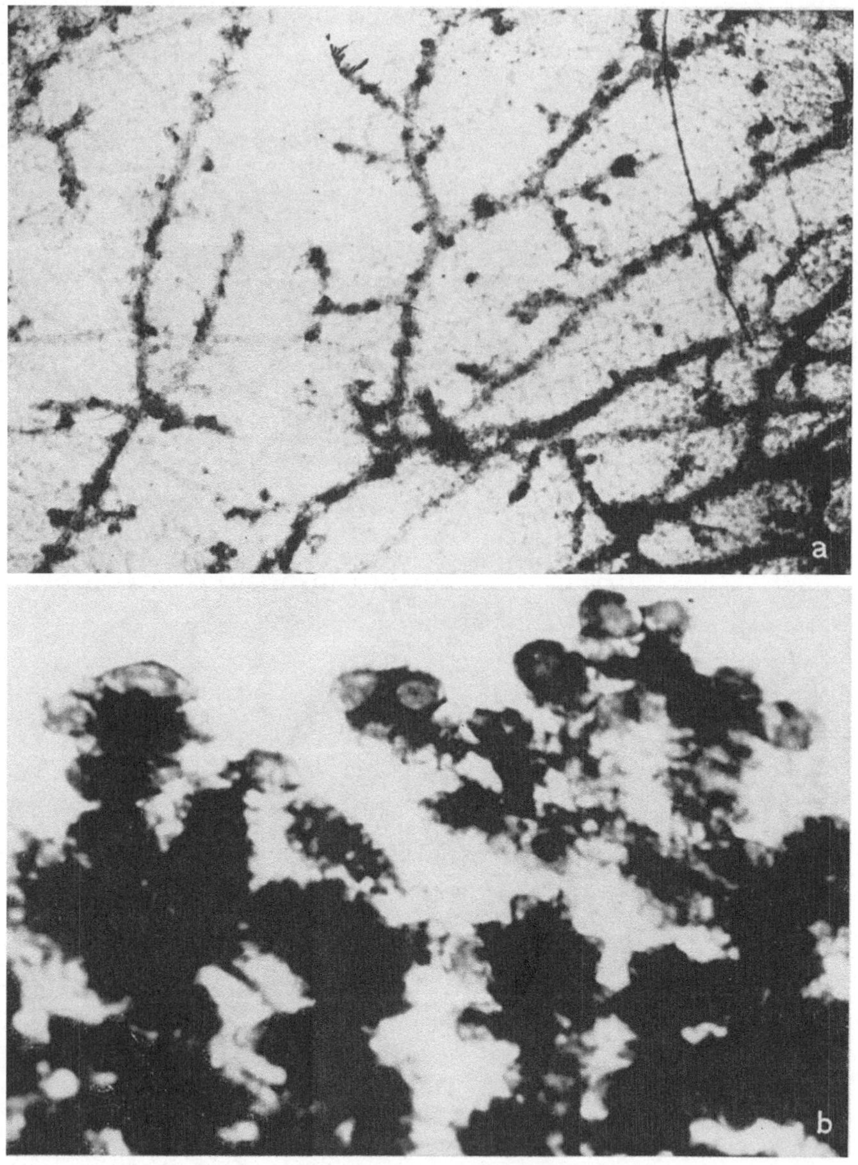

Abb. 3. Legende s. S. 193

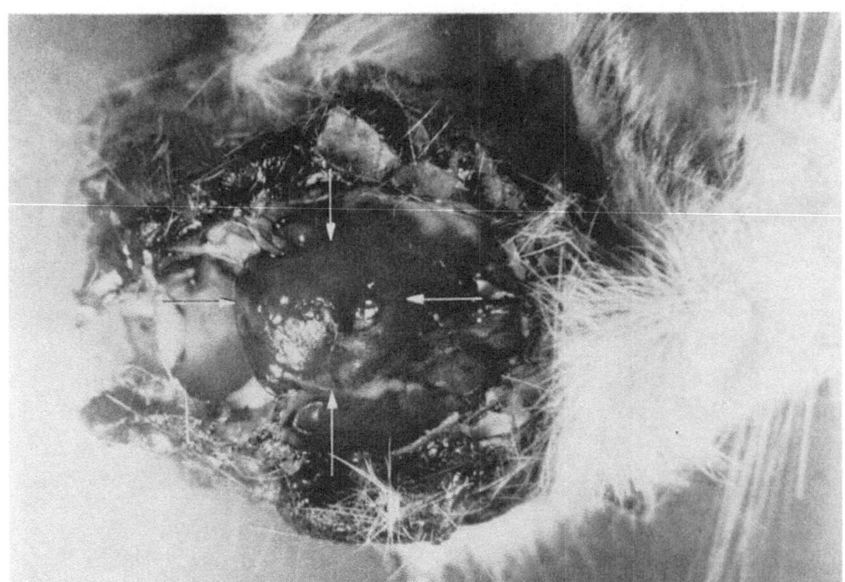

Abb. 4. Hypophysentumor einer Ratte. 30wöchige Behandlung mit täglich 2,0 mg/kg p. o.
Oestradiolvalerianat. Gewicht des Tumors 208 mg (Präparat aus der Abteilung für Toxikologie
der Schering AG)

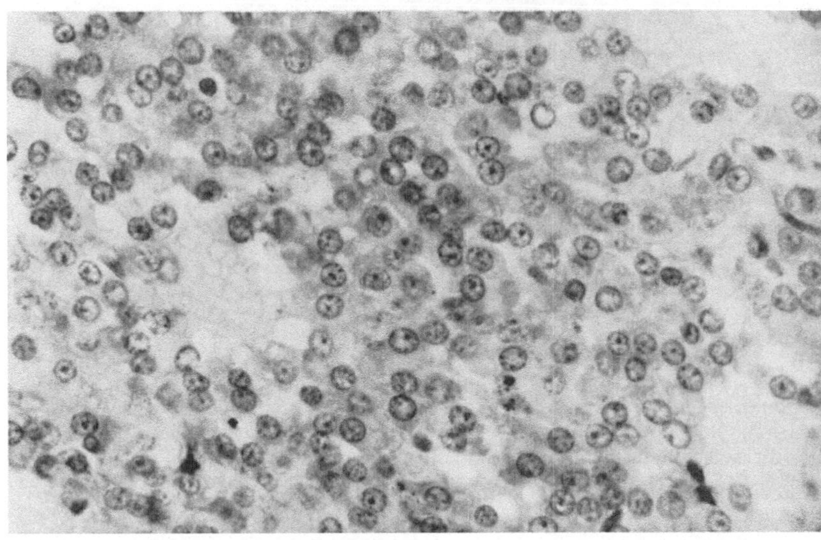

Abb. 5. Histologisches Bild eines Hypophysentumors der Ratte. 52wöchige Behandlung mit
täglich 0,1 mg/kg Äthinyloestradiol p. o. Vergr.: etwa 640mal

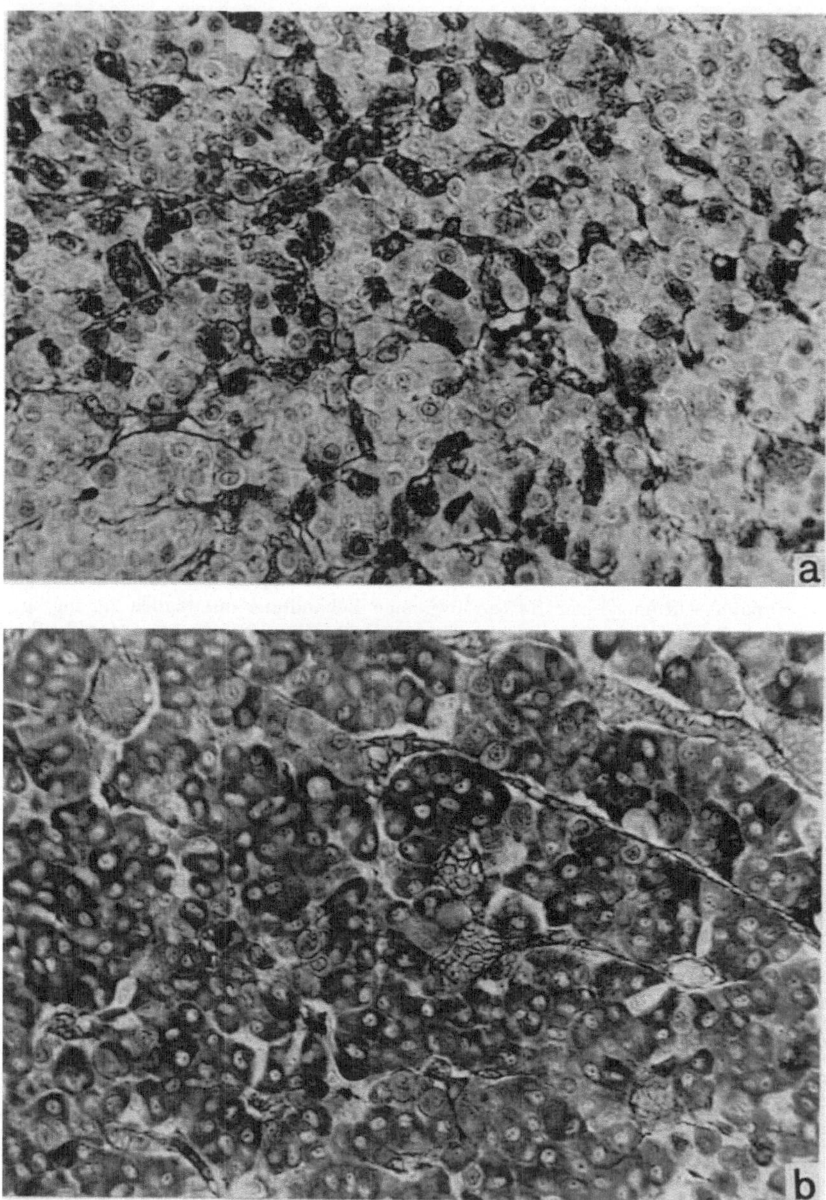

Abb. 6. Hundehypophysen. a) Kontrolle (unbehandeltes Tier); b) nach 30wöchiger Behandlung mit täglich 5,0 mg/kg 4,6-Dichlor-17-acetoxy-16α-methyl-4,6-pregnadien-3,20-dion p. o. Man beachte die starke Vermehrung der acidophilen Zellen. Vergr. etwa 400mal. Färbung: Paraldehydfuchsin (Scott). (Präparate aus der Abteilung für Toxikologie der Schering AG)